THE SARDAR SAROVAR DAM PROJECT

The Sardar Sarovar Dam Project

Selected Documents

Edited by
PHILIPPE CULLET
School of Law, School of Oriental and African Studies (SOAS), University of London, UK
and
International Environmental Law Research Centre (IELRC), Geneva, Switzerland

ASHGATE

Published by
Ashgate Publishing Limited
Gower House
Croft Road
Aldershot
Hampshire GU11 3HR
England

Ashgate Publishing Company
Suite 420
101 Cherry Street
Burlington, VT 05401-4405
USA

Ashgate website: http://www.ashgate.com

British Library Cataloguing in Publication Data
The Sardar Sarovar Dam Project : selected documents
1. Sardar Sarovar (Narmada Project) 2. Water resources development - Law and legislation - India - Cases 3. Dams - Law and legislation - India - Cases 4. Water resources development - India - History - 20th century - Sources
5. Dams - India - History - 20th century - Sources
I. Cullet, Philippe
346.5'404691

Library of Congress Cataloging-in-Publication Data
The Sardar Sarovar Dam Project : selected documents / edited by Philippe Cullet.
p. cm.
Includes bibliographical references and index.
ISBN: 978-0-7546-4910-6
1. Sardar Sarovar (Narmada) Project. Water resources development--Law and legislation--India. 3. Water resources development--Environmental aspects--India. 4. Drinking water--India. 5. Dams--Environmental aspects--India. 6. Irrigation--India. I. Cullet, Philippe.

KNS2532.S27 2007
363.6'10954--dc22

2006100193

ISBN: 978-0-7546-4910-6

Printed and bound in Great Britain by Antony Rowe Ltd, Chippenham, Wiltshire.

Table of Contents

Acknowledgments

The process of gathering the documents reproduced in this volume was only made possible thanks to the help received from many people in many different ways. I would like to thank in particular Prashant Bhushan and Shekhar Singh for providing access to all their files and for providing help in various other ways; S. Muralidhar for access to his files and for advice over many years; Usha Ramanathan for discussions, help in selecting materials and advice over many years; Thomas Berger for allowing the reproduction of selected parts of the 'Report of the Independent Review', Professor Upendra Baxi and Mr Ramaswamy Iyer for kindly accepting to look at the draft manuscript.

At the International Environmental Law Research Centre (IELRC), Shravya Reddy and Dominic Lo are thanked for their extreme patience and dedication to tasks, which were not necessarily the most intellectually challenging; Mohamed Asim for scanning and formatting various legible and illegible documents. Thanks also to Jessy Thomas for patiently reading all the documents with great care and to Thoko Kaime for preparing the index.

A number of people provided significant help in the process of gathering and choosing documents. These include Shripad Dharmadhikary and Nandini Oza, M.S. Ganesh, Rajat Khosla, Radhika Kolluru, Sanjay Parikh, Professor M.H. Qureshi, Jawahar Raja, Dilip D'Souza, Himanshu Thakkar and Umesh.

Finally, I would like to thank the Ashgate team for their support, in particular Alison Kirk for her support through several projects as well as Pauline Beavers, Carolyn Court, Donna Elliott and Neil Jordan.

Abbreviations and Terms Explained

abadi	an inhabited or settled placed
adivasi/aadivasi	tribal (person)
ATR	Action Taken Report
b.cu.m	billion cubic metre
CCA	culturable command area
crore	ten millions (10,000,000). Rs 1 crore is equivalent to €178,571 at exchange rate 56
cumec	cubic metre per second
CWC	Central Water Commission
dharna	peaceful demonstration
EPW	*Economic and Political Weekly*
FMG	Five Member Group
FRL	Full Reservoir Level
FSL	Full Supply Level
ft	foot/feet (1 foot equals 30.48 cm)
GoG	Government of Gujarat
GoI	Government of India
GoM	Government of Maharashtra
GoMP	Government of Madhya Pradesh
GoMs	Group of Ministers
GRA	Grievances Redressal Authority
GWSSB	Gujarat Water Supply and Sewerage Board
ha	hectare(s)
IA	interlocutory application

ILO	International Labour Organisation
Indira Sagar Pariyojana (ISP)	Indira Sagar Project, new name of the Narmada Sagar Project
jan sunwai	public hearing
Kachchh	western district of the state of Gujarat
Kachchh Jal Sankat Nivaran Samiti	Committee for the mitigation of water crisis in Kachchh
kharif	autumn crop
lakh	one hundred thousands (100,000). Rs 1 lakh is equivalent to €1,785.7 at exchange rate 56
m	metre(s)
m.	million(s)
MAF	million acre feet (equal to about 1,233.48 m^3)
m.cu.m	million cubic metre
MH	Maharashtra
MHM	millions of hectare metres
MoEF	Ministry of Environment and Forests
MoWR	Ministry of Water Resources
MP	Madhya Pradesh
MWL	Maximum Water Level
Narmada Sagar Project	Earlier name of the Indira Sagar Project
Narmada Bachao Andolan	Save the Narmada Movement
NBA	Narmada Bachao Andolan
NCA	Narmada Control Authority
nigam	corporation
NPG	Narmada Planning Group
NSP	Narmada Sagar Pariyojana
NSSO	National Sample Survey Organisation
NVDA	Narmada Valley Development Authority

NWDT	Narmada Water Disputes Tribunal
NWDTA	Narmada Water Disputes Tribunal Award
OSG	Sardar Sarovar Project Relief and Rehabilitation Oversight Group (Shunglu Committee)
panchayat	village council
PAF	project affected family
PAP	project affected person
PIL	public interest litigation
rabi	spring crop
R&R	resettlement and rehabilitation
RL	reservoir level
Shri	Mr
Srimati	Ms
SRP	special rehabilitation package
SSNNL	Sardar Sarovar Narmada Nigam Limited
SSP	Sardar Sarovar Project
SSPA	Sardar Sarovar Punarvasvat Agency
taluka/tehsil	Block (sub-division of a district)
ZSI	Zoological Survey of India

Chronology

1946	Planning of different dam projects to harness the Narmada waters starts
1948	Government of India appoints ad hoc committee to scrutinise seven projects, including the Broach Barrage and Canal Project (first name of the Sardar Sarovar Project)
1949	Investigation of Broach project sanctioned by the Government of India
1960	Broach project cleared by the Planning Commission at a height of 160 ft
1961 (April)	Prime Minister Jawahar Lal Nehru inaugurates the project
1963 (November)	Meeting between Union Minister of Irrigation and Power and the chief ministers of Gujarat and Madhya Pradesh leading to the 'Bhopal Agreement' whereby the Navagam dam would be built up to FRL +425 ft and its entire benefits were to be enjoyed by Gujarat
1963 (November)	Madhya Pradesh declining to ratify the Bhopal Agreement, contending that the Navagam dam should not be higher than FRL +162 ft which is the level of the river bed at the Madhya Pradesh border
1964	Government of India appoints Khosla Committee to study the entire Narmada basin and prepare a master plan
1965	Report of the Khosla Committee proposing the construction of a high dam at Navagam at FRL +500 ft. Gujarat broadly endorsed the report, Madhya Pradesh and Maharashtra rejected it
1965	Jalsindhi Agreement between Madhya Pradesh and Maharashtra for the construction of the Jalsindhi dam for power generation (a project opposed by Gujarat that could only have been built if the Sardar Sarovar dam height had been much lower (FRL 210 ft) than eventually decided. Jalsindhi is today in the middle of the submergence zone of the Sardar Sarovar dam)

1968 (July)	Gujarat makes a complaint to the Government of India under Section 3 of the Inter-State Water Disputes Act
1969 (October)	Constitution of the Narmada Water Disputes Tribunal (NWDT) to resolve outstanding issues among concerned states
1972 (February)	NWDT Decision on preliminary issues of law
1972 (July)	Agreement among the chief ministers of Gujarat, Madhya Pradesh, Maharashtra and Rajasthan seeking the intervention of the Prime Minister to allow an expeditious resolution to the conflict
1974	Political agreement among the four concerned states restricting the mandate of the NWDT. Amount of water available in Narmada arbitrarily fixed at 28 MAF
1978 (August)	Final order and Decision of the NWDT (clarifications sought by parties)
1978	First protests in Madhya Pradesh under the banner of the Nimar Bachao Andolan (Save Nimar Movement)
1979 (December)	Final award of the Narmada Water Disputes Tribunal gazetted
1980	Narmada Control Authority set up
1985 (May)	Union of India as well as the states of Gujarat, Madhya Pradesh and Maharashtra sign agreements with the World Bank for loans totalling $450 million
1986	Narmada Dharangrastra Samiti established under leadership of Medha Patkar
1987	Conditional clearances given by the Ministry of Environment and Forests
1987 (April)	Start of work on main dam of Sardar Sarovar Project
1987 (December)	Gujarat Government resolutions modifying its resettlement and rehabilitation package
1988 (October)	Conditional clearance given by the Planning Commission
1989	Narmada Dharangrastra Samiti and several local organisations coming together under the banner of 'Narmada Bachao Andolan'
1990 (June)	Japanese Government cancelling $150 million loan for turbine generators for river bed powerhouse

1992 (June)	Morse & Berger, Report of the Independent Review
1993 (March)	Government of India requesting the World Bank to cancel the remaining loan amount of $181.5 million
1993 (August)	Setting up of the 'Five Member Group' by the Government of India to continue review discussions
1993 (October)	Gujarat High Court Order in effect restraining the publication of the report of the Five Member Group
1994 (24 February)	Sluice gates closed
1994 (25 February)	Gujarat High Court Order prohibiting further work
1994 (April)	Four members of the Five Member Group sign their report accepted by all members. Dr Jayant Patil does not sign in view of the order of 26 October 1993
1994 (May)	NBA filing writ petition in Supreme Court
1994 (July)	Report of Five Member Group submitted to Government of India following Dr Patil's signature but the report is kept sealed
1995 (January)	NCA decides to stop construction work on the dam
1995 (April)	Further report of the Five Member Group
1995 (May)	Supreme Court order restraining further construction
1996 (July)	Decision to first build the dam up to 436 feet and review its operation after five years
1999 (February)	Supreme Court lifts stay on construction
1999 (February)	Grievance Redressal Authority of Gujarat established
1999–	Public interest litigation by Kachchh Jal Sankat Nivaran Samiti against Government of Gujarat requesting higher allocation of water to Kachchh district
2000 (March)	Grievances Redressal Authority of Madhya Pradesh established
2000 (April)	Grievances Redressal Authority of Maharashtra established
2000 (October)	Supreme Court judgment in Writ Petition (Civil) No. 319 of 1994 (*NBA v. Union of India*)
2002 (May)	NCA gives clearance for construction up to 95 metres +3 metre humps
2003 (May)	NCA gives clearance for construction up to 100 metres

2004 (March)	NCA gives clearance for construction up to 110.64 metres
2005 (March)	Supreme Court Judgment in Writ Petition (Civil) No. 328 of 2002 (*NBA v. Union of India*)
2006 (March)	NCA clearance for construction up to 121.92 metres
2006 (9 April)	Note of the Group of Ministers
2006 (15 April)	Review Committee of the NCA split on party lines on the issue of construction. The matter is referred to the Prime Minister, who passes it on to the Supreme Court
2006 (17 April)	Supreme Court does not stop the construction of the dam but indicates that it may do so in the future
2006 (24 April)	Constitution of the Sardar Sarovar Project Relief & Rehabilitation Oversight Group (Shunglu Committee)
2006 (8 May)	Supreme Court order determining that the Court will wait until the receipt of the Shunglu Committee report in July before taking a decision on the construction of the dam
2006 (3 July)	Report of the Shunglu Committee
2006 (10 July)	Supreme Court order confirming that the dam height will stay at 119 metres during the monsoon period

Note on the Documents Reproduced

- The main objective of this compilation is to bring together a selection of some of the main documents concerning the development and implementation of the Sardar Sarovar Project (SSP). It focuses on legal, policy and institutional aspects. This compilation neither seeks to comprehensively address all issues raised by the project nor to bring together all documents pertaining to one specific issue. Instead, it seeks to give readers access to a selection of important documents giving a broad overview of this multi-faceted project, illustrating some of the multiple controversies it has caused and illustrating the evolution of the law and policy framework over the nearly 60 years that have elapsed since a dam was first proposed.

- In this volume, an attempt has been made to reproduce official documents in their original form. They have therefore not been language edited. Nevertheless, the documents reproduced do not always reflect exactly the original text. Thus, typographical and related errors have been corrected; some abbreviations have been standardised throughout the volume for ease of reference; and the transliteration of some Hindi words has been standardised. These changes have not been indicated within the text to enable better readability of the documents. As a result, the documents reproduced here should in no case be deemed to constitute true copies of the original official documents.

- With regard to figures, the following should be noted. Figures have been given in all cases as in the original document. As a result, in original Indian documents, figures are usually given in lakhs (100,000) and crores (10 million). In documents emanating from other agencies, figures are usually given in millions and billions.

- The last part of this book contains a list of all the documents reproduced in chronological order for ease of reference. It also contains a set of references to books, booklets and articles on the SSP. This list has been kept to a minimum and includes only a limited number of relatively easily available sources among the vast literature that has been written on the SSP over the years.

Chapter 1

The Sardar Sarovar Dam Project: An Overview

Philippe Cullet

The Sardar Sarovar Project (SSP) is part of a gigantic scheme seeking to build more than 3,000 dams, including 30 big dams, on the river Narmada, a 1,312 km river flowing westwards from Amarkantak in Madhya Pradesh, touching Maharashtra and ending its course in Gujarat. The SSP is a multi-purpose dam and canal system whose primary rationale is to provide irrigation and drinking water. Power generation is another expected benefit. It is the second biggest of the proposed dams on the Narmada, and its canal network is projected to be the largest in the world. The dam is situated in the state of Gujarat, which will derive most of the benefits of the project, but the submergence – 37,533 hectares in total – is primarily affecting the state of MP (55 per cent) and to a much lesser extent the state of Maharashtra.

The SSP has been one of the most debated development projects of the past several decades in India and at the international level. It is only one of many similar big projects in the Narmada valley and in India generally but it has acquired a symbolic status in development debates. This is due in part to the complexity of such multi-purpose projects and the multiple positive and negative impacts associated with big dams. This is also due to the specificities of this project, which was first proposed nearly 60 years ago.

Firstly, the fact that this project involves four states – Gujarat, MP, Maharashtra and Rajasthan – with the state of Gujarat receiving most of the benefits of the project has repeatedly led to disagreements among the concerned states.

Secondly, the nature of Indian water law, which makes control over water largely a prerogative of the States with some oversight of the Central government in inter-state matters, provided the background for the setting of a complex inter-state institutional machinery to oversee the development of the project. This was also the basis for the setting up of a special tribunal, the Narmada Water Disputes Tribunal (NWDT), to adjudicate the claims of the concerned states.

Thirdly, the involvement of the World Bank in the SSP was also a landmark. It not only led to the commissioning of an independent enquiry of an ongoing project but also led the bank to take the unprecedented step of withdrawing from a funded project. This withdrawal had further significant repercussions at the international level. In particular, it led the World Bank to set up its first internal accountability mechanism, the Inspection Panel. The SSP fiasco also constituted one of the triggers for the setting up of the World Commission on Dams.

Fourthly, the SSP is also a landmark project from the point of view of social movements. While it is now very likely that the dam will be completed as projected by the project promoters, the work undertaken by the Narmada Bachao Andolan (NBA) and other organisations and individuals in the context of the SSP has had important impacts in India and abroad. Thus, to take but one example, before the NBA brought the issue of displacement to the forefront of the policy agenda, oustees had merely been seen as incidental costs of development, so much so that until the early 1990s there were no figures indicating the extent of displacement generated by projects.[1] Overall, regardless of whether oustees are actually resettled or not according to the legal framework put in place and regardless of whether the project eventually fulfils all the environmental and other conditions that have been set over time, the SSP will remain a milestone that has significantly contributed to transforming and redefining people's movements and activism in India and abroad.

This chapter seeks to provide a general framework to understand the documents reproduced in this volume. It addresses some of the main issues relevant in understanding the SSP documents reproduced here. No attempt is made to comprehensively analyse all relevant issues arising from the documents reproduced, something, which would require book-length treatment in itself.

A. Big Dams

The SSP needs to be understood in the broader policy context that has evolved over the nearly 60 years that have elapsed since a dam was first proposed. Throughout the first part of the twentieth century, big dams were seen in the North as one of the solutions to a number of issues. Big dams could, for instance, generate electricity; provide water for irrigation, industrial use and domestic use; store water for use in dry seasons or for transport to water-scarce areas; and contribute to flood prevention. All these perceived benefits led to a construction boom that lasted for several decades. As a result of these efforts, the North has used up 70 per cent of its hydroelectric potential.[2] Dam building in the North declined sharply in the second half of the twentieth century. This was due in part to the realisation that the benefits of large dams were not as important as expected and in part to the growing realisation that big dams had many more negative impacts than had been envisaged

1 Usha Ramanathan, 'Creating Dispensable Citizens', *The Hindu* 14 April 2006, p. 11.

2 Antoinette Hildering, *International Law, Sustainable Development and Water Management* (Delft: Eburon, 2004).

at first. Thus, growing concerns over the environmental impacts of big dams constituted one of the important driving forces behind dams losing their appeal in policy-making circles. However, while dam building may have gone out of fashion in the North during the few decades that coincided with rising environmental consciousness, a new rationale has recently emerged for big dams. Rising global temperatures and the threat of the impact of climate change have led to the search for alternatives to oil-based energy sources. In this context, hydroelectricity is seen as environmentally friendly since it is carbon neutral. While this rationale is noteworthy in the context of the search for climate change mitigation, it does not lessen other existing environmental impacts of big dams.

Dam building on a large scale was also taken up progressively in the South. However, one of the main characteristics of dam building in the South is that it started later than in the North and has largely been taken up without major interruption until this day. Large-scale dam building in big countries like India and China was undertaken in the first place on the basis of the same rationale underlying dam building in the North.[3] One of the main aims that sought to be fulfilled through dams was the promotion of canal irrigation to foster food security. Hydropower was also an important benefit in many cases since electricity generation in the second half of the twentieth century became a central concern, given its pivotal role in fostering economic development. The push for big dam construction was largely nationally driven. However, this has been reinforced by international and bilateral donors promoting big dams as a tool for the economic development of the South. This has not gone without being controversial because the push for big dam building in the South from international actors came more or less at the time when big dams were being sidelined in the North.[4] The emphasis on big dam building in the South went largely unchecked until the World Bank was forced to withdraw from the SSP in 1993. This provided the background for a short decade of reflection on the contribution of big dams to development in the South. This now seems to have been resolved in favour of promoting big dams again. The World Bank thus advocates, for instance, that India still has relatively little capacity to store water and that major investments are required in small and big projects including large dams.[5]

---

3 For a comprehensive analysis of a landmark dam and the controversies around it, *see* Shripad Dharmadhikary, Unravelling Bhakra – Assessing the Temple of Resurgent India (Badwani: Manthan, 2005), available at http://www.manthan-india.org/article19.html.

4 *Cf.* Sanjeev Khagram, *Dams and Development – Transnational Struggles for Water and Power* 178 (Ithaca: Cornell University Press, 2004).

5 John Briscoe & R.P.S. Malik, *India's Water Economy: Bracing for a Turbulent Future* (New Delhi: The World Bank and Oxford University Press, 2006).

B. Development of the Project

The first proposals concerning the construction of dams on the Narmada river for irrigation can be traced back to 1945–46 when the idea was mooted by the Central Waterways, Irrigation and Navigation Commission. In 1948, an ad hoc Committee set up by the Government of India recommended that detailed investigation should be undertaken for seven projects, including the Broach Barrage and Canal project (the first name of the SSP).[6]

The Navagam site was first recommended in 1957.[7] The first project proposal envisaged a dam being built in two stages first up to 160 ft and then to 300 ft. This was revised to 320 ft on the assumption that a higher level canal would be necessary for water to reach Kachchh and Saurashtra.[8] It is this version of the project that Prime Minister Jawaharlal Nehru launched on 5 April 1961.

During the years following the laying of the foundation stone, a number of issues were raised in inter-state proceedings. A 1963 agreement on the height of the dam at Navagam between Gujarat and MP was later repudiated. In the same period, MP and Maharashtra came to an agreement on a dam between Navagam and Harinphal, which would have drastically affected the plan for the dam at Navagam.[9]

The Government of India then decided in 1964 to appoint the Khosla Committee to study the entire Narmada basin and prepare a master plan for the optimum and integrated development of the water resources for irrigation, power generation, navigation, flood control and other aims.[10] The master plan proposed the construction of a high dam for a terminal reservoir at Navagam with FRL +500 ft that was found 'to be the optimum level for providing the maximum storage and reducing to the minimum the amount of water wasted to the sea'.[11] The Khosla Committee thus decided that a high dam at Navagam made more economic sense and gave more benefits than a combination of dams in the Navagam-Hiranphal gorge.

---

6 Khosla Report, reproduced below at page 41 at 1.

7 Dilip D'Souza, *The Narmada Dammed – An Inquiry into the Politics of Development* (New Delhi: Penguin, 2002).

8 *Id.* at 5.

9 P.M. Bakshi, A Background Paper on Article 262 and Inter-State Disputes Relating to Water, *in* Report of the National Commission to Review the Working of the Constitution, Volume II – Book 3 (New Delhi: Ministry of Law, Justice and Company Affairs, 2002).

10 Khosla Report, reproduced below at page 41 at 2.

11 *Id.* at 4.

The proposals of the Khosla Committee were broadly endorsed by Gujarat but opposed by Maharashtra and MP.[12] Continuing disagreements between the states eventually led Gujarat to request the adjudication of the dispute under the Inter-State Water Disputes Act, 1956.[13] As a result, the Central Government constituted the NWDT in October 1969. This was immediately challenged by MP, which argued that the constitution of the tribunal was *ultra vires* of the 1956 Act.[14]

The NWDT first spent a couple of years addressing preliminary issues. This led to its 1972 decision, which confirmed the Tribunal's jurisdiction over the issue and declared that Rajasthan was not entitled to any portion of the Narmada waters given that it was not a riparian state.[15] MP and Rajasthan appealed to the Supreme Court against the decisions of the NWDT. This largely stalled the proceedings of the NWDT. However, political developments in 1972 led to the Congress party (R) being in power in all four concerned states.[16] As a result of this convergence, a political agreement was reached among the chief ministers of the four states in July 1972 concerning the amount of water available and directing the Prime Minister to find a suitable solution.[17]

This was later confirmed in a 1974 agreement among the four states determining the quantity of water in the river that was to be allocated among the states – 28 Million Acre Feet (MAF) – and the specific allocation to Maharashtra (0.25 MAF) and Rajasthan (0.5 MAF).[18] This decision concerning the determination of the amount of water available in the Narmada river, which constitutes the basis for all subsequent planning decisions, has remained controversial. This is due to the fact that the first figure given was computed in 1965 on the basis of relatively few years of observed flow data (1948–62) complemented with 'derived' data or hind-casting

---

12 For MP, *see* Government of MP, Public Works Department, Comments of the State Government on the Report of the Narmada Water Resources Development Committee (Bhopal, January 1966). *See also* Report of the Narmada Water Disputes Tribunal with its Decision in the Matter of Water Disputes Regarding the Inter-State River Narmada and the River Valley Thereof Between the states of Gujarat, MP, Maharashtra and Rajasthan (New Delhi: Government of India, vol. 1, 1979), § 2.4.1.

13 Inter-State Water Disputes Act, 1956, Act 33 of 1956, available at http://www.ielrc.org/content/e5601.pdf.

14 *See* Government of Madhya Pradesh, Demurrer Before the Narmada Water Disputes Tribunal, filed on 24 November 1969 by the State of Madhya Pradesh.

15 Narmada Water Disputes Tribunal, Decision on Preliminary Issues, 23 February 1972.

16 Ranjit Dwivedi, *Conflict and Collective Action – The Sardar Sarovar Project in India* 78 (New Delhi: Routledge, 2006).

17 Agreement of 22 July 1972 Amongst the Chief Ministers of Madhya Pradesh, Gujarat, Maharashtra and Rajasthan, Narmada Development.

18 Agreement of 1974, reproduced below at page 46.

for the period 1915–47.[19] Since then, additional observed flow data puts the figure around 23 MAF.[20] The discrepancy between observed flows and the figure used for planning purposes may pave the way for further conflicts in the Narmada basin once MP makes full use of its water entitlement, should actual water flows be found to be lower than 28 MAF.

The agreement of 1974 led to the resumption of the NWDT proceedings, which went on until 1978. The NWDT gave its decision in August 1978. Following 'references' by all the parties involved,[21] the NWDT's final award was gazetted at the end of 1979. The NWDT took a number of major decisions. On the basis of the 1974 agreement, it allocated 67 per cent of the flow to MP (18.25 MAF) and 33 per cent to Gujarat (9 MAF). The NWDT also fixed the height of the SSP at FRL 138.68 m (455 ft). In other words, the height of the dam was reduced compared to the Khosla proposal but was still higher than MP had suggested. As a result, during the ongoing set of disputes over the dam in the past couple of decades, MP has tried to argue that the dam height should be brought back to 436 ft. The Chief Minister of MP thus tried to persuade the Prime Minister in 1994 that this reduction would save 38,000 persons from being displaced and would save 25,000 acres of land from submergence.[22] The position of MP was taken into account at the highest level. Indeed, the chief ministers of the concerned states and the Prime Minister agreed in 1996 to first plan the dam up to 436 ft and have it operate at that height for five years before moving ahead.[23] This agreement was subsequently abandoned.

The NWDT Award also allocated power benefits among the concerned states. The shares of the three riparian states were put at 57 per cent for MP, 27 per cent for Maharashtra and 16 per cent for Gujarat.[24] The power benefits are meant to ensure that the extensive irrigation water benefits to Gujarat are somewhat counter-balanced with a more favourable allocation of power to the two states facing most of the submergence. This, however, begs the question of the interests of the people

---

19 Report of the Narmada Water Resources Development Committee, Government of India, Ministry of Irrigation and Power (Khosla Report), 1 September 1965 [hereafter Khosla Report].

20 *See, e.g.*, Report of the Five Member Group, partly reproduced below at page 357 [hereafter Five Member Group] indicating at paragraph 3.2.3 that observed data for 1948 to 1988 gives a 75 per cent dependable yield of 22.9 MAF and that actually observed flows, whether for 20 years as at the time of the Tribunal's assessment or for 40 years are only around 23 MAF.

21 According to Section 5(3), Inter-State Water Disputes Act, 1956, above note 13.

22 *Cited in* D'Souza, above note 7 at 155.

23 Brief Record of Decisions Taken in the Meeting of Chief Ministers on Sardar Sarovar Project Convened by the Honourable Prime Minister on 15 July 1996 and 16 July 1996, New Delhi, available at http://www.ielrc.org/content/c9601.pdf.

24 NWDT Award, Clause VIII, reproduced below at page 47.

of MP and Maharashtra. The rationale for the dam, as many times restated is to provide irrigation and drinking water to water-scarce areas of Gujarat. In other words, it is clearly understood that power generation is only an additional benefit of the dam. If irrigation and drinking water are the main expected benefits from the dam, oustees would be expected to get a share of these irrigation and drinking water benefits. The fact that oustees should benefit from the project that is uprooting them is increasingly recognised. Further, it is not out of step with the original planning of the SSP since the Khosla report suggested that oustees should get lands provided with irrigation as well as villages with safe drinking water, roads and schools.[25]

The NWDT Award paved the way for the actual implementation of the project. While construction had officially started in 1961 when Jawaharlal Nehru laid the foundation stone, it is in fact only after the NWDT Award that actual final planning and work on what is today the SSP started. Besides the NWDT Award allowing the Gujarat Government to implement the project, the next event that actually kick-started the project was the World Bank agreeing in 1985 to provide $450 million to finance the construction of the dam and canal network.[26] The involvement of the World Bank was crucial in many respects as it was involved in the final planning for the project during the early 1980s while it was reviewing its suitability for funding.[27] Further, the sanction of a World Bank loan was crucial in imparting a seal of approval on the project that paved the way for the involvement of other transnational actors, such as the Sumitomo Corporation of Japan providing turbines for the project.[28] The involvement of the World Bank also internationalised the project and contributed to the development of a worldwide civil society interest in the SSP and other Narmada dams.

The actual start of the construction also led future oustees to progressively become aware of their destiny. This period coincided with the development of various efforts to ensure that oustees would be at least entitled to the minimum package offered in the NWDT. Among the different groups that started working with

---

25 Khosla Report, reproduced below at page 41 at 4.

26 Various agreements were signed with India and the states. *See, e.g.,* Development Credit Agreement (Narmada River Development (Gujarat) Sardar Sarovar Dam and Power Project) between India and International Development Association, Credit No. 1552 IN, 10 May 1985, reproduced below at page 411 and Gujarat Project Agreement (Narmada River Development (Gujarat) Water Delivery and Drainage Project) between International Development Association and State of Gujarat, 10 May 1985, Credit Number 1553 IN, available at http://www.ielrc.org/content/c8501.pdf.

27 World Bank involvement actually started as soon as the NWDT had crafted an acceptable compromise in 1978. *See* T. Scudder, India's Sardar Sarovar Project (SSP) 10 (unpublished manuscript, 2003).

28 Sardar Sarovar Construction Advisory Committee, Hydropower Complex, available at http://www.sscac.gov.in/m_hydro.html.

oustees, most seem to have at first tried to engage with the project promoters to ensure complete resettlement and rehabilitation. While some organisations like Arch Vahini stuck to this stance,[29] other organisations that eventually coalesced under the single banner of the Narmada Bachao Andolan (NBA) ended up taking a stronger stance against the building of the dam itself, having come to the conclusion that the project could not be realised in such a way as to provide full resettlement and rehabilitation to all oustees.[30] The increasing level of opposition to the dam and the way in which resettlement was being planned and executed was influenced in no small part by the lack of participation of oustees in the development and planning of the project as well as the lack of information regarding the project and ensuing displacement.

By the beginning of the 1990s, the NBA and allied organisations within the country and abroad had made a sufficiently well argued case for their claims. Eventually, the pressure on the World Bank became such that it commissioned the first ever independent review of an ongoing project. The resulting assessment, the Morse Report, was blunt in its assessment of the situation. The two authors opined that:

> [w]e think the Sardar Sarovar Projects as they stand are flawed, that resettlement and rehabilitation of all those displaced by the Projects is not possible under prevailing circumstances and that the environmental impacts of the Projects have not been properly considered or adequately addressed. Moreover, we believe that the Bank shares responsibility with the borrower for the situation that has developed.[31]

At first, the 1992 Morse Report was not well received as witnessed in the reactions of the Government of India and the Government of Gujarat.[32] Nevertheless, it did trigger the eventual pull-out of the World Bank from the project in 1993. World Bank reports commissioned after the pull-out confirmed that the bank accepted that

---

29 The 1987 resolutions of the Government of Gujarat modifying the resettlement and rehabilitation package for oustees in Gujarat led a split in the NGO movement where some organisation like Arch Vahini endorsed it while others did not. *See* Dwivedi, above note 16 at 157.

30 *Cf.* S. Parasuraman, 'The Anti-Dam Movement and Rehabilitation Policy', *in* Jean Drèze et al eds, *The Dam and the Nation – Displacement and Resettlement in the Narmada Valley* 26 (New Delhi: Oxford University Press, 1997).

31 Bradford Morse & Thomas R. Berger, *Sardar Sarovar – Report of the Independent Review* xii (Ottawa: Resource Futures International, 1992), section reproduced below at page 331.

32 *See respectively* Letter from Mr Chitale, reproduced below at page 353 and Government of Gujarat, Comment on the Morse Report, partly reproduced below at page 355.

there had been major shortcomings in this project, including failure to follow its own guidelines.[33]

The involvement and subsequent withdrawal of the World Bank from the project are landmark events from the point of view of the SSP but also from a broader perspective. The involvement in the SSP was, by all accounts, one of the biggest public relations disasters for the World Bank.[34] As a result, a number of initiatives were taken to ensure that similar problems would not resurface. One of the first internal initiatives was the setting up of the Inspection Panel, an internal accountability body.[35] This provides an avenue for individuals and groups to seek redress for breaches by the World Bank of its own operational policies and procedures in the context of a project it finances. Another (indirect) consequence of the SSP fiasco was the setting up of the World Commission on Dams (WCD), an attempt to rebuild a consensus among all concerned actors at the international level to ensure that big dam construction would not be marred by further controversies of the same magnitude. The 2000 WCD report confirmed that the construction of big dams was acceptable but emphasised that project promoters needed to fulfil various obligations, for instance, from the point of view of human displacement and environmental protection.[36] Since 2000, there has indeed been renewed interest in big dams in international policy-making circles. This has led to the reformulation of the rationale for big dams in the World Bank 2004 Water Strategy. The latter provides that of the four effective water resource development mechanisms that play a central role in sustainable growth and poverty reduction, 'major infrastructure such as dams and inter-basin transfers, provide national, regional and local benefits from which all people, including poor people, can gain'.[37]

After the World Bank withdrawal from the project, the development of the SSP took a different turn. While international funding had been withdrawn, it was

---

33 *See, e.g.*, World Bank, Project Completion Report, partly reproduced below at page 422 at § 25.

34 *See, e.g.*, Environment Defense Fund, "Institutional Amnesia": The World Bank's Approach to High-Risk Projects (2003), available at http://www.environmentaldefense.org/documents/3008_Gambling_Amnesia.pdf. The recent World Bank Water Resources Sector Strategy still acknowledges that '[s]ome of the World Bank's greatest and most publicized failures have involved the financing of dams that were planned and built without sufficient attention to social and environmental consequences'. World Bank, Water Resources Sector Strategy – Strategic Directions for World Bank Engagement 8 (Washington, DC: World Bank, 2004) [hereafter Water Resources Strategy].

35 The World Bank Inspection Panel, Resolution No. IBRD 93-10, 34 *International Legal Materials* 503 (1995).

36 World Commission on Dams, *Dams and Development – A New Framework for Decision-Making* (London: Earthscan, 2000).

37 Water Resources Strategy, above note 34 at 2.

decided to pursue the implementation of the project with domestic resources. The remainder of the year 1993, after the withdrawal of the World Bank, was marked by the submergence of Manibeli, the first village in Gujarat to be fully drowned by the dam during the 1993 monsoon. Manibeli's submergence coming within a few months of the World Bank's withdrawal led the NBA to step up its campaign by launching a call for *jal samarpan* or sacrifice by drowning.[38] Eventually, in the face of mounting opposition at home and abroad, the Ministry of Water Resources appointed the Five Member Group (FMG), a group of five experts were asked to 'continue the review discussions initiated during and of June, 1993 on all issues related with the Sardar Sarovar Project'.[39] The FMG process was, on the whole, an attempt to determine how far local experts would concur with the conclusions of the Morse Report. While the terms of reference were relatively narrow and the five members had been chosen by the government, the FMG process led to significant resentment. The constitution of the FMG was thus challenged in the High Court of Gujarat by the Narmada Abhiyan (Narmada Campaign) requesting among other things the setting aside of the Water Ministry's Memorandum and restraining the government from releasing the report to the public.[40] The High Court passed an order in October 1993, which substantially restrained the government from releasing the report to the public until further orders. This was challenged in the Supreme Court, which eventually declared in December 1994 that the report should be published.[41] The report that was eventually released substantially concurred with the Morse Report though in less vigorous terms.[42] A further report by the FMG was submitted in response to a Supreme Court order requesting further views from the group on hydrology, on the height of the dam, on resettlement and rehabilitation of oustees and on the environment.[43] The report submitted in April 1995 by four of the original five members addressed in further detail the requests of the Supreme Court.[44] It could not be adopted unanimously because the members significantly disagreed on issues such as the need to reduce

---

38 *See, e.g.*, The Narmada Sardar Sarovar Project Mass Arrests and Excessive Use of Police Force Against Activists in Central India (Report of the Narmada International Human Rights Panel, October 1993).

39 Ministry of Water Resources, Office Memorandum, No. 6/4/93-PP, 5 August 1993.

40 Special Civil Application No. 9366 of 1993, *Narmada Abhiyan and others v. Government of India and others*.

41 *Narmada Bachao Andolan v. Union of India and Others*, Writ Petition (Civil) No 319 of 1994, Supreme Court of India, Order of 13 December 1994.

42 Five Member Group, above note 20.

43 *Narmada Bachao Andolan v. Union of India and Others*, Writ Petition (Civil) No 319 of 1994, Supreme Court of India, Order of 24 January 1995.

44 Ramaswamy R. Iyer, L.C. Jain, V.C. Kulandaiswamy, Vasant Gowariker, Further Report of the FMG on Certain Issues Relating to the Sardar Sarovar Project, April 1995, available at http://www.ielrc.org/content/c9505.pdf.

the height of the dam and on the need to further review the project in view of the problems identified with resettlement and rehabilitation.[45]

From the point of view of the NBA, the withdrawal of the World Bank was both a boon and a challenge. From March 1993 onwards, much less international pressure could be applied on the Central and State governments. Eventually, in view of the difficulties faced in making itself heard, the NBA decided to have recourse to public interest litigation, which had come to be seen by the mid-1990s as one of the best avenues for ensuring justice and securing the realisation of human rights. As a result, a petition was filed in 1994 in the Supreme Court. Since that day, the Supreme Court has been, in many respects, one of the main actors in the ever unfolding crisis that has been the construction of the SSP over the past couple of decades.

The NBA petition argued among other things that the assumptions on which the NWDT Award had been given in 1979 had significantly changed in the meantime. Further, it asserted that the NWDT had not considered all relevant issues and, in particular, that it had not given project-affected people an opportunity to make representations before it. The petition intimated that these omissions had led to a flawed project with grossly underestimated social and environmental costs, which could not be implemented as per the NWDT Award without serious violations of human rights and damage to the environment. In conclusion, it considered that a review of the project was urgently needed. More specifically, it asked the Court to either order a stoppage of the project and implement proposed alternatives or direct the Union of India to set a new tribunal to review the project which would include participation of project affected people or that the Court should set up an independent team to review the whole project.[46]

Judicial intervention by the Supreme Court has been marked by a series of different stands on the SSP. In May 1994, the Court first declined to stop construction of the dam.[47] A year later, the Court agreed to the suspension of the work on the project and maintained this stance for four years.[48] After 1998, the Court progressively hardened its position. It first authorised the resumption of work on the project in 1999 and eventually, in its main judgment of October 2000, castigated the NBA for having approached the Court too late and gave the

---

45 *Id.* at Section 2.5 and 'A Note by Ramaswamy R. Iyer'.

46 *Narmada Bachao Andolan v. Union of India and Others*, Supreme Court of India, Writ Petition (Civil) No. 319 of 1994.

47 *Narmada Bachao Andolan v. Union of India and Others*, Writ Petition (Civil) No 319 of 1994, Supreme Court of India, Order of 20 May 1994.

48 Order of 5 May 1995, reproduced below at page 127.

government a renewed stamp of approval to complete the project as fast as possible.[49]

While the 2000 judgment only placed minimum constraints on the government to ensure that 'the project is completed as expeditiously as possible',[50] it still imposed that construction should happen in stages and that each stage should only be undertaken after all oustees were effectively resettled (the *pari passu* condition).[51] This has been the basis for further petitions focusing not any more on the whole project but more specifically on the resettlement and rehabilitation of oustees. In particular, petitions have been brought to the Court by individual oustees of MP and by the NBA showing that resettlement and rehabilitation had not taken place as per the NWDT Award and the Court's own 2000 judgment. In a petition of 2002, the Supreme Court first refused to interfere, arguing that issues of resettlement and rehabilitation were best dealt with by the Grievance Redressal Authority (GRA).[52] The case was nevertheless successfully brought back to the Court and following the order of the GRA of MP in September 2004,[53] the Supreme Court gave a further judgment in 2005, which was quite critical of the resettlement and rehabilitation process.[54] In other words, five years after a judgment which was meant to provide for the construction of the dam while ensuring full resettlement for oustees and compliance with environmental conditions, the same Court acknowledged that this had not actually taken place in practice.

Since the 2005 judgment, the pendulum has again swung rapidly towards ensuring that the dam is completed at the earliest, regardless of the various conditions set by the NWDT Award and the Supreme Court earlier. The year 2006 witnessed what will probably be one of the last major confrontations to ensure that the construction of the dam is effected according to all the conditions set in the various relevant legal instruments. The political and legal battle was triggered by the decision of the NCA to authorise the construction of an additional 10 metres of the dam up to 121.92 metres in March 2006, which will lead to the displacement of an estimated 17,255 families. While the NCA gave its clearance on 10 March 2006, the Union Water Resources Minister, Saifuddin Soz, immediately decided to put the decision

---

49 *Narmada Bachao Andolan v. Union of India*, Writ Petition (Civil) No. 319 of 1994, Supreme Court of India, Judgment of 18 October 2000 (Justice Kirpal and Justice Anand), reproduced below at page 138 [hereafter majority judgment].

50 *Id.* § 280.

51 On *pari passu*, see below page 28.

52 *Narmada Bachao Andolan v. Union of India and Others*, Supreme Court, Writ Petition (Civil) No. 328 of 2002, Order of 9 September 2002, reproduced below at page 269.

53 Order of the GRA, reproduced below at page 270.

54 Judgment of 15 March 2005, reproduced below at page 277.

on hold.[55] The decision was put on hold in an attempt to diffuse tension but no specific action was taken to ensure that construction would not take place. As a result, construction was not actually halted.[56] The NBA thus organised from 17 March a dharna in New Delhi, first outside the Water Resources Ministry, then at Jantar Mantar. In view of the lack of response of the government, three NBA activists, Bhagwatibehn, Jamsingh Nargave and Medha Patkar, started an indefinite hunger strike.

This led to a chain reaction in official circles. First, the government sent three ministers to assess the state of resettlement and rehabilitation. Their report given to the government on 9 April was quite critical of the situation on the ground. The Supreme Court was then given opportunities to intervene in the matter since a hearing on fresh petitions by oustees was in any case scheduled. On 17 April, the Court refused to stop the construction of the dam but warned it would do so if rehabilitation was found to be inadequate. The Court, having declined to take action, indicated that it should be an issue for the Prime Minister to resolve, in accordance with some of the directions the Court had given in its 2000 judgment. The Prime Minister was not inclined to take a decision either way and instead requested on 24 April the setting up of yet another committee, the Shunglu Committee, to look into the resettlement and rehabilitation issue. Subsequently, in two further hearings on the SSP on 1 and 8 May, the Supreme Court refused again to suspend the NCA decision. The report of the Shunglu Committee given in early July was sympathetic to the claims of the states that the situation is generally satisfactory with regard to resettlement and rehabilitation. Yet, in the subsequent hearing at the Supreme Court on 10 July, a 'compromise' position was adopted. The dam, which had already been built from 110 metres to 119 metres since March 2006, was not to be built up to 120.92 metres until resettlement and rehabilitation at 119 metres was complete. Without any further approval from the Supreme Court in the meantime, the construction of the dam up to 121.92 metres was completed by December 2006.[57]

C. Rationale and Expected Benefits

The SSP is what is known as a multi-purpose project since it is meant to provide different types of benefits from drinking and irrigation water to electricity

---

55 *See, e.g.*, Gargi Parsai, 'Centre Puts on Hold Decision on Narmada Dam', *The Hindu*, 11 March 2006.

56 *See, e.g.*, Saeed Khan, 'Narmada Dam Construction on Despite Delhi's «Instructions»', 10 March 2006, available at http://www.newkerala.com/news2.php?action=fullnews&id=26733.

57 *See* Advertisement, 'Gujarat Dedicates with Pride 1,450 MW Hydro Power of Sardar Sarovar Narmada Project to the Nation', *The Hindu* 19 January 2007, p. 16.

generation. However, it must be borne in mind that the SSP was, from the start, planned as a source of irrigation water. This is visible in the rationale for the Khosla master plan, which insists that requirements of irrigation should have priority over those of power.[58] This is also confirmed by the fact that the provision of drinking water only became a significant component of the project in later years. In other words, the SSP was from the start planned as a project that would help solve the water problems of water-scarce areas of Gujarat and Rajasthan.[59]

The general rationale to bring water to water-scarce areas can be looked at from different angles. Thus, the first justification was the perceived need after independence to extend irrigation so as to ensure aggregate food security for the country.[60] According to Justice Kirpal, this remains a primary rationale for big dams since 'large-scale river valley projects per se all over the country have made India more than self-sufficient in food'.[61] Another justification was an attempt to settle farmers in regions bordering Pakistan, an objective that was at least as much based on military strategy objectives as on food security.[62] To this are added the generation of electricity and the provision of drinking water to a large part of Gujarat. A number of other reasons have been added over time. Justice Kirpal argues, for instance, that the displacement of oustees should be seen as a favour made to them since '[i]t is not fair that tribals and the people in undeveloped villages should continue in the same condition without ever enjoying the fruits of science and technology for better health and have a higher quality of life style'.[63] Similarly, he argues that while a dam displaces more people than a thermal power plant, the cost of electricity generation is lower and it is ecologically friendly from the perspective of global warming.[64] The SSP has also been justified on its potential for fostering tourism in the Narmada valley. Not only is the dam itself, a new 'temple of engineering', seen as a tourist attraction but it is argued that the water released by the dams will improve the flow of 'clean running water' at the bathing ghats for pilgrims.[65]

---

58 Khosla Report, above note 19 at 210.

59 *See, e.g.*, R. Rangachari et al., 'Some Agreed Conclusions', *in* Shekhar Singh and Pranab Banerji eds, *Large Dams in India – Environmental, Social & Economic Impacts* 273 (New Delhi: Indian Institute of Public Administration, 2002).

60 Khosla Report, above note 19 at 210.

61 Majority judgment, above note 49 at § 265.

62 Note that this objective finds an echo in the majority judgment, above note 49 at § 262.

63 Majority judgment, above note 49 at § 267.

64 *Id.* § 278.

65 *See* Narmada Planning Agency, Government of Madhya Pradesh, Development of Recreational, Tourist, Sports and other Social and Community Facilities in the Basin (1984).

Before turning to expected benefits, it is apt to briefly scrutinise project costs. While an overall assessment of the project should go beyond economic costs and benefits, these are still fundamental as the project is meant to be financially viable. The various figures given by different analysts over time may not necessarily be fully comparable. Nevertheless, they clearly point out that one of the main characteristics of the overall costs of the SSP is that they have grown exponentially over time. In 1983, costs were estimated at Rs 4,877 crore by Tate Economic Consultancy Services.[66] In 1985, the World Bank put the total cost at Rs 13,640 crore. By 1992, according to estimates of the Gujarat Government it was Rs 20,470 crore.[67] In 1994, other figures based on World Bank estimates put the total cost at Rs 34,000 crore.[68] By 2006, the Government of Gujarat had spent Rs 21,411.81 crore on the project.[69] Two brief comments can be made on these figures. Firstly, aggregate figures by themselves point to problems in the overall planning of the project. The fast increasing costs of the project not only threaten its overall economic viability but have also led, in the words of an analyst to the 'economic insolvency that plagues the project'.[70] Secondly, even if the cost estimates made in the 1980s were accurate and made the project economically viable at the time of the World Bank assessment, this has in fact always been a fiction. Estimates of the total costs of the project presented here do not include the huge expenditure on the drinking water component, which has, for all practical purposes, been planned only in the past 15 years.[71]

While it is impossible today to provide accurate estimates of the total benefits that will accrue during the project lifetime, it is certain that the project was conceived as being financially viable on the basis of earlier estimates that are completely out of step with actual costs. This therefore begs the question of alternatives to the SSP. Could similar benefits have been achieved through smaller projects rather than a gigantic scheme? The answer is that without any doubt there are alternatives which could have been considered and that could have delivered similar benefits at a much lower cost. One such alternative is the proposal by Paranjape and Joy that would ensure the delivery of the same amount of water authorised by the NWDT

66 Sanjay Sangvai, *The River and Life – People's Struggle in the Narmada Valley* 13 (Mumbai: Earthcare Books, revised ed. 2002).

67 Amita Baviskar, *In the Belly of the River – Tribal Conflicts Over Development in the Narmada Valley* (New Delhi: Oxford University Press, 2nd ed. 2004).

68 Sangvai, above note 66 at 27.

69 Rajya Sabha, Starred Question No 558, Construction of Narmada Dam near Kevadia. Response of the Minister of Water Resources (Prof. Saif-Ud-Din Soz) of 23 May 2006.

70 Himanshu Upadhyaya, 'Sardar Sarovar Project – Dubious Record', 23/10 *Frontline* 42 (2006).

71 For more details on the drinking water component, see below at page 18.

but with a much lower dam that would reduce submergence by two-thirds.[72] While the NWDT did not, as such, consider alternatives to the SSP, it had to address the claims of MP and Maharashtra, which wanted to build a dam at Jalsindhi. The construction of the Jalsindhi dam would have drastically limited the size of the SSP but, at the same time, the area coming under submergence would have been reduced from an estimated 36,400 ha for the SSP to 14,600 ha with the two smaller dams and the submergence of cultivable area reduced from 12,000 ha to 1,440 ha.[73] The ultimate power generation of the two dams would also have been higher than for the single SSP.[74] Despite finding the two smaller dams attractive from a submergence and power generation point of view, the NWDT found that both reservoirs were too small and that all excess inflows of surplus years from the entire catchment area below the Indira Sagar Pariyojana (ISP) would go to waste to the sea. Further, the NWDT determined that with a lower dam, irrigation benefits would be substantially reduced and that irrigation in Rajasthan could only be through lift irrigation.[75] This again confirms the importance of irrigation in the decision to build the high SSP dam that is being built today and that these benefits were put above other considerations such as displacement and submergence of arable land.

The benefits of the project are meant to be so large for Gujarat and her people that the project has been nicknamed ‘Gujarat’s lifeline’.[76] In fact, the extent of benefits and their distribution within Gujarat remains a matter of intense debate. While the project has always been premised on the benefits it would bring to water-scarce areas,[77] an unavoidable aspect of the project is that the main canal first runs through fertile and highly developed parts of the state. It has not remained unnoticed to people living in areas close to the dam that large-scale benefits could be derived from the additional water coming from the Narmada river. This probably explains why the setting up of several sugar factories has been reported in the command area of the first part of the main canal.[78] More generally, industries

---

72 *See, e.g.*, Suhas Paranjape & K.J. Joy, ‘Alternative Restructuring of the Sardar Sarovar: Breaking the Deadlock’, 41/7 *EPW* 601 (2006).

73 Report of the Narmada Water Disputes Tribunal with its Decision in the Matter of Water Disputes Regarding the Inter-State River Narmada and the River Valley Thereof Between the States of Gujarat, Madhya Pradesh, Maharashtra and Rajasthan (New Delhi: Government of India, vol. 2, 1979).

74 *Id.* § 13.5.1.

75 *Id.*

76 *See, e.g.*, Dionne Bunsha, ‘Heights of Intolerance’, 23/11 *Frontline* 108 (2006).

77 *See, e.g.*, Lyla Mehta, ‘Manufacture of Popular Perceptions of Scarcity: Dams and Water-related Narratives in Gujarat’, 29/12 *World Dev.* 2025, 2028 (2001) indicating that the 1972 Irrigation Commission on Narmada Waters stressed the importance of providing Narmada water to scarcity areas of Gujarat irrespective of what share of Narmada waters would be allocated to Gujarat.

78 Sangvai, above note 66 at 114.

are expected to be major beneficiaries of the project.[79] The tension between the benefits that people in water-scarce areas expect from Narmada waters and the lure of additional benefits for people living in already prosperous areas benefiting from relatively adequate water availability has been a long-standing issue. This is, for instance, reflected in the ongoing public interest litigation initiated by the Kachchh Jal Sankat Nivaran Samiti, an organisation representing the interests of the people of the Kachchh district of Gujarat to request additional water allocation for the district which constitutes on its own about a quarter of the size of the state and is 100 per cent drought-prone and whose relief has been consistently used as a justification for the SSP.[80] Yet, the planned allocation of Narmada waters for Kachchh is only 2 per cent.[81] It is a telling fact that more than 40 years after the Khosla report and a couple of decades after implementation of the project started, the issue of who within the beneficiary state will benefit is still the object of severe disagreements. The fact that the government does not seem ready to allocate more water to the poorer and more drought-prone areas of the state seems to indicate that concerns over the possibility that most water will be appropriated by people and industries living in the first part of the command area rather than preferentially channelled to poorer and more water-scarce areas of the state have not been unfounded.

Moving beyond Gujarat, two main inter-state aspects need to be considered with regard to the benefits of the SSP. Firstly, while MP and Maharashtra get no or limited water allocation out of the SSP, they are meant to be partly compensated for the submergence of parts of their territory with a higher allocation of power benefits. Keeping in mind that the main rationale for the SSP is the provision of water to foster food security through irrigation, it appears that neither of the two states benefit at all. While the concerned states themselves seem to have generally accepted this, it is an element which needs to be borne in mind when analysing the SSP on the whole. MP, the state that sees a significant part of its territory submerged by the SSP is rewarded by power benefits, which will, as intended in the planning of the project itself, rapidly diminish over the lifetime of the project. In view of the Khosla report's insistence on the need to make the national interest prevail, it can be argued that MP is doing this for the broader common good. Yet, it is unsure whether the people of MP, not just people who are ousted by the dam, can ever make sense of this decision. It is indeed extremely unlikely that the food security and economic benefits that will accrue to Gujarat will ever trickle down to benefit the broader population of MP. In the case of the SSP, there is a clear inter-state dimension to the distribution of benefits and costs, which contributes to highlighting this issue. However, the same would be true of other dams on the Narmada situated wholly within MP or other dams elsewhere. The costs borne by

---

79 Dwivedi, above note 16 at 123.

80 *Kachchh Jal Sankat Nivaran Samiti v. State of Gujarat*, reproduced below at page 105.

81 Five Member Group, above note 20 at paragraph 3.3.3.

one part of a region, state or country are usually justified in terms of the benefits that this brings to the broader polity. However, as further analysed below with regard to oustees, it is relatively difficult to justify 'costs' like displacement where the project planners do not ensure that oustees are among the first to benefit from the project, for instance, through relocation in the command area of the project.

The second inter-state element refers to the broader plan for the SSP. From the start, it was always planned that the SSP would only deliver its expected benefits if it was built alongside the similarly very large ISP.[82] There are therefore physical links between the two projects since the non-completion of ISP would have had important consequences on the expected benefits of the SSP. This explains why Gujarat was made to pay 17.63 per cent of the costs of ISP.[83]

There are three main benefits expected from the SSP. These are irrigation water, drinking water and power generation. The irrigation benefits constitute, as mentioned above, the main justification for the dam. If everything goes according to plan, the dam will feed a network of canals comprising a 458 km long main canal up to Gujarat-Rajasthan border, 42 branch canals, 7,500 km of distributaries, and nearly 30,000 km of minors and sub-minors. The network distribution system is planned to eventually have an aggregate length of about 75,000 km. In keeping with the project rationale, which is to bring irrigation water to water-scarce areas of Gujarat as well as to some areas of southern Rajasthan, the main canal is planned to extend for another 74 km into Rajasthan. The reservoir of the SSP is planned as having a gross storage capacity of 0.95 million hectare metres and a live storage of 0.58 million hectare metres. This is meant to provide irrigation to about 1.80 million hectares in about 3,400 villages in Gujarat, 75,000 ha in arid areas of Rajasthan and 37,500 ha (by lift irrigation) in Maharashtra. Overall, this would make it the largest irrigation project in the world.[84] At this juncture, it is not possible to make a full assessment of the irrigation component of the SSP. This is because the canal network is not complete. Further, the irrigation component has been marred by institutional delays, for instance, in the setting up of water user associations at the local level, which are a prerequisite for the availability of Narmada water.

The provision of drinking water to villages and towns in Gujarat has become another central component of the project. There is no doubt that the provision of drinking water to water-scarce parts of Gujarat is a major benefit from a project like the SSP. Nevertheless, this part of the project also raises a number of questions. Firstly, the drinking water plan appears to be an afterthought. In fact, it

---

82 The ISP was commissioned during the financial year 2004-2005.

83 NWDT Award, Clause X, reproduced below at page 47.

84 Scudder, above note 27 at 8.

was only effectively taken up after 1990.[85] Indeed the Morse Commission still found that the plans for the delivery of water to drought-prone areas of Gujarat were in the earliest stages of development.[86] Even by 1994, as indicated by the Five Member Group, there was very little clarity on the drinking water component and its implementation.[87] Secondly, as mentioned above, the huge drinking water scheme, which is being implemented at present to bring Narmada water to villages all the way to coastal villages in Saurashtra and elsewhere, is not actually part of the cost of the SSP. This is in fact a separate project whose cost should be added to the overall cost of the SSP when attempting to calculate costs and benefits.[88] These are very substantial costs since the capital costs of the water supply scheme are now estimated at Rs 8,026 crores.[89] Thirdly, the lack of clarity on the drinking water component is not only because planning started very late but also that the scheme itself seems to be regularly changing. Thus, the number of villages and towns to be served by the drinking water component seems to be regularly growing. This increased from no allocation in the 1970s to 4,270 villages in 1983–84, 7,235 in 1990, 8,215 during the 1990s to reach the current figure of 9,633 villages and 131 towns.[90] However, what looks like ever increasing benefits of the SSP for the people of Gujarat is not free from controversy. Indeed, as witnessed through the ongoing litigation concerning drinking water benefits for the Kachchh region, it is not necessarily the regions that gave the project its rationale in the first place that get preferential allocation of project benefits.[91]

The third main, though ancillary, benefit of the SSP is electricity generation. This is achieved through an underground River Bed Power House, which should provide a peak of 1,200 megawatts capacity, and a surface Canal Head Power House providing a maximum of 250 megawatts capacity. With regard to hydropower generation, the main question that arises is that of the actual benefits that will accrue to MP and Maharashtra. Indeed, while there will be 1,450 megawatts of installed capacity, it has been estimated that actual power generation will only be 425 megawatts in the early stages of the project, diminishing to 50 megawatts by the time all the water allocation for Gujarat is used for irrigation.[92]

---

85 Indira Hirway & Subhrangsu Goswami, Concurrent Monitoring of the World's Largest Drinking Water Pipeline Project – A Study of Narmada Based Project in Gujarat (Ahmedabad: Centre for Development Alternatives, 2006).

86 Morse & Berger, above note 31 at xxiii.

87 Five Member Group, above note 20 at Section 3.4.

88 *See, e.g.*, Himanshu Upadhyaya, 'Sardar Sarovar Project – Dam Minus Drinking Water', *The Hindu*, 16 April 2006.

89 Padmaparna Ghosh & S.V. Suresh Babu, '907 KM from Parliament', *Down to Earth* 15 May 2006, p. 23.

90 *See* Sangvai, above note 66 at 119 and Hirway & Goswami, above note 85 at 2.

91 *Kachchh Jal Sankat Nivaran Samiti v. State of Gujarat*, reproduced below at page 105.

92 Sangvai, above note 66 at 121.

D. Issues Raised

The SSP is not only a vast project, it is also a project that creates an extensive submergence zone. In fact, the reservoir is of such proportions that it will extend up to 214 km upstream from the dam. This can be explained by the fact that while the dam itself and the first part of the submergence zone is in a comparatively hilly area, the tail end of the reservoir is by and large a plain area. This explains that each additional metre of dam construction at this juncture has such important consequences in terms of displacement. It also explains why there could have been heated arguments as late as the mid-1990s over the reduction in height of the dam by only 19 ft since this would have made a big difference in terms of submergence and displacement.

1. Displacement

The submergence created by the dam has a number of direct and indirect impacts. From a human development point of view, there is, however, no greater impact than the ousting of hundreds of thousands of people. The fact that ousted people should be resettled by the state has not been a real issue over the past few decades. However, behind what may look like a consensual position on the basic moral and legal obligation of the state, there are a host of mostly unresolved problems. These include a variety of issues ranging from the definition of who is an oustee to the type of resettlement and rehabilitation package.

The first issue, which is still the object of debate, is the most basic one, the number of people who are displaced by the dam. When the NWDT Award was given, only estimates of the number of oustees were available. It was estimated at that time that only 6,147 families were to be displaced.[93] This is, therefore, the basis on which the economic viability of the project was determined. In the mid-1980s, the Department of Environment and Forests put the total at 66,675 people (or about 13,335 families).[94] By the early 1990s, the Five Member Group gave a figure of 40,245 families affected by the SSP.[95] In 2000, the Supreme Court acknowledged a figure of 40,827.[96] Today, estimates range from about 40,000 (about 200,000 people) to about 100,000 families displaced (about 500,000 people).[97]

---

93 NWDT Award, Clause IV(1), reproduced below at page 47.

94 Department of Environment & Forests, Environmental Aspects of Narmada Sagar and Sardar Sarovar Multi-purpose Projects (1986), available at http://www.ielrc.org/content/c8601.pdf.

95 Five Member Group, above note 20 at paragraph 5.3.

96 Majority judgment, above note 49 at § 83.

97 Scudder, above note 27 at 3.

Different reasons explain the differences in human count. Firstly, earlier estimates were just that, estimates that were not backed by actual door-to-door surveys.[98] Secondly, not all people ousted by the SSP are actually recognised as oustees. Thus, people ousted by the construction of the canal network do not fall under the protection offered by the NWDT Award.[99] Yet, this is a very substantial number of people, with estimates ranging from 140,000 to 170,000 farmers affected.[100] Thirdly, there are debates over the definition of oustees since it is only people who can show that they own land who get the full benefit of the resettlement and rehabilitation package. This has proved immensely problematic in the adivasi areas where people have technically been encroachers for generations. Further, this excludes all families whose main livelihood is not agriculture and therefore do not occupy land, such as fisherpeople, river bed farmers or boat people.[101] Fourthly, the NWDT accepted that every major son should be treated as a separate family. This was earlier contested by MP and Maharashtra.[102] More recently, the main issue has been the time lag between surveys and resettlement, which led to a situation where many people who are today major sons may not have been adults when the surveys were first undertaken. Fifthly, the fate of people whose livelihood depends on the water of the Narmada downstream of the dam has not been effectively addressed. This includes, for instance, an estimated 5,000 families of the fishing community who are likely to be deprived of their means of livelihood once the project is completed.[103]

It is not only the huge number of people displaced that attracts attention. Underlying concepts and principles concerning displacement also warrant analysis. As such, the very fact of displacing people in the name of economic and social development benefits may be an unacceptable proposition unless very strict safeguards are introduced. Thus, as acknowledged even in the early Khosla Report, oustees should at the very least be provided with lands falling under the command area of the project that is displacing them so that they actually benefit from the project that is severely affecting their lives.[104]

---

98 *See also* Morse & Berger, above note 31 at xv indicating that even in 1985, 'when the credit and loan agreements were signed, no basis for designing, implementing, and assessing resettlement and rehabilitation was in place. The numbers of people to be affected were not known; the range of likely impacts had never been considered; the canal had been overlooked'.

99 *See, e.g.*, Majority judgment, above note 49 at § 196.

100 *See* Tata Institute of Social Sciences, The Sardar Sarovar Project: Experiences with Resettlement and Rehabilitation' (Mumbai: TISS, 1993) and Dwivedi, above note 16 at 131.

101 Sangvai, above note 66 at 125.

102 Morse & Berger, above note 31 at xvii.

103 Dwivedi, above note 16 at 132.

104 Khosla Report, reproduced below at page 41 at 4.

The NWDT Award offers a resettlement and rehabilitation package, which is more progressive than what would have been available under the Land Acquisition Act, 1894. Yet, this does not mean that it is an appropriate package that meets the needs of all people affected by the project. In fact, this is not something that could have been expected to come out of the NWDT process since it was nothing more than an inter-state procedure, which did not seek the participation of affected people.[105] Regardless of the merits and shortcomings of the NWDT Award, the real question today is whether the concerned authorities will comply with the clear terms of the Award. Compliance with the terms concerning the time frame for rehabilitation before submergence is one critical issue. Following the Award, the Supreme Court affirmed in 1992 that rehabilitation should be complete six months before actual submergence.[106] However, the orders of the Supreme Court of April and May 2006 seem to imply that this condition has all but been cast away, insofar as the Court allowed the construction to proceed despite affidavits and the report of the Group of Ministers showing that resettlement and rehabilitation was not complete.[107]

Compliance with the terms of the NWDT Award has been the object of significant debates in recent years with the introduction of the so-called special rehabilitation package. Under the gloss of the word 'special', what the Government of MP has tried to do since 2001 is to subvert the terms of the NWDT that provide land for land compensation.[108] The special rehabilitation package can be read as a desperate attempt by a government that has not been able to find land to resettle oustees to show at least minimal compliance with the terms of the NWDT Award. As proposed since 2001, the state acknowledges that it is unable to find enough land to resettle all oustees and that compliance with the NWDT Award would considerably slow down the project. As a result, it proposes to withdraw from the provision of land and give oustees money with which they can purchase land. The special rehabilitation package is in direct contravention to the NWDT Award directing land for land compensation. It also goes against the 2000 judgment that specifically commended the NWDT Award for going beyond the limited compensation scheme of the Land Acquisition Act.[109] In fact, even the Group of

---

105 *Cf.* Morse & Berger, above note 31 at xv.

106 *B.D. Sharma v. Union of India*, Writ Petition (Civil) No. 1201 of 1990, Order of 9 August 1991, reproduced below at page 265.

107 *See, e.g.*, Affidavit of the State of MP, reproduced below at page 130 and Note of the Group of Ministers, reproduced below at page 314.

108 *See* Narmada Valley Development Authority, Proposal for Amending the Terms of Resettlement and Rehabilitation under NWDT, reproduced below at page 299, Narmada Control Authority, Consideration of Proposal for Special Rehabilitation Package, reproduced below at page 301 and Narmada Valley Development Department, Terms of Special Rehabilitation Package, reproduced below at page 302.

109 Majority judgment, above note 49 at § 179.

Ministers called for its cessation.[110] Nevertheless, the Supreme Court has failed to challenge its validity in recent orders. As a result, the Government of MP has made wide-ranging use of the special rehabilitation package in recent times.[111]

On the whole, the story of resettlement and rehabilitation in the context of the SSP is a failure from the perspective of the conditions that were laid down in the NWDT Award and elsewhere. While resettlement and rehabilitation in each of the three states where submergence is taking place has proved problematic, the situation in MP is the most problematic. This has been acknowledged in various documents. Justice Kirpal noted, for instance, that in the State of MP there 'seems to be no hurry in taking steps to effectively rehabilitate Madhya Pradesh PAFs in their home State'.[112] In fact, it has been directly or indirectly acknowledged at various points since 2000 that the Government of MP neither has enough cultivable land to resettle all oustees nor finds itself capable of purchasing land.[113] The Shunglu Committee referred to this problem and the report specifically stated that:

> the quality of land available in the Land Bank was by and large average; it was not irrigable and cultivable. Considerable efforts would be needed to bring it to standard of cultivable and irrigated land.[114]

It also indicated that the 'offer of government land in place of land acquired did not meet the requirements of PAFs'.[115]

A related issue, which has been problematic for many years, is the attempt to distinguish between temporarily affected (during the period of the monsoon only) and permanently affected oustees. This artificial distinction has been made to try to delay the point at which oustees need to be resettled. This has been condemned for many years.[116] In 2005, the Supreme Court clearly confirmed that no distinction

---

110 Note of the Group of Ministers, General observation No. 4 reproduced below at page 314.

111 Thus, according to one monitoring report given in 2006, out of the 4,262 oustees entitled to land compensation covered in that report, 3,834 oustees (or 90 per cent of this group) have opted for the special rehabilitation package and 428 oustees have been allotted government land. *See* Narmada Control Authority, Resettlement and Rehabilitation Sub-group, Review of Status of R&R at Dam Height EL 121.92 m of the Sardar Sarovar Project (SSP), reproduced below at page 304.

112 Majority judgment, above note 49 at § 247.

113 *See, e.g.*, Narmada Valley Development Authority, above note 108 at § 2.

114 Report of the Sardar Sarovar Project Relief and Rehabilitation Oversight Group on the Status of Rehabilitation of Project Affected Families in Madhya Pradesh (Shri V.K. Shunglu Chairman) (2006) at § 2.19, available at http://www.ielrc.org/content/c0608.pdf, partly reproduced below at page 325.

115 *Id.* § 4.7.

116 *See, e.g.*, Report of the Fact Finding Team that Visited Narmada Valley and Relocation Sites in July 1994 (Usha Ramanathan, Rita Manchanda and Krishan Mahajan), available at http://www.ielrc.org/content/c9401.pdf.

should be made.[117] Yet, this does not seem to have been accepted by all concerned parties.

The failure of resettlement and rehabilitation also goes beyond the terms of the NWDT Award. This is the case, for instance, with regard to information and participation, as in situations where oustees were never informed about displacement until surveyors placing stone markers to indicate submergence level came to their village.[118]

More generally, it is noteworthy that strong criticism of the process of resettlement and rehabilitation in the SSP has come from a variety of agencies and quarters, including official sources and committees appointed by the government as well as a number of independent reports, commissions and committees. The overall picture, which emerges, is that apart from concerned authorities themselves asserting that they are complying with all their commitments, a majority of other sources, in addition to the NBA, contradict the official view.

Another issue that has attracted significant attention is the quality of the resettlement and rehabilitation package offered to oustees. Again, while concerned authorities tend to claim that all is well on this front, various official and independent reports have found major shortcomings with regard to proposed and existing resettlement sites. With regard to resettlement sites in MP, for instance, the Group of Ministers noted that:

> Dharampuri had been shown to the GoMs as a success story by the Madhya Pradesh Government and it turned out to be the worst example of not doing anything by way of settlement when there was apparently no difficulty in respect of resources. The people there showed to the GoMs two dry water pumps and a heap of stones that had been dumped there a day before the GoMs' arrival indicating that roads would be built soon.[119]

An independent team led by Professor Chenoy that visited several resettlement sites later in 2006 gives a similar picture.[120] Earlier, the Daud Committee that surveyed rehabilitation sites in Maharashtra found that:

> [e]xcepting the resettlement village of Simamli in Gujarat, which offers a little satisfaction, rest of the resettlement villages from Maharashtra in particular, visited by us, lack almost all the basic facilities required for habitation, specially quality and availability of suitable agricultural land. One cannot ignore the enormous number of complaints that the Committee came to hear from the aggrieved people about having been shifted to the new sites without being

---

117 Judgment of 15 March 2005, reproduced below at page 277 at § 58.

118 Baviskar, above note 67 at 201.

119 Note of the Group of Ministers, Section on Dharampuri, reproduced below at page 315.

120 Chenoy, K.M et al., *Narmada Rehabilitation – Fact and Fiction* (New Delhi: Indian Social Institute, 2006).

> provided with compensatory agricultural land. One of the greatest shortcomings is that of non-availability of water even for domestic purposes like cooking and drinking. Even in Simamli, it is not as if everything is as it should be.[121]

Resettlement and rehabilitation also needs to be understood in a broader context. From a human rights perspective, displacement is a 'problem' and everything needs to be done to ensure that oustees' lives are affected as little as possible in a situation where there is a major upheaval in their lives. As noted above, even the 1965 Khosla Committee report noted that displacement was a 'major human problem'. Today, it is accepted that displacement is a human right issue.[122] However, this does not seem to have been accepted in the context of the SSP. The Secretary, Ministry of Water Resources thus asserted in 1992 that:

> [t]he basic right of human individuals is the one of development and that is what is being ensured through different development programmes including the water resources development programmes such as the Narmada Development Projects. There is, therefore, *no question of any violation of the human rights* when such projects are undertaken or implemented (emphasis added).[123]

In 2000, Justice Kirpal confirmed in his judgment that:

> the displacement of the tribals and other persons would not per se result in the violation of their fundamental or other rights. The effect is to see that their rehabilitation at new locations they are better off than what they were. At the rehabilitation sites they will have more and better amenities than which they enjoyed in their tribal hamlets. The gradual assimilation in the mainstream of the society will lead to betterment and progress.[124]

The implication is that displacement contributes to the realisation of oustees' human rights rather than causing a major upheaval. While a discussion of this position is beyond the scope of this chapter, it indicates that even in the twenty-first century there is in fact no consensus on what constitutes a human right and its realisation. While oustees' lives may in fact be improved from a socio-economic point of view *if* they are resettled in such a way that their overall lifestyle is improved, this does not lessen the trauma of displacement and should in fact be seen as the 'price' that the agency proposing displacement should be made to pay in return for deriving benefits from a project rather than as a 'benefit' for oustees. Further, it is inappropriate to talk about displacement as a favour done to adivasis who can henceforth benefit from new government services such as health and

---

121 Justice S.M. Daud Report, partly reproduced below at page 377.

122 *See, e.g.*, Guiding Principles on Internal Displacement, *in* Report of the Representative of the Secretary General, Mr. Francis M. Deng, submitted pursuant to Commission resolution 1997/39, UN Doc. E/CN.4/1998/53/Add.2 (1998).

123 Letter from Mr Chitale, Secretary, Ministry of Water Resources to Mr Vergin, World Bank concerning the Morse Report, 1992, partly reproduced below at page 355 at § 4.

124 Majority judgment, above note 49 at § 91.

education.[125] All these services are in fact due to all citizens and should have been available to oustees earlier. It is therefore impossible to justify displacement in this way.

2. *Environmental aspects*

Environmental aspects of the SSP have also been controversial. The history of compliance with environmental conditions laid down by various agencies has been a story of controversies, which have in many cases not been resolved to this day.

In the context of the World Bank loan appraisal, limited consideration was given to environmental issues.[126] Even though environmental conditions were much weaker two decades ago than they would be today,[127] the Morse Report still found that '[m]ost of the Bank's 1985 legal requirements for the environment have not been met'.[128]

At the domestic level, one of the issues that has dominated the environmental agenda in the SSP is that of environmental impact assessment which has been recurrently brought back to the table. In 1983, clearance was rejected.[129] In the following years, the Ministry of Environment and Forests (MoEF) attempted to maintain pressure on project proponents to comply with environmental conditions and remained unwilling to grant clearance.[130] Eventually, in 1987 it agreed to give a 'conditional' clearance.[131] Since this clearance was only conditional, its validity was conditional on the intervention of the Prime Minister who is deemed to have sanctioned the project.

This is analysed in detail in the majority and minority judgments of the Supreme Court in 2000. Justice Kirpal speaking for the majority thought that the Prime Minister's note constituted an endorsement of the environmental clearance. On the contrary, Justice Bharucha writing the minority judgment thought that the decision should have been left to the MoEF. Further, his understanding based on the documents of the relevant ministries was that 'the requisite data for assessment of

---

125 *See* Tata Institute of Social Sciences, above note 100 at 24.

126 Morse & Berger, above note 31 at 223.

127 *See* Section 9(b), World Bank, Operational Manual Statement: Environmental Aspects of Bank Work, OMS No. 2.36, May 1984.

128 Morse & Berger, above note 31 at 233.

129 *Id.* at 222.

130 *See, e.g.*, Department of Environment & Forests, above note 94.

131 Ministry of Environment and Forests Environmental Clearance, reproduced below at page 77.

the environmental impact of the project was not available when the environmental clearance thereof was granted'.[132]

In other words, without going beyond the analysis undertaken by Supreme Court judges, it becomes apparent that the actual clearance of the SSP from an environmental point of view is in question. This is in fact an issue which was taken up by MPs.[133] It should also be noted in this context that one of the reasons for the confusion surrounding the clearance can be partly traced back to the fact that the project was sanctioned for funding by the World Bank before environmental clearance was considered at the national level. This eventually put pressure on the MoEF to tone down its stand on the project.[134]

The question of environmental clearance becomes even more intricate once it is considered that the 1994 Environmental Impact Assessment Notification provided that any clearance was valid only for five years.[135] While the SSP was 'cleared' before the coming into force of the Notification, the fact that it has been an ongoing project for so many years at the very least raises the question of the need for a fresh clearance once the Notification was issued. The question of the lapse of any clearance given in the 1980s was in fact taken up by the NCA Environment Sub-group, which opined already in 1990 that the conditional clearance should be deemed to have lapsed.[136] In this context, recent changes to the environmental impact assessment framework are striking. A new Notification that supersedes the 1994 document was adopted in 2006.[137] It generally weakens the framework adopted a decade earlier. With regard to the validity of clearances granted for river valley projects, it not only lengthens the period of validity to ten years but also provides that it can be extended by another five years.[138] Regardless of the problematic aspects of this Notification that weakens existing environmental protection standards, this would not shield the SSP from the necessity to apply for a fresh clearance.

One of the critical conditions that was put down in the 1987 conditional clearance is the condition that remedial measures should be undertaken *pari passu*, or in

---

132 *Narmada Bachao Andolan v. Union of India*, Writ Petition (Civil) No. 319 of 1994, Supreme Court of India, Judgment of 18 October 2000 (Justice Bharucha), reproduced below at page 228 [hereafter minority judgment].

133 *See* Parliamentary Questions on Environmental Impact Assessment, 1993, reproduced below at page 80.

134 *See, e.g.*, Dwivedi, above note 16 at 135 and Sangvai, above note 66 at 24.

135 Section III(c), Environmental Impact Assessment Notification, 1994.

136 Morse & Berger, above note 31 at 230.

137 Government of India, The Environmental Impact Assessment Notification, 14 September 2006, Gazette of India, Extraordinary, Part-II, and Section 3, Sub-section (ii) Ministry of Environment and Forests.

138 *Id.* at section 9.

other words at the same pace as construction work. Over time, this condition has become one of the most controversial issues of the legal framework covering the SSP.

The *pari passu* condition has been at the centre of debates at various stages of the project. Thus, when it was proposed to close the sluice gates of the dam in 1994, the NCA Environment Sub-group recommended that works should be halted because it was concerned over the slow pace of resettlement and unfinished environmental studies.[139] Another type of concern emerged in the context of the development of the Command Area Development Action Plan. It was drafted in 2003, years after it should have been completed, therefore again raising concerns about the *pari passu* condition.

3. *The pari passu condition*

Pari passu construction implies that all mitigating work that needs to be done with regard to the construction up to the next level has to be undertaken before construction of that phase is allowed. In other words, *pari passu* construction implies that the pace at which the dam is built cannot outpace rehabilitation work.

The introduction of the concept of *pari passu* in the mid-1980s was the result of debates between the water and environment ministries. While the latter thought that the project was not fit for environmental clearance, the former were keen to implement it. The *pari passu* condition was meant to allow construction work to proceed soon while not affecting the implementation of environmental mitigation measures.

Even assuming that the principle of *pari passu* was implemented in letter and spirit, it is unacceptable from an environmental policy and planning point of view.[140] The very idea of undertaking environmental impact assessments of a proposed project is undermined by the *pari passu* principle. Indeed, internationally agreed standards for environmental impact assessment provide that it is a proposed activity, which is assessed, not an activity that is already being implemented.[141]

Regardless of theoretical controversies over the principle of *pari passu*, in the initial years after the conditional environmental clearance was given, the content of the *pari passu* condition was rapidly eroded to the point where it was the pace of

---

139 Patrick McCully (compiling), Sardar Sarovar Project (SSP), An Overview (Berkeley, CA: International Rivers Network, 1994).

140 *Cf.* Morse & Berger, above note 31 at xxi.

141 *See* Article 2(3), Convention on Environmental Impact Assessment in a Transboundary Context, Espoo, 25 February 1991, 30 *International Legal Materials* 800 (1991).

construction that determined the pace of remedial work rather than the contrary.[142] Thus, in 1993, it was suggested, in an attempt to ensure the early closure of the sluice gates, that a 'temporary waiver' of the *pari passu* could be sought from the MoEF.[143]

The concept of *pari passu* was again given a prominent role in the context of the Supreme Court judgment in 2000 which provided that dam construction should go on as fast as possible but that this should be done in stages to ensure that remedial work was effectively completed before construction of the next stage. In other words, while the majority judgment sought the completion of the project at the earliest, the *pari passu* condition was the one compromise that was made in favour of the oustees and the environment.

This remains the main leverage that individuals and groups seeking effective remedial measures can use to try to ensure that the project authorities abide by the conditions that have been put forth. Ineffective implementation of the *pari passu* conditions posed in the 2000 judgment led to the filing of numerous complaints at the level of the Grievance Redressal Authority, as well as to the filing of new petitions in the Supreme Court. One instance of the complete failure to apply the principle of *pari passu* was noted by the Court in the context of Picchodi and Jalsindhi villages.[144] This concerned the fact that applicants were not only clearly project affected families (PAFs) but also happened to be PAFs at the height of 95 and 100 metres when the dam stood at 110 metres. It is this complete failure to implement the *pari passu* principle that the Supreme Court condemned in 2005.[145]

However, while the 2005 judgment seemed to uphold the notion of *pari passu* proposed five years earlier, the principle has been further watered down in the meantime. Events in 2006 seem to indicate that for all practical purposes *pari passu* now means that remedial work is to follow construction work as well as possible, but construction work is clearly given priority. This stands out, for instance, in the decision of the Resettlement and Rehabilitation Sub-group of March 2006 that grants clearance while specifically noting the findings of the GRA that rehabilitation work is expected to be completed by June 2006, or more or less when the dam was expected to reach 121.92 metres, in complete contravention of the terms of the NWDT Award.[146] The orders of the Supreme Court in the spring

---

142 *See, e.g.*, Ramaswamy R. Iyer, 'Narmada: The Cost of Delaying Rehabilitation', *Hindu*, 1 May 2006, page 11. *See also,* Morse & Berger, above note 31 at 230.

143 NCA, Minutes of the 48th Meeting, New Delhi 31 December 1993, Item XLVIII-2 (252), pages 5-9, Closure of Construction Sluices at EL +18 m of SSP.

144 Judgment of 15 March 2005, § 21, reproduced below at page 277.

145 *Id.*

146 Narmada Control Authority, Resettlement and Rehabilitation Sub-group, Minutes of the 63rd Meeting, Item No. LXIII-2 (345), 8 March 2006, reproduced below at page 304.

of 2006 where construction was not stopped despite the evidence presented, for instance, by the Group of Ministers that resettlement and rehabilitation at 121.92 metres was not complete confirm this stand.

E. Sardar Sarovar and Beyond

The Sardar Sarovar Project is not only a project of superlatives, it is also a project that will leave its imprint in development debates, whether expected benefits are eventually delivered or not. Firstly, debates around the SSP have made an important contribution to rethinking the development contribution that can be expected from dams. Dams have always been premised on their contribution to economic development, whether in the form of irrigation water or power benefits. Given the focus on the economic development contribution of dams, any remedial measure required by the construction of a dam such as resettlement and rehabilitation of oustees and environmental remedial measures used to be seen largely as costs. The fact that oustees and their resettlement was seen as a 'cost' in the search for greater economic benefits explains why it has been difficult to move towards a position where displacement is seen not as a cost but as an important human right problem. From this perspective, one major contribution of the debates around the SSP and other dams has been to ensure that it is today very difficult to consider oustees only as 'costs' to be tabulated in the search for greater economic good. Thus, the whole 2000 judgment is, on the surface at least, concerned with the plight of oustees.[147] Further, in another judgment concerning another big dam, the Tehri dam in Uttarakhand, the judges specifically indicated that the overall project benefits from the dam cannot be used as an excuse to deprive oustees from their fundamental rights. Rehabilitation is a corollary of the rights recognised at Article 21 (right to life) of the Constitution and includes not only the provision of food, clothes and shelter but also support to rebuild lives.[148] This changed paradigm neither means that oustees' rights are always enforced, as witnessed in the case of the SSP, nor that the state is ready to recognise their rights without struggle. Recent developments in the case of the SSP tend to show that the 'cost' mentality still seems to predominate and the state is ready to shirk its responsibilities to oustees to ensure successful implementation of the physical works.

Secondly, the debate over big dams remains largely mired in a dichotomy that conceives big dams as modern and therefore in principle superior to other solutions that could bring similar benefits in terms of economic development. As a result, while the SSP controversy at the national and international levels led to what can historically be seen as a policy breather in the mid-1990s, this period is now over.

---

147 Majority judgment, above note 49 at § 80.

148 *N.D. Jayal v. Union of India*, Judgment of 1 September 2003, 2003(7) *SCALE* 54 at § 60.

At the international level, the lessons learnt from the SSP in terms of resettlement and rehabilitation or environmental remedial measures have hopefully been learnt and not forgotten. In any case, what is apparent is that in the long run, the SSP crisis will only have been a temporary crisis of confidence in big dams. In the twenty-first century, the threat of global warming that has become one of the most politically sensitive environmental problems – while not the most severe environmental issue for most developing countries at present – has provided a new rationale for dam building since dams provide greenhouse gas-free power. Big dams are thus being proposed as modern solutions to a question of economic development (power generation) and an environmental problem (climate change). In other words, the big dams of tomorrow will be premised on being not only good for development but also good for the environment. This obviously obfuscates the fact that the negative environmental side effects of dams will be no less important whether they can be justified as carbon-neutral energy-producing projects or not.

Thirdly, big dams have been projected as symbols of development for decades. The symbolism can be used for various purposes. As noted in the case of the SSP, big dams are first premised on their potential for fostering irrigation and therefore contributing to the food security of the country. Over time and in view of the increasing problems faced by people in getting access to water for domestic use, the latter has also been added as a basic justification for dam building. Further, big dams are seen as potent signs of the overall economic development of a country. In other words, dams can also be used politically as rallying points for strengthening the idea of nationhood, as indicated in the case of the SSP where one of the justifications put forward from the time of the Khosla report was the introduction of irrigation in zones bordering Pakistan to ensure a permanent population presence in those areas.

Fourthly, the SSP represents one specific type of development endeavour, which is overwhelmingly driven by the state. This is visible not only in the fact that the state has initiated and implemented the project but also in the heavy use of the state apparatus. Thus, a whole administrative machinery was set up from 1979 onwards to plan, oversee and implement the project, starting with the Narmada Control Authority (NCA) established to provide the overall coordination and direction of the implementation of the SSP.[149] The NCA is supplemented by a number of other bodies, including the important sub-groups on environment and resettlement and rehabilitation. Further, a review committee composed of union ministers and the chief ministers of the relevant states was also set up with the power to review any decision of the NCA. While this review committee is meant to take decisions that are final and binding on all states, the situation since the 2000 judgment is that the

149 Constitution of the Narmada Control Authority, reproduced below at page 82 at § 9.

final authority on the SSP is in fact the Prime Minister rather than the review committee.[150]

1. Legal aspects

Legal issues have played an important role in the evolution and implementation of the SSP. This is due in part to the specific conditions under which the project was taken up. The fact that MP and Maharashtra saw their land submerged while the water benefits of the SSP went to Gujarat implied from the start that such an inter-state project might be mired in disagreements among the concerned states. Further, the planning of the project without participation or information to the oustees provided the basis for future conflicts over their rights. Several points can be noted.

Firstly, over the course of the nearly 60 years that have elapsed since a proposal for a dam was first mooted, the law that regulates remedial measures concerning dam and canal building has dramatically evolved. This is true at the national and international levels. From the point of view of environmental remedial measures, a clear and rapid evolution can be seen insofar as most of what is today's national and international environmental law has evolved over the project lifetime. As a result, there are a host of norms, which did not exist at certain earlier points in the project planning and implementation that have become central parts of environmental law. As noted above, in the case of environmental impact assessment, the norms that have been adopted imply that a project like the SSP should be subjected to further assessment, even if the procedure followed at the outset had been uncontroversially comprehensive.

From the point of view of resettlement and rehabilitation, the situation is much more complex. In principle, the main legislative framework is the Land Acquisition Act, 1894. This is supplemented in the case of the SSP by the NWDT Award, which provides a resettlement and rehabilitation package that goes beyond the colonial act. While the NWDT Award has been repeatedly deemed to be unchangeable, it needs to be read alongside the string of Supreme Court judgments and orders that have eventually come to condone the weakening of resettlement and rehabilitation provisions. This is, for instance, visible in the implied acceptance of the special rehabilitation package.[151]

Besides legal provisions that are binding in the case of the SSP, efforts to adopt a national rehabilitation policy are also noteworthy. While such a policy would have no direct implication for the SSP, it would likely influence the implementation of the conditions posed in the NWDT Award. After nearly two decades of debates, a

---

150 Majority judgment, above note 49 at § 280.

151 On the special rehabilitation package, see above at page 22.

policy was eventually put forward in 2004.[152] This policy has bee criticised for not going far enough in strengthening the protection oustees. As a result, a more progressive rehabilitation policy was drafted by the National Advisory Council (NAC) in 2005.[153] Yet, a new draft policy suggested by the Government in 2006 seems to again backtrack on the NAC proposal.[154] The overall result as far as the SSP is concerned is that beyond the terms of the NWDT Award, there does not seem to be a strong consensus yet to strengthen the rights of oustees in development projects. This may in part explain, but not justify, the relative ease with which the terms of NWDT Award are being weakened.

Secondly, courts and tribunals have played a central role in the overall debate concerning the SSP. The inter-state nature of the SSP paved the way for the setting up of the NWDT under the Inter-State River Water Disputes Act, 1956. The NWDT Award has played a central role in the shaping of the project. Further, various organisations and movements have used the courts in an attempt to achieve goals that could not be reached through political means. In particular, since 1994 the NBA has used the courts on repeated occasions to try and obtain what could not be obtained directly from the government. As noted above, apart from the petition of 1994 leading to the main SSP judgment of 2000, the NBA has repeatedly gone back to courts to try and ensure fairer outcomes for oustees. This includes various strategies from seeking a review of the 2000 judgment to trying to ensure compliance with the conditions posed therein.[155] The NBA has by no means been the only organisation or movement seeking to use the courts to achieve its aims. Thus, in Gujarat, the courts have been approached on various occasions. In some cases, courts did attempt to provide relief to oustees as in the case of the High Court Order of 25 February 1994 that enjoined the authorities not to take any further action towards the closure of the sluice gates.[156] More recently, the High Court was approached concerning the allocation of water for the Kachchh

---

152 National Policy on Resettlement and Rehabilitation for Project Affected Families, *Gazette of India*, Extraordinary Part-I, Section 1, No- 46, 17th February 2004.

153 Draft National Development, Displacement and Rehabilitation Policy, May 2005, available at http://nac.nic.in/concept%20papers/rehabilitationmay2005.pdf. The National Advisory Council was set up by the government in 2004 to oversee the implementation of its National Common Minimum Programme, a set of governance principles adopted by coalition partners. *See* Order constituting a National Advisory Council, Government of India, Cabinet Secretariat, 31 May 2004, No, 631/2/1/2004-Cab.

154 *See* Comparative Statement of National Policy for Resettlement and Rehabilitation of Project Affected Families (NPRR–2003) and National Rehabilitation Policy (NRP-2006), available at http://www.dolr.nic.in/NRP2006-draft.pdf.

155 *See, e.g.*, Review Petition and Review Order, reproduced below respectively at page 244 and page 262.

156 *Lok Adhikar Sangh v. State of Gujarat*, Civil Application 522 of 1994 in Special Civil Application No. 4285 of 1991, Order of 25 February 1994.

district.[157] Further, courts have also been used by groups that seek to ensure the speedy completion of the project. Thus, as noted above the Narmada Abhiyan challenged the setting up of the Five Member Group in an attempt to weaken it.[158] Courts have also been a forum used to try and stifle dissent. To date, the most debated case has involved Ms Arundhati Roy who was accused of having 'led a huge crowd and held a dharna in front of this court and shouted abusive slogans against the court including slogans ascribing lack of integrity and dishonesty to this institution' in the context of the aftermath of the 2000 Supreme Court judgment.[159] The 2002 judgment in these contempt proceedings found that Arundhati Roy had 'committed the criminal contempt of this court by scandalising its authority with malafide intentions'. It went on to find that:

> [a]s the respondent has not shown any repentance or regret or remorse, no lenient view should be taken in the matter. However, showing the magnanimity of law by keeping in mind that the respondent is a woman, and hoping that better sense and wisdom shall dawn upon the respondent in the future to serve the cause of art and literature by her creative skill and imagination, we feel that the ends of justice would be met if she is sentenced to symbolic imprisonment besides paying a fine of Rs. 2,000.[160]

The judgment was widely condemned, not only because it sought to stifle dissent in the specific case of the SSP but also because the law of contempt is not uniformly applied and because the notion that citizens cannot criticise judges in itself is extremely outdated.[161] This is not the only case where attempts have been made to use the courts to stifle dissent. On at least two occasions, the latter in mid-2006, the NBA was faced with allegations concerning the alleged foreign source of its funding and the seditious nature of its activities.[162] While such allegations led to an apology in 1990 and will probably fizzle away as well in the latest case, they conform to the pattern of Strategic Litigation Against Public Participation (SLAPP) which have also been used in other cases, for instance, in the context of the Maheshwar dam, another big dam on the Narmada river.[163]

---

157 *Kachchh Jal Sankat Nivaran Samiti v. State of Gujarat*, reproduced below at page 105.

158 *See* above at page 10.

159 Re: Arundhati Roy (Contemner), Judgment of 6 March 2002, Contempt Petition (Crl.) 10 of 2001, available at www.ielrc.org/content/c0201.pdf.

160 *Id.* at 14.

161 *See, e.g.*, Rajeev Dhavan, 'Arundhati Roy's Contempt', *The Hindu* 5 April 2002.

162 The National Council for Civil Liberties is the same organisation that was attacking the NBA on the same grounds after the 2000 Supreme Court judgment. *See* Ashish Kothari, 'Against a People's Movement', 18/15 *Frontline* 3 August 2001.

163 *See, e.g.*, *Shree Maheshwar Hydel Power Corporation Ltd v. Chittaroopa Palit*, Bombay City Civil Court, Order of 23 October 2001. *See also* Rajeev Dhavan, 'Corporates versus the People', *The Hindu* 16 May 2003.

Thirdly, the role that courts have played in the context of the SSP is o relevance for the development of public interest litigation. The reason w were approached by the NBA was linked to the perception that the Supreme Court was likely to respond favourably to their claims. This reasoning seemed appropriate in the initial years. However, in the 2000 judgment, the court made strong strictures against the NBA, which have broader implications for public interest litigation. In the course of the discussion of the NBA's petition, the judges repeatedly stressed that the NBA should have approached the Court much earlier since it was in effect challenging a clearance obtained seven years earlier. Thus, after the 2000 judgment:

> [w]hen such projects are undertaken and hundreds of crores of public money is spent, individual or organisations in the garb of PIL cannot be permitted to challenge the policy decision taken after a lapse of time. It is against the national interest and contrary to the established principles of law that decisions to undertake developmental projects are permitted to be challenged after a number of years during which period public money has been spent in the execution of the project.[164]

In the specific SSP case, the principle of laches was used by judges to justify their refusal to look into issues raised in the petition relating to the height of the dam and the extent of submergence, environment studies and clearance, hydrology, seismicity and other issues.[165] As a result, the judges asserted that they could only look into the issue of resettlement and rehabilitation. Interestingly, the judges still decided to address other issues even though they determined that it is 'not necessary to do so'.[166]

The same type of argument was subsequently used in another dam-related case. In the 2003 Tehri dam judgment, the judges again determined that they could not entertain a petition asking them to think about the decision to construct the project when implementation was nearly complete.[167] The issue of laches raises broader questions. While courts have not hesitated to take a strong stance in projects or issues of relatively minor importance, they have shied away from taking the same type of decisions in major projects. Laches thus becomes an excuse used by courts not to address certain issues in the case of major development projects. The NBA was told in clear terms that its pleas would not be entertained because it had come to the Court too late for the Court to review the project itself. The Court thus emphasised the finality and binding nature of the NWDT Award.[168] Yet, as mentioned above, with regard to the introduction of the special rehabilitation

164 Majority judgment, above note 49 at § 78.

165 On laches, *see* majority judgment, above note 49 at § 76ff.

166 Majority judgment, above note 49 at § 80.

167 *N.D. Jayal v. Union of India*, Judgment of 1 September 2003, 2003(7) *SCALE* 54 at § 12.

168 Majority judgment, above note 49 at § 204.

package, which goes against the terms of the Award, the same Court does not seem to think that this is cause for major concern. Further, while some level of certainty is indeed required in planning small and large projects, the amount of money spent on an ongoing project or the length of time that has elapsed since the initiation of a project can never be used as an excuse for not reviewing it. What matters at the end of the day is not whether the dam is built but whether all the conditions posed have been fulfilled. In practice, arguing that the amount of money already spent on the project precludes further debate implies that the fate of oustees is made subsidiary to the completion of the construction work. This is not acceptable from either a human rights or an environmental perspective. This has been a recurrent issue since the Department of Environment and Forests already argued in the mid-1980s that the amount of money invested until then on the SSP should not preclude project planners from modifying some parameters to minimise environmental damage while at the same time ensuring optimal utilisation of water resources.[169] In fact, leniency towards organisations or individuals seeking to ensure compliance with environmental or human rights conditions should be granted in a case like the SSP where even in 1992 it was acknowledged by a government official that benefits to the water-scarce areas of Gujarat may only be felt by 2025.[170]

Fourthly, international law has played a subsidiary but noteworthy role in SSP debates. The World Bank loans brought to the SSP the whole set of policies that provide the framework for the implementation of bank-funded projects. These internal documents are of great relevance because they are binding for all projects and because they generally closely follow existing international norms in their respective fields. In the case of the SSP, they constituted the basis for the assessment of the independent review chaired by Bradford Morse.

International treaties signed by India have also been important signposts in the overall law-related discussions. This has in particular been the case of Convention 107 of the ILO. Convention 107 is a relatively old treaty of 1957, which seeks to protect the rights of tribal people and in particular provides land for land compensation.[171] It has been superseded by Convention 169 of 1989, which updates Convention 107 to strengthen the rights of tribal people.[172] India has only ratified the older convention. Nevertheless, compliance with the relatively non-

---

169 Department of Environment & Forests, above note 94 at conclusion § 5.

170 D'Souza, above note 7 at xiii.

171 Article 12, Convention 107 concerning the Protection and Integration of Indigenous and Other Tribal and Semi-Tribal Populations in Independent Countries, 26 June 1957, partly reproduced at page 443.

172 Convention 169 concerning Indigenous and Tribal Peoples in Independent Countries, 27 June 1989, 28 *International Legal Materials* 1382 (1989).

demanding commitments under Convention 107 has proved difficult as visible in the references made concerning the SSP at the ILO.[173]

2. *Sardar Sarovar legacy*

As of early 2007, it appears likely that the Sardar Sarovar Dam will be completed. In other words, the main component of the SSP is likely to be implemented as planned. This constitutes a major achievement from an engineering point of view. Nevertheless, in the twenty-first century, the successful conclusion of the physical works is not sufficient to declare the project a success. Indeed, the fact that from the outset measures were put in place to ensure resettlement and rehabilitation of oustees and remedial measures for negative environmental impacts implies that a comprehensive assessment of the project must take these dimensions into account. From this point of view, it is unlikely that the SSP will be remembered as a successful project. Whereas the conditions expressed in the NWDT Award of 1979 cannot be said to be extremely onerous, these conditions have not been fulfilled and may, or may not, be fulfilled in the future. The most severe failure concerns the situation of oustees that are affected by non-compliance with the terms of the Award with reference to the timeline for resettlement, by changes in the terms of the Award when they are offered the 'special rehabilitation package' and by resettlement and rehabilitation in sites which do not provide them even with the same basic amenities they had in their original villages.

The completion of the dam is one important part of the SSP. However, a comprehensive assessment of the project needs to also examine other elements such as the gigantic canal network and the massive drinking water project. It is only when these infrastructure projects are completed that it will be possible to say whether the SSP has fully delivered on its promises. Indeed, it is irrigation for water-scarce areas of Gujarat and some parts of Rajasthan as well as more recently drinking water for a large part of Gujarat which provide the underlying justification for the SSP. Knowing that water is flowing through the main canal in the already relatively well-watered areas of Southern and Central Gujarat is a positive outcome but is insufficient from the point of view of the overall rationale given to justify such a huge project.

While the fate of oustees and the actual benefits delivered by the SSP should be the primary concern of all actors involved, it is also necessary to look beyond the project itself to examine its broader impacts. From the perspective of resettlement and rehabilitation, the outcome of the SSP is likely to be both significant and minuscule. The inter-state nature of the project led – without the participation of

---

173 *See, e.g.*, the two documents reproduced below concerning SSP in the context of Convention 107 at page 444. A number of other references to SSP have been made to the Committee of Experts.

the oustees in the NWDT – to a decision, which was more progressive than the conditions prevailing at that point under the Land Acquisition Act, 1894. This is not to say that an ideal resettlement and rehabilitation law should adopt the parameters of the NWDT Award given its limitations in terms of the restricted definition of 'oustee' and the limitation of its application to dam-displaced people as opposed to SSP-wide displaced people including oustees from the canal network. Further, the NWDT Award is fundamentally vitiated because there was neither information nor participation of affected people in the planning or implementation of the project. Nevertheless, the NWDT Award provided at least the basis for land for land compensation, which was a step forward. While this was to be welcomed, it is uncertain what is left of these progressive features today. On the one hand, as noted above, within the SSP the principle of land for land compensation has for all practical purposes been abandoned since the special rehabilitation package authorises cash compensation instead. On the other hand, while efforts to develop a national policy on resettlement and rehabilitation were undertaken from the mid-1980s, the document eventually adopted in 2004 in effect does no more than promising to attempt to provide land for land where possible.[174] Even if a stronger policy modelled after the NAC proposal was adopted as official national policy, this would still fall short of being binding on the government. A resettlement and rehabilitation act, which supersedes the Land Acquisition Act, is what is required to ensure that oustees' rights effectively become rights rather than privileges granted by the state.[175]

The visibility of the SSP for the past two decades has largely been due to the presence of the NBA and other movements, organisations and individuals that have ensured, though not without much difficulty, that the SSP has remained a politically live issue. While the SSP is by far not the only project where civil society has played a big role in trying to ensure that projects are not implemented without full implementation of remedial measures, it is a special case because of the visibility that the NBA managed to give to the project at the national and international levels. The success of the NBA in making the SSP a national as well as an international issue have been key elements in the way the SSP has been implemented over the past couple of decades. The NBA has ensured through its work that generations of people in India and abroad who would not have known about the impacts of big dams earlier are today aware of the type of problems that arise. Overall, it is remarkably difficult to judge the actual impacts of the NBA and associated organisations on the SSP and big dam policy. The NBA was more than instrumental in forcing the World Bank to call for what became the Morse Report, which in turn led to the bank's withdrawal, the setting up of the Inspection Panel

---

174 National Policy on Resettlement and Rehabilitation, above note 152.

175 One proposal going in that direction is the Bill introduced by Rajya Sabha MP, E.M. Sudarsana Natchiappan, The Multi Purpose Inter-State River Projects and Setting up of Large Enterprises (Rehabilitation of Oustees) Bill, 2006.

and was instrumental in the setting up of the WCD. The NBA was also successful in forcing the government to respond to its claims, for instance, by setting up the FMG, managed to obtain a four-year pause in the dam construction when it first approached the Supreme Court and was instrumental in ensuring the setting up of a GRA in each of the three states.

However, from a longer term perspective, the NBA's achievements seem to be partly dwarfed by the fact that the dam is inexorably being built and resettlement and rehabilitation is not being achieved as per the terms of the NWDT Award. Overall, the use of existing legal mechanisms to ensure the realisation of oustees' rights has brought some relief to the oustees but not full success. When the analysis is taken one step further, it becomes apparent that the use of courts to achieve justice has been a double-edged sword in the case of the SSP. While even in the 1990s the idea of approaching the Supreme Court was not an obvious choice, it was eventually done because it seemed, on the basis of the existing case law, that relief could be obtained. The partly unforeseeable problem is that over the past 12 years, public interest litigation has suffered a series of setbacks and the wave of optimism which led people to expect courts to provide full justice in all environmental and human rights cases in the early 1990s has proved misplaced. This general trend has been more marked in the case of big development projects where courts have been much less keen to challenge the government than on smaller issues.

Among the Narmada dams built or planned, the SSP is the one that has received most attention and been subjected to most scrutiny. However, this does not imply that other projects are proceeding without controversies. In fact, significant problems with regard to resettlement and rehabilitation have been highlighted in various cases ranging from the relatively older Bargi dam to recent issues in the case of the ISP.[176] In the case of the Maheshwar dam, an additional controversy occurred because it was the first big dam project to be handed over to the private sector.[177] Apart from the Bargi dam, which was substantially completed before the NBA started large-scale work on Narmada dams, other big dams in the valley have also been the focus of close attention by the NBA. In fact, for the past many years, some NBA activists have devoted themselves entirely to issues and concerns related to dams upstream from Sardar Sarovar, such as Maheshwar, Omkareshwar

---

176 On the Bargi dam, *see, e.g.*, The Human Cost of the Bargi Dam (Jabalpur: Bargi Bandh Visthapit Avam Prabhavit Sangh, 1998). On the Indira Sagar Pariyojana, *see, e.g.*, Arundhati Roy, 'The Road To Harsud', *Outlook* 26 July 2004.

177 On the Maheshwar dam, *see, e.g.*, Richard E. Bissell, Shekhar Singh & Hermann Warth, Report of an Independent Review – Maheshwar Hydroelectric Project: Resettlement and Rehabilitation (Report conducted for the Ministry of Economic Cooperation and Development (BMZ), Government of Germany, 2000).

and ISP.[178] As a result, the success or failure of the SSP from the point of view of project construction or from the point of view of remedial measures cannot be analysed in isolation from what has happened and is happening in the case of other dams. The SSP offers many lessons and some hope for people affected by other big dams on the Narmada or elsewhere.

---

178 In terms of recent litigation outcomes, *see, e.g.*, concerning ISP, *Narmada Bachao Andolan v. Narmada Hydro-Electric Development Corporation*, High Court of Madhya Pradesh at Jabalpur, Writ petition No. 3022/2005, Order of 8 September 2006, available at http://www.ielrc.org/content/c0613.pdf.

Chapter 2

Early Documents and Decisions

Khosla Report, 1965

Source: Report of the Narmada Water Resources Development Committee, Government of India, Ministry of Irrigation and Power (Khosla Report), 1 September 1965 (extracts).

Introduction

Past Investigations

The idea of the integrated development of the Narmada basin for irrigation, power, flood control, navigation, fisheries, etc. was first mooted in 1945–46 by Shri A.N. Khosla, Chairman, Central Waterways, Irrigation and Navigation Commission. In 1946, at the request of the States concerned, a study of topography and preliminary reconnaissance was undertaken and seven sites were considered suitable for detailed investigation and preparation of projects.

In 1948, an ad hoc Committee under the Chairmanship of Shri A.N. Khosla, was appointed by the Government of India to scrutinise the estimates prepared for the investigation of the above seven projects, and to recommend priorities.

This Committee, after studies and discussions with officers concerned, recommended that detailed investigation should be done for Bargi, Punasa and Tawa projects in Madhya Pradesh and Broach Barrage and Canal project in the former Bombay State.

This work was taken up by the Central Waterways, Irrigation and Navigation Commission (now the Central Water & Power Commission) and project reports for Punasa, Tawa and Broach projects were prepared. Work on the Bargi had to be suspended for want of funds but this was restarted in 1960 and completed in 1963.

The Tawa project was sanctioned and work on it was started about three years back but progress has been slow on account of paucity of funds. No further progress has been made on the other three projects.

The site and scope of the Broach Barrage and Canal project have since undergone radical change. Investigations on the finally revised proposal are in an advanced stage.

The Committee

The present Committee was appointed by the Government of India in September 1964, to study the entire Narmada basin and to prepare, with the help of the Central Water & Power Commission, a Master Plan for the optimum and integrated development of the water resources of the Narmada river for irrigation, power generation, navigation, flood control, etc.

Shortly after the appointment of the Committee, the work of organising its Secretariat and collection of reports and data from the Central Water & Power Commission, and other sources was taken up.

Shri K.S.S. Murthy, Deputy Secretary, Ministry of Irrigation & Power, acted as the Secretary of the Committee in addition to his own duties. The services of Sarvashri R.L. Mohan, Under-Secretary, Ministry of Irrigation & Power; T.A. Deodas and H.J. Desai, Deputy Directors in the Central Water & Power Commission, were, placed whole time at the disposal of the Committee in September, October and November 1964, respectively.

On 14 June 1965, Shri K.S.S. Murthy had to go abroad. Thereafter the work of the Committee had to be performed by the other three officers.

Analysis of Data Made Available

All the available project reports for Narmada basin were collected and in response to the Committee's request, the States of Madhya Pradesh, Maharashtra, Gujarat and Rajasthan sent numerous reports, notes, memoranda and other literature explaining their viewpoints and giving details of their projects and the benefits they expected to derive from Narmada basin development. A large number of comments were also received from the States dealing with the project reports, memoranda and views expressed by the other States. Comments were also received from all the States on the hydrology of the Narmada originally prepared by the Central Water & Power Commission.

All these papers were carefully seen and analysed in the Committee's Secretariat and numerous additional studies were made (with the help of the Central Water & Power Commission, where necessary) and a large number of graphs, tables and statements were prepared to enable the Committee to make a realistic assessment of the view-points and needs of the various States for the preparation of the Master Plan.

The literature received from the States showed that in a few cases fairly detailed investigation had been done, in others only preliminary work had been carried out, while in the majority of cases, no investigation whatsoever had been undertaken.

The Master Plan

Owing to inadequacy of data and general lack of field investigations and project reports, considerable difficulty was experienced in coming to conclusions. However, there was enough material to enable the Committee, with the help of personal discussions with the representatives of the States concerned, to draw up a Master Plan in broad outline, which it is hoped will satisfy, in large measure, the just demands and needs of the States concerned.

In drawing up the Master Plan, the Committee was guided by the following basic considerations:

(1) National interest should have over-riding priority;

irrespective of State boundaries, the Plan should, therefore, provide for maximum benefits in respect of irrigation, power generation, flood control, navigation, etc.

(2) Rights and interests of States concerned should by fully safeguarded subject to (1) above;

(3) Requirements of irrigation should have priority over those of power;

subject to the provision that the suitable apportionment of water between irrigation and power may have to be considered, should it be found that, with the full development of irrigation, power production is unduly affected;

(4) Irrigation should be extended, with the waters available, to the maximum area within physical limits of command irrespective of State boundaries;

in particular, irrigation should be extended to the arid areas along the international border with Pakistan both in Gujarat and Rajasthan, to encourage sturdy peasants to settle in these border areas – later events have confirmed the imperative need for this – and;

(5) All available water should be utilised to the maximum extent possible for irrigation and power generation;

and, where no irrigation is possible, for power generation; and

the quantity going to waste to the sea without being used for irrigation or generating power should be kept to the unavoidable minimum.

In accordance with the above, it is proposed to extend irrigation to areas of Gujarat (Rann of Kachchh) and Rajasthan (Barmer and Jalore desert) which border on Pakistan and to settle on them sturdy peasants on a permanent basis. These areas have no source of irrigation, except from the Narmada river and that with a canal taking off from Navagam at +300 FSL. This is an essential component of the Master Plan both from the point of view of national security and food production in these rainless areas.

Fortunately, the available supplies in the Narmada are enough to meet all the irrigation requirements of Madhya Pradesh and Gujarat and, for want of command, only to a limited extent, those of Rajasthan. The irrigation requirements of Maharashtra from the Narmada are insignificant – only 0.10 MAF. The remaining areas of Rajasthan desert bordering on Pakistan, not included for irrigation from the Narmada waters, are proposed to be irrigated from the Mahi-Sabarmati complex in integration with the Narmada irrigation system in Gujarat.

The irrigation requirements at full development of Madhya Pradesh have been assessed at 15.6 MAF for 6.5 million acres of irrigation and that of Gujarat 10.65 MAF and Rajasthan 0.25 MAF for 4.68 million acres of irrigation. These requirements, according to Committee's analysis, will be met in 95 per cent of the years while in 5 per cent of the years, there will be a small shortage in Madhya Pradesh and a somewhat bigger shortage in Gujarat.

Stress has been laid in the Plan on adequate drainage facilities and resort to large-scale pumping to obviate the hazards of waterlogging resulting from irrigation on a big scale. The greater the consumptive use for irrigation, the greater will be the regeneration from surface and sub-surface drainage, thereby adding to the quantity of water available for generation of power. Large scale pumping in areas of high subsoil water levels, will similarly release corresponding surface reservoir supplies for generation of additional power.

The Plan provides for the construction of a high dam for a terminal reservoir at Navagam with FRL +500, which has been found to be the optimum level for providing the maximum storage and reducing to the minimum the amount of water wasted to the sea. This will also result in the maximum generation of power and maximum flood control. The relative benefits and economics of a dam or combination of dams at other sites in the Navagam-Hiranphal gorge have been considered. The Navagam site and the construction of a dam at this site with FRL +500, have been accepted as being the most economical, giving the maximum benefits.

The construction of the terminal and other storage dams would involve large scale submersion of lands of which between 40 to 50 per cent may be cultivable and, therefore, pose a major human problem. But if this problem is dealt with in a human way as set forth in Chapter XIV, the displaced persons will move from their original rainfed lands to lands provided with irrigation, and to model new villages with safe drinking water, electricity, roads, schools, etc. and small industries which, between them, will transform the present exclusively agricultural economy of these persons to an agro-industrial one.

The amount of power generated under the Master Plan, 20 years from the start of construction, will be 2,014 MW, of which 951 MW will be at the terminal dam at Navagam. This power development will be very large in the earlier years, when irrigation is still developing, as the consumptive use of water will be small and the bulk of the river supplies will be available for power generation.

Assuming that authorization to start essential preliminaries on the key projects of the Plan is given in 1965, the power picture, on the assumption that irrigation development will take place according to plan, will be somewhat as given below. If irrigation development is slower, the power development will be correspondingly higher.

Power development in MW on Narmada system
(Implementation of plan assumed to start in 1965)

	1975	1980	1985	1990	1995
At Navagam	1,054	1,140	951	696	511
Above Navagam	554	1,198	1,063	843	793
Total	1,608	2,338	2,014	1,539	1,304
Water flowing to sea without generating power (MAF)	7.15	6.11	6.02	5.95	5.60

The figures of power generation given in this table will be higher if allowance is made (a) for the regeneration water increasingly coming back into the river with increasing irrigation and (b) for extensive supplemental supplies pumped from the subsoil water reservoirs created by increasing irrigation. Both these additional supplies will add increasingly to power generation (indicated in the table) with increasing consumptive use for irrigation.

The load forecasts of Maharashtra, Gujarat and Madhya Pradesh show that the power will be used up as fast as it is generated. For that reason, it has been proposed to install the ultimate capacity at each dam simultaneously with its completion so that the large amount of water available from storage, while irrigation is still in the early stages of development, does the maximum amount of generation of power before being allowed to flow into the sea.

It may be mentioned here that the full cost of the Navagam dam allocated to power and of the electrical installations, which latter include the cost of 400 kV transmission lines from Navagam to Hiranphal, will have been recovered in the twenty-sixth year after start of construction and thereafter there will be a net profit to the project, after paying for depreciation and working expenses (there will be no interest charge after the twenty-sixth year as the full investment will have been recovered) of Rs 4.71 crores per year, assuming a sale rate of 3 paise per unit (kWh) at the bus-bars. This unit rate of 3 paise allows for the extra cost of 400 kV transmission lines from Navagam to Hiranphal.

Navigation would be feasible from the mouth of the Narmada in the Gulf of Cambay and from the Kandla port, to the Navagam reservoir and beyond to the Bargi reservoir and possibly further up. It is proposed that navigation works should be constructed simultaneously with the canal works and dams so that the facilities of canals and reservoirs can be used for navigation as soon as these are completed.

The periodic floods, which devastate the Narmada valley in Gujarat, would cease, as the storage in the several dams proposed on the river, which will be up to 28.2 MAF, would moderate the peak discharge to safe limits.

The reservoirs formed by the major and medium projects should, with proper management, produce vast quantities of fish, a valuable food that is in short supply and has to be imported at considerable cost in foreign exchange.

The more important dams, such as Punasa and Navagam, with small extra expense, can be developed as attractive tourist resorts.

Narmada Water Dispute, Agreement between Madhya Pradesh, Maharashtra, Gujarat and Rajasthan, 1974

Source: Narmada Water Dispute, Agreement between Madhya Pradesh, Maharashtra, Gujarat and Rajasthan, 12 July 1974. Signed by: Chief Minister of Madhya Pradesh, Chief Minister of Maharashtra, Adviser to the Governor for the State of Gujarat, Chief Minister of Rajasthan.

It is agreed:

1) that the water dispute referred to the Narmada Water Disputes Tribunal be determined by the Tribunal on the basis of this agreement between the States of Madhya Pradesh, Maharashtra, Gujarat and Rajasthan (hereinafter referred to as 'Madhya Pradesh', 'Maharashtra', 'Gujarat' and 'Rajasthan' respectively);

2) that development of Narmada should no longer be delayed in the best regional and national interests;

3) that the quantity of water in Narmada available for 75 per cent of the years be assessed at 28 MAF and that the Tribunal in determining the disputes referred to it do proceed on the basis of that assessment;

4) that the requirements of Maharashtra and Rajasthan for use in their territories are 0.25 and 0.5 MAF, respectively and that the Tribunal in determining the disputes referred to it do proceed on the basis that the requirements of Maharashtra for use in its territories are 0.25 MAF and that Rajasthan will get for use in its territories 0.5 MAF without prejudice to the level of the canal;

5) that the net available quantity of water for use in Madhya Pradesh and Gujarat is 27.25 MAF and that the Tribunal in determining the disputes referred to it do proceed on the basis that the net available quantity of water for use in Madhya Pradesh and Gujarat is 27.25 MAF;

6) that the Tribunal do allocate this balance of water namely, 27.25 MAF, between Madhya Pradesh and Gujarat after taking into consideration various contentions and submissions of the parties hereto;

7) that the height of Navagam Dam be fixed by the Tribunal after taking into consideration various contentions and submissions of the parties hereto;

8) that the level of the canal be fixed by the Tribunal after taking into consideration various contentions and submissions of the parties hereto;

9) that in the light of this agreement, issue No. 4, 5, 7, 7(a), 7(c), 7(d), 7(e), 7(f), 8, 10, 11, 12 and 20 framed by the Tribunal on 28 January 1971 may be deleted and that issue No. 6, 7(b), 13 and 17 may be suitably modified as in the Annexure to this Agreement. All other issues may be determined by the Tribunal after taking into consideration the various contentions and submissions of the parties hereto;

10) that for the limited purpose of effectuating the terms of this agreement Madhya Pradesh do withdraw the proceedings filed by it before the Hon'ble Supreme Court and arising out of the decision of the Tribunal dated 23 February 1972 on the preliminary issues of law;

11) that for the limited purpose of effectuating the terms of this agreement, Rajasthan do withdraw the proceedings filed by it before the Hon'ble Supreme Court and arising out of the decision of the Tribunal dated 23 February 1972 on the preliminary issues of law; and

12) that Rajasthan shall be a party to the further proceedings before the Tribunal, without prejudice to the legal position regarding the rights of a non-riparian State.

Narmada Water Disputes Tribunal, Final Order and Decision of the Tribunal, 1979

Source: Narmada Water Disputes Tribunal, Final Order and Decision of the Tribunal, 12 December 1979.

Clause I: Date of Coming into Operation of the Order

This Order shall come into operation on the date of publication of the Decision of this Tribunal in the Official Gazette under Section 6 of the Inter-State Water Disputes Act, 1956.

Clause II: Determination of the Utilisable Quantum of Narmada Waters

The Tribunal hereby determines that the utilisable quantum of waters of the Narmada at Sardar Sarovar Dam site on the basis of 75 per cent dependability should be assessed at 28 MAF (34,537.44 M.cu.m.)

Clause III: Apportionment of the Utilisable Quantum of Narmada Waters

1) The Tribunal hereby orders that out of the utilisable quantum of Narmada waters, (a) Madhya Pradesh is entitled to a share of 18.25 MAF (22,511.01 M.cu.m.), (b) Gujarat is entitled to a share of 9 MAF (11,101.32 M.cu.m.), (c) Rajasthan is entitled to a share of 0.5 MAF (616.74 M.cu.m.) and (d) Maharashtra is entitled to a share of 0.25 MAF (308.37 M.cu.m.);

2) Further, it is clarified that the apportionment relates to actual withdrawals and not consumptive use;

3) Within its share of water, each party State is free to make such changes in the pattern of water use and in the areas to be benefited within or outside the Narmada basin in its territory, as it may consider necessary.

Clause IV: Order with Regard to Excess Waters and Sharing of Distress

1) The utilisable flow of Narmada in excess of the 28 MAF (34,537.44 M.cu.m.) of utilisable flow in any water year, i.e., from 1 July to 30 June of next calendar year is apportioned in the following ratios of allocation, i.e., 73 for Madhya Pradesh, 36 for Gujarat, 1 for Maharashtra and 2 for Rajasthan;

2) In the event of the available utilisable waters for allocation in any water year from 1 July to 30 June of the next calendar year falling short of 28 MAF (34,537.44 M.cu.m.), the shortage should be shared between the various States in the ratio of 73 for Madhya Pradesh, 36 for Gujarat, 1 for Maharashtra and 2 for Rajasthan;

3) The available utilisable waters in a water year will include the waters carried over from the previous water year as assessed on 1 July on the basis of stored waters available on that date;

4) The available utilisable waters on any date will be inclusive of return flows and exclusive of losses due to evaporation of the various reservoirs;

5) It may be mentioned that in many years there will be surplus water in the filling period after meeting the storage requirements and withdrawals during the period. This will flow down to sea. Only a portion of it will be utilisable for generating power at Sardar Sarovar river bed power house and the rest will go waste. It is desirable that water, which would go waste without even generating power at the last river bed power house, should be allowed to be utilised by the party States to the extent they can. Gujarat is, therefore, directed that whenever water starts going to waste to sea without generating power, or based on the information received from upstream gauging stations,

it anticipates that water would so go to waste, it shall inform the Narmada Control Authority (hereinafter referred to as the Authority) and designated representatives of all the concerned States. Gujarat shall also inform them when such flows cease. During the period of such flows, the party States, whose reservoirs are spilling and the spill water cannot be stored elsewhere, may utilise such flows from the said reservoirs as they like and such utilisation by the party States will not count towards allotment of supplies to them, but, use of such water will not establish any prescriptive rights.

Clause V: Period of Operation of the Order of Apportionment

Our Orders with regard to the equitable allocation in Clauses III and IV are made subject to review at any time after a period of 45 years from the date of publication of the Decision of the Tribunal in the Official Gazette.

Clause VI: Full Supply Level of the Navagam Canal

The Tribunal hereby determines that the Full Supply Level of Navagam Canal off-taking from Sardar Sarovar should be fixed at 91.44 m (+300 ft) at its head regulator with a bed gradient of 1 in 12,000 from head to 290 km (mile 180), that is, up to the offtake of Saurashtra branch. From that point to Rajasthan border the bed gradient should be 1 in 10,000. These bed gradients may be changed by Gujarat and Rajasthan by mutual agreement. Gujarat and Rajasthan shall be at liberty to decide the canal capacity required by each in the light of water, which would be expected to be available within their share.

Clause VII: Full Reservoir Level and Maximum Water Level of the Sardar Sarovar Dam

The Tribunal hereby determines that the height of the Sardar Sarovar Dam should be fixed for Full Reservoir Level +138.68 m (+455 ft) and Maximum Water Level at +140.21 m (+460 ft). Gujarat shall take up and complete the construction of the dam accordingly.

Clause VIII: Sharing of Costs and Benefits

(1)(i) The Tribunal hereby determines that out of the net power produced at Navagam at Canal Head and river bed power houses on any day the share of Madhya Pradesh will be 57 per cent; Maharashtra's share will be 27 per cent and Gujarat's share will be 16 per cent.

(ii) The party States shall make available in annual instalments their share of funds required according to approved construction programme and take all the necessary steps to complete the Sardar Sarovar Dam within ten years of the date of publication of the Final Order and Decision of the Tribunal in the Official Gazette.

(2) The Tribunal makes the following further Orders:

(i) The power generated in the river bed and canal power houses at Navagam will be integrated in a common switchyard.

(ii) Madhya Pradesh and Maharashtra will be entitled to get 57 per cent and 27 per cent respectively of the power available at bus bar in the switchyard after allowing for station auxiliaries.

(iii) The above entitlement applies both to availability of machine capacity for peak loads and to the total energy produced in any day.

(iv) The entitlement of power and energy for any day can be utilised fully or partly by the concerned States or sold to another participating State under mutual agreement. It cannot, however, be carried forward except under a separate agreement or working arrangement entered into among the affected parties.

(v) Gujarat will construct and maintain the transmission lines needed to supply the allotted quantum of power to Madhya Pradesh and Maharashtra up to Gujarat State border, along an alignment as agreed to between the parties and if there is no agreement along such alignment as may be decided by the Narmada Control Authority. The transmission lines beyond Gujarat State border shall be constructed and maintained by Madhya Pradesh and Maharashtra in their respective States.

(vi) The power houses and appurtenant works including the machinery and all installations as well as the transmission lines in Gujarat State will be constructed, maintained and operated by Gujarat State or an authority nominated by the State.

(vii) The authority in control of the power houses shall follow the directions of the Narmada Control Authority in so far as use of water is concerned.

(viii) The scheme of operation of the power houses including the power required and the load to be catered for the different party States during different parts of the day shall be settled between the States at least one week before the commencement of every month and shall not be altered during the month except under agreement among the States or under emergencies. If and when Sardar Sarovar power complex gets linked with the regional or national power grid, the operation of the Sardar Sarovar power complex will be governed by such altered system conditions. But in that event the Narmada Control Authority should arrange to take such steps as are necessary to enable the three States of Gujarat, Madhya Pradesh and Maharashtra to get their entitlement of power and energy from the Sardar Sarovar power complex according to these orders.

(ix) The capital cost of the power portion of Navagam complex shall comprise the following:

(a) Full cost of Unit III electrical works and control works pertaining thereto, up to and including the switchyard.

(b) Full cost of transmission lines in Gujarat State constructed for supplying power to Madhya Pradesh and Maharashtra.

(c) 56.1 per cent of the net cost of common facilities such as dam and appurtenant works i.e. Unit I of Sardar Sarovar Project, after allowing for credits, if any;

(d) 56.1 per cent of the credit given to Madhya Pradesh for the downstream benefits derived from Narmada Sagar Dam.

(x) Madhya Pradesh and Maharashtra shall respectively pay to Gujarat 57 per cent and 27 per cent of the capital cost of the power portion of the Sardar Sarovar headworks worked out vide (ix) above. This amount shall be paid in annual instalments until the capital works are completed. Each instalment will be worked out on the basis of the budgeted figures of the concerned works at the commencement of each financial year and shall be set off and adjusted against actual figures at the end of the financial year.

(xi) In addition to the payments vide (x) above, Madhya Pradesh and Maharashtra shall also pay to Gujarat 57 per cent and 27 per cent respectively of the operation and maintenance costs of the Sardar Sarovar power complex each year. These payments are also to be based on budgeted figures at the commencement of each financial year and adjusted against actual cost at the end of the year.

(xii) Notwithstanding the directions contained hereinabove, the party States may, by mutual agreement, alter, amend, or modify any of the directions in respect of sharing of power and payment for it.

Clause IX: Regulated Releases to be Made by Madhya Pradesh for the Requirement of Sardar Sarovar Project

With regard to the quantum and pattern of regulated releases, the Tribunal makes the following Order:

It has been agreed by the party States and decided by the Tribunal in its Order dated 8 October 1974, that the utilisable quantity of water of 75 per cent dependability in the Narmada at Sardar Sarovar Dam site should be assessed at 28 MAF (34,537.44 M.cu.m.). The actual inflow of 75 per cent dependability, however, is only 33,316.29 M.cu.m (27.01 MAF) and this is brought up to utilisable quantity of 28 MAF (34,537.44 M.cu.m) by means of carryover in various reservoirs allowing for evaporation losses and regeneration. Out of 28 MAF (34,537.44 M.cu.m), 11,101.32 M.cu.m (9 MAF) has to be provided for Gujarat and 0.5 MAF (616.74 M.cu.m) for Rajasthan at Sardar Sarovar. The requirements at Sardar Sarovar have to be met by releases by Madhya Pradesh and by inflows from the intermediate catchment, surplus to the requirements of Madhya Pradesh below Narmada Sagar and Maharashtra. The releases from Maheshwar work out to 10,015.86 M.cu.m (8.12 MAF). Making uniform monthly releases the amount of water to be released by Madhya Pradesh per month would be 834.65 M.cu.m (0.677 MAF). The actual inflow in the river system, however, would vary from year to year and, therefore, the releases by Madhya Pradesh would also vary.

The inflow during the filling period, July to October, cannot be predicted at the beginning of the season. It is only in October that it would be fully known whether the particular year is a normal year or the extent to which it is a surplus or deficit year. Normally the releases by Madhya Pradesh during the filling period, therefore, would have to be more or less on the basis of the year yielding 28 MAF (34,537.44 M.cu.m.) utilisable quantity. The month of July and early part of August are crucial for kharif sowing. It is important that during this period regulatory arrangements should ensure that due share of water is made available to all parties.

Having regard to the facts mentioned in the preceding two paragraphs, we order that detailed rules of regulation and water accounting shall be framed by Narmada Control Authority in accordance with the guidelines given below. These guidelines may, however, be altered, amended or modified by agreement between the States concerned.

(i) The 28 MAF (34,537.44 M.cu.m.) utilisable supplies of 75 per cent dependability in a water year (1 July to 30 June next year) shall be shared by the party States as under:

Madhya Pradesh	18.25 MAF	(22,511.01 M.cu.m.)
Gujarat	9.00 MAF	(11,101.32 M.cu.m.)
Rajasthan	0.50 MAF	(616.74 M.cu.m.)
Maharashtra	0.25 MAF	(308.37 M.cu.m.)
	28.00 MAF	(34,537.44 M.cu.m.)

(ii) Surplus or deficit utilisable supplies in a water year shall be shared to the extent feasible by the party States in the same proportion as their allotted shares in (i) above. The surplus water shall first be utilised for filling up the reservoirs to capacity and surplus water shall be utilised for irrigation and other purposes only after that has been ensured.

(iii) The water available in the live storages of the various reservoirs on 30 June shall be reckoned as an inflow to be shared in the next water year.

(iv) The releases necessary to ensure Gujarat and Rajasthan's share of water in a water year shall be let down by Madhya Pradesh at a reasonably uniform rate, permitting only such variation as the Authority may direct or approve and keeping in view the directions for regulated releases.

(v) The Authority shall ensure by so directing the releases by Madhya Pradesh that there is at all times sufficient utilisable water in Sardar Sarovar to meet the requirements of the next ten days subject to water being available in the storages in Madhya Pradesh after taking into account the proportionate requirements of Madhya Pradesh. For this purpose, Gujarat and Rajasthan would intimate their requirements of the ten-day period well in advance.

(vi) Utilisation in a water year by each party State shall be figured out on the basis of actual daily discharge at canal head on every major and medium project. For minor works, it shall be on the basis of area irrigated under different crops, the delta for each crop being approved by the Authority. For pumping schemes, drawing directly from the river, its tributaries or reservoirs, whether for irrigation, domestic or industrial use, water drawn shall be reckoned on the basis of the rated capacity of pumps and the number of hours they run. For a cross check, the seasonwise and cropwise area irrigated by each pumping scheme shall also be recorded, and if the figures of water drawn as worked out by the two aforesaid methods differ, the decision of the Authority as regards water drawn shall be final.

(vii) Withdrawals from Sardar Sarovar for Navagam Canal for Gujarat and Rajasthan shall be measured at the head of Navagam Canal. The supply to Rajasthan shall be measured at Gujarat-Rajasthan border. The loss in the canal in carrying the supply for Rajasthan shall be determined by the Authority after the canal has been constructed and shall reckon against the share of Rajasthan.

Water let down into the river from Sardar Sarovar through power house turbines shall be measured on the basis of power generated by it and that escaped through the spillway by measurement at the spillway.

Gujarat may let down water from Sardar Sarovar for its downstream use by making specific indent for it and such releases shall reckon against its share. Such releases for downstream use shall be made through the turbines and the power so generated shared between Madhya Pradesh, Maharashtra and Gujarat in the prescribed ratio.

Water let down into the river from Sardar Sarovar except at the specific indent of Gujarat shall not reckon against the share of Gujarat. The water drawn from Sardar Sarovar for use in Madhya Pradesh and Maharashtra, as the case may be, shall reckon against the share of water of that State.

(viii) For major and medium projects, water account shall be kept by ten-day period. The last ten-day period of a month may have 11 days, ten days or less, depending upon the number of days in the month. For minor schemes water accounts shall be kept by crop seasons, kharif (July to October) rabi (November to March) and hot weather (April to June). For pumping schemes and domestic and industrial uses it shall be monthly.

(ix) The water use by minor and pumping schemes in any ten-day period may provisionally be taken to be the same as in the corresponding period in the previous year on the basis of average use during the crop period. For final water account, however, it will be determined as in (vi) above.

(x) Each State shall furnish to the Authority and make available to any party State desiring the same, such data and information as the Authority may require and ask for.

(xi) The Authority shall arrange the review of the ten day releases made by Madhya Pradesh at least once a month and oftener as considered necessary for directing any change in the releases. It may designate a person for doing so.

(xii) The Authority shall direct final adjustment to be made in the following water year of the use in excess of the authorised use, if any, by any State or States during the preceding water year by curtailing the share(s) of the State or States concerned which have used water in excess and make over the same to the State or States which have received short supplies. Water supplied to Rajasthan on any day in excess of 10 per cent over and above its indent shall reckon against use by Gujarat.

(xiii) The Authority shall furnish the annual water account for the water year to the Governments of the party States by the end of August of the next water year. Each State may make any observation on the account and/or point out corrections in it, if any, within one month of its receipt. After making the necessary modifications, the Authority shall furnish to each party State the final annual water account for the water year by 31 October. The Authority shall cause the annual water account to be published each year.

Clause X: Payment to be Made by Gujarat to Madhya Pradesh for such Regulated Releases

(1) Madhya Pradesh shall take up and complete the construction of Narmada Sagar dam with FRL 262.13 m (860 ft) concurrently with or earlier than the construction of Sardar Sarovar Dam.

(2) The Tribunal further orders that Gujarat should credit to Madhya Pradesh each year 17.63 per cent of the expenditure on account of Narmada Sagar dam in the financial year commencing from the year of taking up of the construction of Narmada Sagar dam. This will be initially credited on the basis of budget allotment to be adjusted at the end of the year on actual expenditure. The post construction expenditure on maintenance is not to be considered as cost of construction.

Clause XI: Directions Regarding Submergence Land Acquisition and Rehabilitation of Displaced Persons

Sub-Clause I: Definitions

I(1): 'Land' The expression 'land' shall have the same meaning as defined in the Land Acquisition Act, 1894 (hereinafter referred to as the Act) which states 'the expression 'land' includes benefits to arise out of land, and things attached to the earth or permanently fastened to anything attached to the earth'.

I(2): 'Oustee' – An 'oustee' shall mean any person who since at least one year prior to the date of publication of the notification under Section 4 of the Act, has been ordinarily residing or cultivating land or carrying on any trade, occupation, or calling or working for gain in the area likely to be submerged permanently or temporarily.

I(3): 'Family' – (i) A family shall include husband, wife and minor children and other persons dependent on the head of the family, e.g. widowed mother.

(ii) Every major son will be treated as a separate family.

Sub-Clause II: Lands Which are to be Compulsorily Acquired

II(1): Madhya Pradesh and Maharashtra shall acquire for the SSP under the provisions of the Land Acquisition Act, 1894, all lands of private ownership situated below the FRL +138.68 m (455 ft) of Sardar Sarovar and all interests therein not belonging to the respective States. If on the basis of the aforesaid, 75 per cent or more land of a contiguous holding of any person is required to be compulsorily acquired, such person shall have the option to compel compulsory acquisition of the entire contiguous holding.

II(2): Madhya Pradesh and Maharashtra shall also acquire for the SSP, under the provisions of the Land Acquisition Act, 1894, all buildings with their appurtenant land situated between FRL +138.68 m (455 ft) and MWL +141.21 m (460 ft) as also those affected by the backwater effect resulting from MWL +141.21 m (460 ft).

II(3): The backwater level at the highest flood level in Sardar Sarovar shall be worked out by the Central Water Commission in consultation with Madhya Pradesh and Gujarat.

Sub-Clause III: Liability of Gujarat to Pay Compensation for Land Acquisition and Rehabilitation etc.

III(1): Gujarat shall pay to Madhya Pradesh and Maharashtra all costs including compensation, charges and expenses incurred by them for or in respect of the compulsory acquisition of lands required to be acquired as aforesaid.

III(2): Gujarat shall pay to Madhya Pradesh and Maharashtra and the Union of India compensation for the respective government lands and structures on principles similar to those underlying the Land Acquisition Act, 1894. Where any dispute or difference arises between Gujarat, Madhya Pradesh, Maharashtra and the Union of India with respect to the compensation payable as aforesaid, any of the three States of Gujarat, Madhya Pradesh and Maharashtra or the Union of India may refer the matter in dispute to arbitration. The State of Gujarat on the one hand and the States of Madhya Pradesh, Maharashtra or the Union of India (as the case may be) on the other hand shall respectively nominate one arbitrator each. In the event of disagreement between the arbitrators, such dispute or difference shall be referred to an umpire who shall be a person appointed in that behalf by the Chief Justice of India from among persons who are, or have been judges of the Supreme Court. The decision of the arbitrators or, as the case may be, of the umpire shall be final and binding on the parties and shall be given effect to by them.

III(3): Gujarat shall pay to Madhya Pradesh and Maharashtra land revenue in accordance with the respective Land Revenue Codes of Madhya Pradesh and Maharashtra in respect of all lands in their respective territories acquired for Gujarat or conveyed to it.

III(4): Gujarat shall pay to Madhya Pradesh and Maharashtra all costs, charges and expenses incurred by Madhya Pradesh and Maharashtra for the purpose of removal and reinstallation of any ancient or historical monuments, archaeological remains, religious place of worship or idols likely to be affected by submergence under Sardar Sarovar and that in the event of such payment being made, no separate compensation as hereinbefore provided shall be required to be paid in respect of the same having been affected by the submergence.

III(5): Gujarat shall pay to Madhya Pradesh and Maharashtra all costs, charges and expenses required to be incurred by them for rehabilitation of oustees and oustee families in their respective territories in accordance with the directions hereinafter contained.

III(6): Gujarat shall pay to Madhya Pradesh and Maharashtra costs on account of establishment charges for land acquisition and rehabilitation and other departmental staff which Madhya Pradesh and Maharashtra may consider necessary for the purpose of such acquisition and rehabilitation.

<u>Sub-Clause IV: Provision for Rehabilitation</u>

IV(1): According to the present estimates the number of oustee families would be 6,147 spread over 158 villages in Madhya Pradesh, 456 families spread over 27 villages in Maharashtra. Gujarat shall establish rehabilitation villages in Gujarat in the irrigation command of the Sardar Sarovar Project on the norms hereinafter mentioned for rehabilitation of the families who are willing to migrate to Gujarat. For oustee families who are unwilling to migrate to Gujarat, Gujarat shall pay to Madhya Pradesh and Maharashtra the cost, charges and expenses for establishment of such villages in their respective territories on the norms as hereinafter provided.

IV(2)(i): According to the present estimates the number of oustee families below RL 106.68 metres (RL 350 ft) would be 30 spread over 20 villages in Madhya Pradesh and 250 families spread over 20 villages in Maharashtra. Within six months of the publication of the decision of the Tribunal in the Official Gazette, Gujarat, Madhya Pradesh and Maharashtra shall determine by mutual consultation the location of one or two rehabilitation villages in Gujarat to rehabilitate oustees from areas below RL 106.68 metres (RL 350 ft). Gujarat shall acquire necessary lands for the rehabilitation villages and make available the same within two years of the decision of the Tribunal. Within six months of the decision of the location of the rehabilitation villages in Gujarat, Madhya Pradesh and Maharashtra shall intimate to Gujarat the number of oustee families from areas below RL 106.68 metres (RL 350 ft) willing to migrate to Gujarat. For the remaining oustee families, Madhya Pradesh and Maharashtra shall arrange to acquire lands for rehabilitation within the respective States.

IV(2)(ii): Madhya Pradesh and Maharashtra shall set up adequate establishments for land acquisition and rehabilitation of oustee families. Gujarat shall deposit within three months of the decision of the Tribunal Rupees 10 lakhs each with

Madhya Pradesh and Maharashtra in advance towards cost of establishment and rehabilitation in these States to be adjusted after actual costs are determined. Madhya Pradesh and Maharashtra shall start land acquisition proceedings for areas below RL 106.68 metres (RL +350 ft) within six months of the decision of the Tribunal and convey the lands to Gujarat for project purposes within three years of the decision of the Tribunal. Within 18 months of the decision of the Tribunal, Gujarat shall make an advance payment of Rs 70 lakhs to Madhya Pradesh and Rs 100 lakhs to Maharashtra towards the compensation of land, to be adjusted after actual costs are determined.

IV(2)(iii): Regarding the oustee families from areas above RL 106.68 metres (RL +350 ft), Gujarat shall intimate to Madhya Pradesh and Maharashtra within six months of publication of the decision of the Tribunal in the Official Gazette the number and general location of rehabilitation villages proposed to be established by Gujarat in accordance with the decision of the Tribunal. Within one year of the receipt of proposal of Gujarat, both Madhya Pradesh and Maharashtra shall intimate to Gujarat the number of oustee families willing to migrate to Gujarat. The three States by mutual consultation shall determine within two years of the decision of the Tribunal, the number and general location of rehabilitation villages required to be established by Gujarat in its own territory. Madhya Pradesh and Maharashtra shall intimate to Gujarat the number of such villages to be established in Madhya Pradesh and Maharashtra and for which Gujarat would be required to make payments to Madhya Pradesh and Maharashtra respectively.

IV(2)(iv): Gujarat shall acquire and make available a year in advance of the submergence before each successive stage, irrigable lands and house sites for rehabilitation of the oustee families from Madhya Pradesh and Maharashtra who are willing to migrate to Gujarat. Gujarat shall in the first instance offer to rehabilitate the oustees in its own territory.

IV(3): Gujarat shall also provide the following grants and amenities to the oustees:

(a) Resettlement Grants (Rehabilitation Grant) - Gujarat shall pay per family a sum of Rs 750 inclusive of transportation charges as resettlement grant.

(b) Grant-in-aid

In addition, Gujarat shall pay per family grant-in-aid in the following scale:

Where total compensation is received	Grand-in-aid
Above Rs 2000	Nil
Between Rs 2000 and Rs 500	Rs 500 less an amount equal to one-third of the compensation in excess of Rs 500

Less than Rs 500	Rs 500

(c) Civic amenities

1. One primary school (3 rooms) for 100 families.

2. One Panchayat Ghar for every 500 families.

3. One dispensary for every 500 families.

4. One seed store for every 500 families.

5. One children's park for every 500 families.

6. One village pond for every 500 families.

7. Drinking water well with trough for every 50 families.

8. Each colony should be linked to main road by roads of appropriate standard.

9. One platform for every 50 families.

10. Every oustee family shall be entitled to and allotted a house site i.e. a plot of land measuring 18.29 * 27.43m (60 ft * 90 ft) free of cost. In addition, a provision of 30 per cent additional area for roads, government buildings, open space etc. shall be made by Gujarat under civic amenities.

11. The State of Gujarat shall make the following provision for rehabilitation in Madhya Pradesh and Maharashtra:

(a) Resettlement	Rs 750 per family
(b) Grant-in-aid	Rs 500 per family
(c) Acquisition of land for resettlement of families affected – 0.40 hectares (one acre) for 6 families	Rs 1,500 per acre
(d) Civic amenities:	
1. One primary school for100 families	Rs 30,000 each
2. One Community hall-cum-panchayat bhavan for 500 families	Rs 20,000 each
3. One dispensary for 500 families	Rs 25,000 each
4. One seed store for 500 families	Rs 10,000 each
5. One children's park for 500 families	Rs 6,000 each
6. One well with trough for 50 families	Rs 10,000 each
7. One pond for 500 families	Rs 20,000 each

8. One tree platform for 50 families	Rs 1,500 each
9. One religious place of worship for 100 families	Rs 1,000 each
10. Construction of approach roads and link roads for Abadies – 3 km per every new Abadi	Rs 30,000 per km
11. Electrical distribution lines and street lights – 2 km per 100 families	Rs 11,000 per km
12. Social amenities for each municipal town going under submergence, viz, water supply and sanitary arrangements layout, levelling of site etc.	Rs 5,00,000 each town

IV(4)(i): Gujarat is directed to provide for rehabilitation and civic amenities as per directions contained hereinabove in Sub-clause IV(3) in its estimate for B-Land compensation and rehabilitation.

IV(4)(ii): Notwithstanding the provisions hereinbefore contained, Gujarat shall not be liable to pay any compensation for the loss of public properties, facilities or amenities such as drinking water wells, primary school buildings, internal roads, village sites, approach roads, dispensaries, panchayat buildings, rural electrification, highway, bridges, telegraph lines, power lines etc. if corresponding alternative properties, facilities or amenities are to be provided at the cost of the Sardar Sarovar Project. The party owning the facility shall have the option to accept compensation for utilities as existing or ask for their replacement or relocation at the cost of Gujarat.

IV(5): It is made clear that the monetary values in Clause IV(3)(c) are liable to be changed at the time of actual rehabilitation. Where any dispute or difference arises as regards the changed valuation, the matter shall be determined by arbitration in the manner provided in Clause III(2) above and Gujarat's liability shall stand altered accordingly.

IV(6)(i): In the event of Gujarat being unable to resettle the oustees or the oustees being unwilling to occupy the area offered by Gujarat, Madhya Pradesh and Maharashtra shall make such provisions for rehabilitation, civic amenities etc. on the lines mentioned in Clauses IV(1) to (4) above. Gujarat shall, in that event, be liable to pay all such expenses, costs etc., arising out of or in connection with rehabilitation and provision of civic amenities for the oustees including the cost of all acquisition proceedings and payment of compensation etc., as per the Land Acquisition Act, for the land allotted to oustees for cultivation and habitation.

IV(6)(ii): In no event shall any areas in Madhya Pradesh and Maharashtra be submerged under the Sardar Sarovar unless all payment of compensation, expenses and costs as aforesaid is made for the acquisition of land and properties and arrangements are made for the rehabilitation of the oustees therefrom in accordance with these directions and intimated to the oustees.

IV(7): Allotment of Agricultural Lands: Every displaced family, from whom more than 25 per cent of its land holding is acquired, shall be entitled to and be allotted irrigable land to the extent of land acquired from it subject to the prescribed ceiling in the State concerned and a minimum of 2 hectares (5 acres) per family, the irrigation facilities being provided by the State in whose territory the allotted land is situated. This land shall be transferred to the oustee family if it agrees to take it. The price charged for it would be as mutually agreed between Gujarat and the concerned State. Of the price to be paid for the land a sum equal to 50 per cent of the compensation payable to the oustee family for the land acquired from it will be set off as an initial instalment of payment. The balance cost of the allotted land shall be recovered from the allottee in 20 yearly instalments free of interest. Where land is allotted in Madhya Pradesh or Maharashtra, Gujarat having paid for it vide Clause IV(6)(i) supra, all recoveries for the allotted land shall be credited to Gujarat.

IV(8): Any dispute between the States in respect of Clauses IV(1) to (7) of these directions shall be referred to and determined by arbitration in the manner provided in Clause III(2) of these directions.

Sub-Clause V: Programme for Payment to be Made by Gujarat to Madhya Pradesh and Maharashtra

V(1): As soon as practicable after the publication of the decision of the Tribunal in the Official Gazette, Gujarat shall prepare and furnish to the other party States, a fresh estimate of sub-head B-Land for the Sardar Sarovar Project as permitted by the Tribunal including in particular, costs of acquisition of lands in Madhya Pradesh and Maharashtra and of rehabilitation of oustee families in Madhya Pradesh and Maharashtra.

V(2)(i): As soon as practicable after the publication of the decision of the Tribunal in the Official Gazette and in any case before expiry of three months thereafter, both Madhya Pradesh and Maharashtra shall furnish to Gujarat three sets of Majmuli/Taluka maps of all talukas in their respective territories likely to be submerged wholly or partly under Sardar Sarovar. These maps shall indicate village boundaries. Within three months after the receipt of the Majmuli/Taluka maps Gujarat shall mark thereon the boundary of the area situated below the FRL as also that between FRL and MWL including the area affected by back water resulting from MWL and shall return one respective set so marked to Madhya Pradesh and Maharashtra.

V(2)(ii): As soon as practicable after the receipt of one set of the Majmuli/Taluka maps marked as aforesaid and in any case within six months thereof, the Governments of Madhya Pradesh and Maharashtra shall publish notifications under Sub-section (1) of Section 4 of the Act notifying that the lands in their respective territories situated below the FRL and buildings with their appurtenant lands between FRL and MWL, as also those affected by the back water effect resulting from MWL (to be specified in the notifications) are likely to be needed for the Sardar Sarovar Project.

V(2)(iii): As soon as practicable after publication of the decision of the Tribunal in the Official Gazette as hereinbefore referred to and in any case within one year thereof, Gujarat shall intimate to Madhya Pradesh and Maharashtra yearwise programme of construction of the dam.

V(2)(iv): Objections, if any, received against the proposed acquisition of lands as notified under Section 4 of the Act shall be heard and disposed of and any reports to the State governments as contemplated by Sub-section (2) of Section 5A of the Act shall be made with utmost expedition. The Governments of Madhya Pradesh and Maharashtra shall issue requisite notifications under Section 6 of the Act with utmost expedition and in any case before the expiry of three years from the dates of publication of the respective notifications under Sub-section (1) of Section 4 of the Act.

V(2)(v): As soon as practicable after receipt of the yearwise programme of construction of the Sardar Sarovar dam from Gujarat both Madhya Pradesh and Maharashtra in consultation with Gujarat shall finalise their respective yearwise programme of completing the proceedings for compulsory acquisition of lands in their respective territories up to the stages of making awards under Section 11 of the Act of taking possession of the lands under Section 16 of the Act.

V(3)(i): Gujarat is required to pay to Madhya Pradesh and Maharashtra compensation for compulsory acquisition of lands, market value of government lands to be conveyed to Gujarat and expenditure to be incurred in connection with the rehabilitation of oustee families to be rehabilitated in Madhya Pradesh and Maharashtra as hereinbefore provided. Madhya Pradesh and Maharashtra shall on or before 30 September of each year intimate to Gujarat the amounts required to be paid by Gujarat to Madhya Pradesh and Maharashtra respectively having regard to (a) the extent of lands in Madhya Pradesh and Maharashtra in respect of which awards are likely to be made under Section 11 of the Act (b) the extent of government lands likely to be conveyed by Madhya Pradesh and Maharashtra to Gujarat during the next financial year and (c) the expenditure likely to be incurred by Madhya Pradesh and Maharashtra in connection with rehabilitation of oustee families in Madhya Pradesh and Maharashtra during the next financial year. In arriving at these estimates for the next financial year, Madhya Pradesh and Maharashtra shall also take into account the differences, if any, between the payments made by Gujarat in pursuance of this clause for the current financial year and the amount actually payable during the said financial year.

V(3)(ii): On the basis of these estimates, Gujarat shall on or before 31 May of the following financial year make payments to Madhya Pradesh and Maharashtra of the amounts estimated as provided in Clause V(3)(i) above.

V(3)(iii): Gujarat shall at each successive stage of submergence intimate to Madhya Pradesh and Maharashtra the area coming under submergence at least 18 months in advance. The inhabitants of the area coming under the respective stages of submergence will be entitled to occupy or use their properties without being required to pay anything for such occupation and use until a date to be notified by the State concerned, which shall not be less than six months before submergence. They must vacate the area by the notified date.

V(4)(i): On payment of the amounts to be paid each year by Gujarat as compensation for compulsory acquisition of lands as aforesaid, Madhya Pradesh and Maharashtra shall, as expeditiously as possible, complete the acquisition and transfer such lands to Gujarat so as to vest the lands in Gujarat to be used only for the purpose of submergence and subject to Clauses V(5) to (8) of these directions.

V(4)(ii): On payment of the market value of government lands by Gujarat as hereinbefore provided Madhya Pradesh and Maharashtra and the Union of India shall convey such lands to Gujarat so as to vest in Gujarat to be used only for the purpose of submergence and subject to clauses V(5) to (8) of these directions.

V(5): Gujarat shall pay to Madhya Pradesh and Maharashtra in accordance with the respective Land Revenue Codes, the amount of land revenue payable every year for the lands in their respective territories acquired for Gujarat or conveyed to it, at the rates prevailing in Madhya Pradesh and Maharashtra respectively from time to time.

V(6): Madhya Pradesh and Maharashtra, as the case may be, shall remit, each year to Gujarat any revenue which they may derive from the cultivation of lands which get periodically exposed in Sardar Sarovar, after deducting collection charges for the same.

V(7): Notwithstanding vesting in Gujarat of the lands coming under submergence, Madhya Pradesh and Maharashtra shall continue to enjoy all rights of sovereignty intact over the submerged area in the respective States.

V(8): Madhya Pradesh and Maharashtra respectively shall be exclusively entitled to all rights of fishing, boating and water transportation over the part of lake over the submerged land within Madhya Pradesh and Maharashtra respectively, provided, however, that such right is not exercised to the prejudice of any utilities of the Sardar Sarovar Project or cause hindrance in the legitimate performance of their duties by the project personnel.

V(9): All residual rights not specifically transferred to Gujarat in respect of the lands coming under submergence shall continue to vest in the government in whose territory they are situated.

V(10): In the event of the said lands not being used for the purpose of submergence for which it is acquired, the State of Gujarat shall retransfer such land to Madhya Pradesh or Maharashtra as the case may be, subject to the condition that Madhya

Pradesh and Maharashtra refund to Gujarat the amount of compensation received from Gujarat in respect of such land.

V(11): In the event of any land acquired for rehabilitation of oustee families is not used for the purpose, it shall be returned to the original owner on payment, where feasible or otherwise disposed of and due credit given to Gujarat.

V(12): All costs incurred by Gujarat on acquisition of land and rehabilitation of oustees in respect of Sardar Sarovar shall be charged to Sardar Sarovar Project estimate, Unit I – Dam and Appurtenant Works.

<u>Sub-clause VI</u>

Nothing contained in Clause XI shall prevent the alteration, amendment and modification of all or any of the foregoing clauses by agreement between all the party States.

Clause XII: Allocation of Cost of Sardar Sarovar Project between Irrigation and Power

We determine that the cost of Unit I – Dam and Appurtenant Works – should be apportioned between Irrigation and Power as follows:

- Irrigation 43.9 per cent
- Power 56.1 per cent

Clause XIII: Allocation of Irrigation Component of Cost of Sardar Sarovar Project and between Gujarat and Rajasthan

(a)(i) The irrigation component of the cost of Unit I of Sardar Sarovar Project (Dam and Appurtenant Works) should be shared by Gujarat and Rajasthan in the ratio of 18:1.

(a)(ii) Madhya Pradesh and Maharashtra shall contribute a pro rata share to the irrigation component of the cost of Sardar Sarovar Dam as also towards its operation and annual maintenance, for water drawn from Sardar Sarovar for use in their territory. The pro rata share shall be in proportion the quantity of water so drawn to 9.5 MAF. The amount so contributed shall be credited to Gujarat and Rajasthan in the ratio of 18:1.

(b) The cost of Navagam Canal with its design approved by Narmada Control Authority shall be shared by the two States as under:

(i) The cost differential in respect of land, earth work and lining for the gradients proposed by Gujarat and that now prescribed, to be borne by Rajasthan in full.

(ii) The actual cost of the canal less (i) above to be shared on cusec-mile basis.

The actual cost should be shared by Gujarat and Rajasthan on cusec-mile basis on the first instance and on completion of the work the share cost shall be adjusted as indicated above. Rajasthan shall credit its share cost each year initially on the basis of budget allotment. This should then be adjusted at the end of the year to actual expenditure. The post-construction expenditure on maintenance is not to be considered as cost of construction.

Should any difference arise between Rajasthan and Gujarat on figures of cost in respect of Navagam Main Canal for purposes of sharing the cost, the matter shall be referred to the Narmada Control Authority and on such a reference its decision shall be final and binding.

Clause XIV: Setting up of Machinery for Implementing the Decision of the Tribunal

We make the following orders with regard to setting up of machinery for implementing the decision of the Tribunal:

Sub-clause I: Constitution of the Authority.

1(1): An inter-State, administrative authority to be called Narmada Control Authority (hereinafter referred to as the 'Authority') shall be established for the purpose of securing compliance with and implementation of the decision and directions of the Narmada Water Disputes Tribunal (herein referred as the 'Orders').

1(2): The Authority shall consist of seven high-ranking engineer members, of whom one each shall be of the rank of Engineer-in-Chief, Chief Engineer, or Additional Chief Engineer of the Irrigation Department, Power Department or the State Electricity Board appointed by the government of each of the States of Madhya Pradesh, Gujarat, Maharashtra and Rajasthan and three other eminent engineers of a rank not less than that of a Chief Engineer to be appointed by the Central Government in consultation with the party States. One of the three independent members shall be nominated by the Central Government, as the Chairman of the Authority with a deliberative vote at meetings where decisions are taken on any matter affecting the interest of more than one State and he will be in charge of the administrative work of the Authority. The Central or State government, as the case may be, shall have the power to remove or suspend from the Authority any member who, in its opinion, is not suitable to continue as member.

1(3): Each independent member shall be a full-time member and be appointed for a term not exceeding five years. The members appointed by the State governments shall be part-time members. The appointing authority for independent member or that for part-time member, as the case may be, shall determine the terms and conditions of appointment in each case. As far as practicable, the first appointment of the seven members of the Authority shall be made within three months from the date of publication of the decision of the Tribunal in the Official Gazette.

1(4): Vacancies of members – On any vacancy occurring in the offices of the three independent members, the Central Government shall appoint a person to such vacant office, and on any vacancy occurring in the office of the four members other than the independent members, the State government by whom the member whose office falls vacant was appointed shall appoint a person to the vacant office.

In case of illness or absence for any cause whatever of a member, the Central Government or State government by whom he was appointed (as the case may be) may appoint a person as an acting member during such illness or absence and such acting member shall, while so acting, have all the powers and perform all the duties and be entitled to the indemnities of the member (vide Sub-clause 5) in whose stead he so acts, save and except that the next senior independent member appointed by the Central Government and not the acting member shall act as Chairman at business meeting of the Authority or as the Chairman of the Authority in the event of illness or absence of the Chairman of the Authority.

Sub-clause 2: Secretary of the Authority

The Authority shall employ a Secretary, who shall be an Engineer. He shall not be a member of the Authority.

Sub-clause 3: Quorum and Voting

Five members shall be a quorum and the concurrence of the majority shall be necessary for the transaction of the business of the Authority except such business as the Authority may from time to time prescribe as routine. The Authority shall not prescribe as routine any business in which the interests of any two of the States are likely to be in conflict. For the transaction of routine business three members shall be a quorum and in the absence of the Chairman of the Authority, the Chairman elected at the meeting shall have a deliberative vote and in the event of an equality of votes a casting vote also. Subject as aforesaid the members shall have equal powers.

Sub-clause 4: Disposal of Business by the Authority

4(1): Subject to the provisions of Sub-clause 4(2) below, the Authority may dispose of any matter before it either by circulation or by holding a meeting. However, it will be open to any member of the Authority to require that a matter shall not be disposed of by circulation but at a meeting.

4(2): On the following matters the Authority shall record its decision by a Resolution at a meeting in which the Chairman and all the members from the party States are present:

(i) Framing of Rules of Business;

(ii) Delegation of functions to a member or Secretary or any official of the Authority;

(iii) Categorising any part of the business of the Authority as of a formal or routine nature;

(iv) Any other matter which any of the four party States require that it shall be decided at a meeting where all the members from the party States are present.

However, if any particular item under this Sub-clause cannot be disposed of at two successive meetings owing to the absence of one or more members from the party States, it shall be disposed of under Sub-clause 3 of Clause XIV.

4(3): Subject to the foregoing provisions, the Authority shall frame its own Rules for the conduct of its business.

4(4): The Authority shall cause proper minutes or records of all its proceedings to be kept as a permanent record.

Sub-clause 5: Indemnity of Members

No member, officer or employee of the Authority shall be liable for loss, injury or damages resulting from (a) action taken by such member, officer or employee in good faith and without malice under the apparent authority of the Orders, even though such action is later determined to be unauthorised, or (b) the negligent or wrongful act of omission of any other person, employed by the Authority and serving under such member, officer or employee unless such member, officer or employee failed to exercise due care in the appointment of such other person or the supervision of his work.

Sub-clause 6: Officers and Servants of the Authority

The Authority may from time to time appoint or employ such and so many officers and servants as it thinks fit and remove or dismiss them, under the rules and regulations applicable to the appointment, removal and dismissal of the Central Government officers and servants. All such officers and servants shall as such be subject to the sole control of the Authority. The scales of pay and other service conditions shall be as applicable to Central Government employees. Persons employed in the services of the four States may be appointed or employed by the Authority in such proportions as the Authority may deem fit. The Authority shall arrange with the State governments to spare the services of the persons employed in the State governments for whole-time employment with the Authority, or for the performance of any work or services for the Authority. The Authority may also make direct recruitment of any personnel or obtain the same from the Centre or other source as considered appropriate.

Sub-clause 7: Administrative & Field Organisation Costs

(1) All expenses of the Authority (including the salary and expenses of the independent members) shall be borne by the State governments of Madhya Pradesh, Gujarat, Maharashtra and Rajasthan in equal shares. The expenses pertaining to a member representing a State shall be borne by the State concerned. The cost of maintaining, operating and controlling the gauging and other

hydrological stations in each State and the telecommunication systems for communicating the data shall be borne by the State concerned.

(2) The costs of construction of the storages, power installations, diversion works, headworks and canal networks shall be borne wholly by the State government in whose territory the work is located except for works whose cost has been ordered by the Tribunal to be shared between two or more party States. Where the capital cost is thus shared, the operation and maintenance cost shall also be shared in the same proportion.

Sub-clause 8: Powers, Functions & Duties of the Authority

8(1): The role of the Authority will mainly comprise co-ordination and direction. Normally all bilateral matters should be dealt with mutually by the States concerned and referred to the Authority only if there is a dispute.

8(2): The Authority shall be charged with the power and shall be under a duty to do any or all things necessary, sufficient and expedient for the implementation of the Orders with respect to:

(i) the storage, apportionment, regulation and control of the Narmada waters;

(ii) sharing of power benefits from Sardar Sarovar project;

(iii) regulated releases by Madhya Pradesh;

(iv) acquisition by the concerned State for Sardar Sarovar project of lands and properties likely to be submerged under Sardar Sarovar;

(v) compensation and rehabilitation and settlement of oustees; and

(vi) sharing of costs.

8(3): In particular and without prejudice to the generality of the foregoing functions, the Authority shall perform *inter alia* the following functions:

(i) Madhya Pradesh or Gujarat, as the case may be, shall submit to the Authority the SSP Report, the NSP Report, the Omkareshwar Project Report and the Maheshwar Project Report. The Authority shall point out to the States concerned, the Central Water Commission, the Central Electricity Authority and Planning Commission any features of these projects that may conflict with the implementation of the Orders of the Tribunal. Any subsequent changes in the salient features or substantial increase in cost in respect of dams, power houses and canal headworks shall be reported to the Authority for taking appropriate action in the matter.

(ii) The Authority shall decide the phasing and shall co-ordinate construction programmes of the NSP and Sardar Sarovar Unit II – Canals with a view to obtaining expeditiously optimum benefits during and after the completion of the construction of the projects, having due regard to the availability of funds.

(iii) The Authority shall obtain from the concerned States periodical progress reports both as to works and expenditure, and shall on receipt of such reports review the progress of construction of different units of the projects and whenever necessary advise the State concerned on the steps to be taken to expedite the work, except in respect of Unit I – Dam and Appurtenant Works and Unit III – Power Complex of Sardar Sarovar Project. The States shall submit, in respect of projects in Sub-clause 8(3)(i), completion reports to the Authority.

(iv) The Authority shall issue appropriate directions whenever necessary for timely and full compliance by the concerned States with the Orders of the Tribunal in the matter of acquisition for and making available to Gujarat lands and properties likely to be submerged under the Sardar Sarovar Project and in the matter of compensation and rehabilitation of oustees thereunder.

(v) The Authority shall cause to be established, maintained and operated by the State governments concerned or any one or more of them, such stream and other gauging stations, equipped with automatic recorders where necessary, discharge, silt and evaporation observation stations and measuring devices as may be necessary from time to time for securing the records required for carrying out the provisions of the Orders. If deemed necessary, the Authority may require the installation, maintenance and operation by the State concerned of measuring devices of approved type at the head of main canals as also at the offtake of the canal for Rajasthan for measuring amount of water diverted from the Narmada river system.

(vi) Concurrent records shall be kept of the flow of the Narmada at all stations considered necessary by the Authority and the records correlated.

(vii) The Authority shall frame rules of regulation and water accounting as per guidelines given in Clause IX. It shall determine the share of water of each State for every ten-day period for purposes of regulation and water accounting.

(viii) The Authority shall ensure implementation of the Orders of the Tribunal in respect of (a) quantum and pattern of regulated releases by Madhya Pradesh; (b) payment for such regulated releases/sharing of costs.

(ix) The Authority shall collect from the State concerned data of the areas irrigated by Narmada waters in each season, of power generated at each hydro-electric power station at and downstream of Narmada Sagar, of withdrawals for domestic, municipal and industrial or any other purposes and of waters going down the river from Sardar Sarovar Project.

(x) The Authority shall determine the volume of water flowing in the river Narmada and its tributaries in a water year (1 July to 30 June next year).

(xi) The Authority shall determine from time to time the volume of water stored by each State in reservoirs and other storages and may for that purpose adopt any device or method.

(xii) The Authority shall determine at appropriate periodic intervals the use of Narmada waters made by the States, or such of them as necessary, at any place or in any area at any time and for that purpose it may take note of all diversions or obstructions, whether natural or artificial or partly natural and partly artificial, from the river Narmada and its Tributaries and measure such use by any method as it deems fit.

(xiii) The Authority or any of its duly authorised representative shall have power to enter upon any land and property upon which any project or development of any project, or any work of gauging evaporation or other hydrological station or measuring device has been or is being constructed, operated or maintained by any State for the use of Narmada water. Each State through its appropriate departments shall render all co-operation and assistance to the Authority and its authorised representatives in this behalf.

(xiv) The Authority shall meet as often as necessary and decide on a proper management of waters including in particular the manner and details of withdrawal of waters from the storages on the Narmada river system in accordance with the orders. In particular, the Authority shall meet at the end of filling season, and review the availability of water in the storages on the Narmada river system and decide upon the pattern of their regulation for the next irrigation season, taking into account the carryover storages.

(xv) The Authority shall give directions for a phased programme of construction for generation and transmission of power in fulfilment of the shares of power allocated to the three States of Madhya Pradesh, Maharashtra and Gujarat from Sardar Sarovar and for payments therefore in accordance with the Orders of the Tribunal. The Authority shall also ensure that generation and transmission of power from Sardar Sarovar complex are in accordance with the Orders.

(xvi) The Authority shall issue appropriate directions for the establishment, maintenance and operation of an effective system of flood forecasting and flood control, including reporting of heavy precipitation and telecommunication systems. The safety of a structure shall primarily be the responsibility of the Chief Engineer in charge of the structure and no decision or order shall be binding on him if in his opinion the safety of the structure will be endangered thereby. The Authority shall publish annually and make available to party States the data regarding operation of reservoirs during floods.

8(4): In the light of its experience, the Authority may modify or add to the functions enumerated hereinabove in Sub-clauses 8(3)(i) to (xvi) by a resolution.

8(5): All the concerned States shall submit to the Authority all the relevant information called for by the Authority in connection with the Narmada Valley Development expeditiously.

Sub-clause 9: Annual Report of the Authority

The Authority shall prepare and transmit to each of the four States as early as possible and in any case before the end of the current water year (1 July to 30 June) an annual report covering the activities of the Authority for the preceding year and to make available to each State on its request any information within its possession any time and always provide access to its record to the States and their representatives.

Sub-clause 10: Records of the Authority and their Location

The Authority shall keep a record of all meetings and proceedings, maintain regular accounts, and have a suitable office where documents, records, accounts and gauging data shall be kept open for inspection by the four States or their representatives at such times and under such regulations as the Authority may determine. The location of the central, regional and sub-regional offices of the Narmada Control Authority shall be determined by the Authority. The headquarters of the Authority shall be at New Delhi until such time as it decides on its permanent location.

Sub-clause 11: Contracts and Agreements

The Authority shall enter into such contracts and agreements as may be necessary and essential for the full and proper performance of the functions and duties conferred or imposed on it.

Sub-Clause 12: Financial Provisions

(1) All the capital and revenue expenditure required to be incurred by the Authority shall be borne by the four States of Madhya Pradesh, Gujarat, Maharashtra and Rajasthan equally. The governments of the four States shall provide the necessary funds to the Authority to meet all capital and revenue expenditure required to be incurred by the Authority for the discharge of its functions.

(2) On the constitution of the Authority each of the governments of the four States shall contribute Rs 5,00,000 (Rupees five lakh) to the fund of the Authority in the first instance.

(3) The Authority shall in the month of September of each year prepare detailed estimates of the amounts of money required during the 12 months from the first day of April of the ensuing year, showing the manner in which it is proposed to expend such money. The Authority shall on or before 15 October forward a copy of such detailed estimate to the concerned Chief Engineers of the four States and indicate the amount required to be contributed by each State for the ensuing financial year. Each of the State governments shall pay to the Authority its contribution as indicated by the Authority on or before 30 April of the ensuing year.

(4) The Authority shall maintain detailed and accurate accounts of all receipts and disbursements and shall after the close of each financial year prepare an annual statement of accounts and send copies thereof to the Accountants General as well as the concerned Chief Engineers of the four States. The form of the annual statement of accounts shall be such as may be prescribed by rules. The accounts maintained by the Authority shall be open for inspection at all reasonable times by the four States through their duly authorised representative or representatives.

(5) Disbursement shall be made from the fund of the Authority only in such manner as may be prescribed by the Authority. The Authority may incur such expenditure as it may think fit to meet any emergency in the discharge of its functions.

(6) The accounts maintained by the Authority shall be audited by the Comptroller & Auditor General of India or his nominee, who shall certify subject to such observations as he may wish to make on the annual accounts of the Authority. The Authority shall forward to the Accountants General and the concerned Chief Engineers of the four States copies of the Report of the Comptroller & Auditor General of India and shall include the same in its Annual Report.

Sub-clause 13: Decision of the Authority

The decisions of the Authority on all matters covered under Sub-clause 8 shall be final and binding on the four party States. However, there shall be a Review Committee which may *suo motu* or on the application of any party State review any decision of the Authority. In urgent cases the Chairman of the Review Committee may, on the application of the party State, grant stay of any order of the Authority pending final decision on review.

Sub-clause 14: Review Committee

14(1): The Review Committee shall consist of five members including a Chairman as under:

(i) Union Minister for Irrigation as the	Chairman
(ii) Chief Minister of Madhya Pradesh	Member
(iii) Chief Minister of Gujarat	Member
(iv) Chief Minister of Maharashtra	Member
(v) Chief Minister of Rajasthan	Member

The Secretary of the Union Ministry of Agriculture and Irrigation, Department of Irrigation shall be the Convenor of the Review Committee but shall not have any voting right. In case there is President's rule in any of the States, the Governor of that State or his authorised representative will act as member of the Review Committee.

14(2): The Chief Ministers of the four States may nominate the respective Irrigation Ministers either generally or specially as the alternate member with full powers of voting, taking decisions etc.

14(3): The Review Committee may review the decision of the Authority at a meeting at which the Chairman and all the members of the Review Committee are present. It is expected that the decisions of the Review Committee will be by consensus. Failing consensus it shall be by majority of votes of members including the Chairman.

14(4): Advance notice of the proposed meeting of the Review Committee, its agenda and agenda notes will be forwarded by the Convenor to the party States.

14(5): The decision of the Review Committee shall be recorded in writing and shall be final and binding on all the States.

Sub-clause 15: Construction Outside Jurisdiction of the Authority

The planning and construction of the projects will be carried out by each State through its own agencies, save and except to the extent prescribed in Sub-clause 16 of Clause XIV.

Sub-clause 16: Supervisory Function of the Authority over Construction of Sardar Sarovar Project

(1) The four party States have financial commitment in respect of Unit I – Dam and Appurtenant Works of the Sardar Sarovar Project and three of them namely, Gujarat, Maharashtra and Madhya Pradesh have such commitment in respect of Unit III – Power Complex of the Project. With a view to ensuring efficient, economical and early execution of these Units of the Project, and taking into account the financial commitments of the party States, it is desirable and necessary that a Construction Advisory Committee should be constituted for the purpose. We, therefore, order that such an Advisory Committee to be called Sardar Sarovar Construction Advisory Committee should be set up within three months from the date of publication of the Decision of the Tribunal in the Official Gazette.

(2) The construction Advisory Committee shall have a whole-time Secretary of the rank of Chief Engineer to be appointed by Union of India and such other staff as may be necessary.

(3) The Committee shall comprise:

(i) The Secretary to the Government of India, in charge of Irrigation - Chairman.

(ii) Chairman, Central Water Commission (CWC), or a member of the CWC representing him in case the Chairman is unable to attend a meeting.

(iii) Chairman, Central Electricity Authority (CEA), or a member of the CEA representing him in case the Chairman is unable to attend a meeting.

(iv) Chairman, Narmada Control Authority (NCA) or an independent member of NCA representing him in case the Chairman is unable to attend a meeting.

(v) Joint Secretary (Financial Adviser) in the Union Ministry of Agriculture & Irrigation (Department of Irrigation).

(vi) Secretaries in charge of Finance Department of governments of Madhya Pradesh, Gujarat, Maharashtra and Rajasthan.

(vii) Secretaries in charge of Irrigation Department of governments of Gujarat and Rajasthan.

(viii) Secretaries in charge of Power Department of Madhya Pradesh, Maharashtra and Gujarat.

(ix) Secretaries in charge of Revenue Department or any other Department concerned with land acquisition of Madhya Pradesh, Maharashtra and Gujarat.

(x) General Manager or Chief Engineers of Gujarat in charge of the project and Chief Engineers of Madhya Pradesh, Maharashtra and Rajasthan concerned with the Project.

(xi) Chairman, State Electricity Boards of Madhya Pradesh, Maharashtra and Gujarat.

(xii) Financial Adviser, Sardar Sarovar Project.

The Chairman may co-opt any other member for any particular meeting.

(4) The Sardar Sarovar Construction Advisory Committee shall:

(i) scrutinise the project estimates prepared for these works, advise necessary modifications and recommend the estimates for the administrative approval of the concerned governments;

(ii) examine and make recommendations on all proposals pertaining to technical features and designs as may be referred to it by any of the party States and where necessary consult experts for the purpose;

(iii) examine and make recommendations on the programme of construction of different parts of the project in a co-ordinated manner, keeping in view the funds available, the economics of the project and the desirability of obtaining quick results;

(iv) examine the requirement of funds for the construction of works and other purposes according to the approved programme and make necessary recommendations;

(v) examine and recommend, from time to time, the delegation of such powers, both technical and financial, as it may deem necessary for the efficient execution of the project, to the General Manager/Chief Engineers, Superintending Engineers,

Executive Engineers and Sub-Divisional Officers engaged in the execution of the project;

(vi) examine and, where necessary, recommend specification for various classes of work;

(vii) examine and make recommendation on all sub-estimates and contracts, the cost of which exceeds the powers of sanction of the General Manager/Chief Engineers;

(viii) review progress reports, both for works and expenditure from the General Manager/Chief Engineers and recommend, where necessary, steps to be taken to expedite the work.

(5) The headquarters of the Construction Advisory Committee will be fixed by the Committee.

(6) The Construction Advisory Committee will frame rules regarding procedure and delegation of power for the purpose of carrying out its business.

(7) The recommendations of the Construction Advisory Committee shall be conveyed to the governments concerned by the Committee and copies sent to the Review Committee and Narmada Control Authority for information.

(8) The recommendations of the Construction Advisory Committee shall normally be accepted by the State governments concerned. In the event of any disagreement, the matter shall be referred to the Review Committee and the decision of the Review Committee shall be final and binding on all the concerned States.

In all matters relating to the construction of the Sardar Sarovar Dam and appurtenant works (Unit I), Power House and generating machinery (Unit III) and transmission lines to feed power to Madhya Pradesh, Maharashtra and Gujarat up to the next sub-station in each case, the Narmada Control Authority will carry out only such functions as do not specifically devolve upon the Construction Advisory Committee set up under Sub-clause 16, Clause XIV.

(9) The Construction Advisory Committee will be dissolved after three years of the completion of construction of Units I and III of the Sardar Sarovar Project. The post-construction management of Units I and III will be by Gujarat under the supervision of the Narmada Control Authority.

(10) The expenditure of the Construction Advisory committee will be borne by the four States of Madhya Pradesh, Gujarat, Maharashtra and Rajasthan equally.

Sub-clause 17

Nothing contained in this Order shall prevent the alteration, amendment or modification of all or any of the foregoing clauses by agreement between all the States concerned.

Sub-clause 18

The Union of India has consented to participate in the machinery to be established by the Order of the Tribunal, if so directed and to do its best to implement the decision of the Tribunal. Accordingly, we direct the Union of India to participate in the machinery set up by the Order of the Tribunal to implement the directions of the Tribunal specifically under Clauses 1(2), 4, 12(6), 13, 14 and generally to implement all the other directions so far as the Union of India is concerned.

Clause XV: Order as to Costs of Proceedings

(i) The States of Gujarat, Madhya Pradesh, Maharashtra and Rajasthan shall bear their own costs of appearing before the Tribunal. The expenses of the Tribunal shall be borne and paid by the aforesaid four States in equal shares. These directions relate to the references under Section 5(1) of the Inter-State Water Disputes Act, 1956.

(ii) The States of Gujarat, Madhya Pradesh, Maharashtra and Rajasthan shall bear their own costs of appearing before the Tribunal in the references under Section 5(3) of the said Act. The expenses of the Tribunal in respect of the aforesaid references shall be borne and paid by the States of Gujarat, Madhya Pradesh, Maharashtra and Rajasthan in equal shares.

Clause XVI: Period of Operation of Certain Clauses of the Final Order

In addition to Clauses III and IV (mentioned in Clause V), our orders in Clause VII with regard to Full Reservoir Level and Maximum Water Level of the Sardar Sarovar Dam, Clause VIII with regard to Sharing of Costs and Benefits, Clause IX with regard to Regulated Releases to be made by Madhya Pradesh for the Requirement of Sardar Sarovar Project, Clause X with regard to payment to be made by Gujarat to Madhya Pradesh for such Regulated Releases, Clause XII with regard to Allocation of Costs of Sardar Sarovar Project between Irrigation and Power, Clause XIII with regard to Allocation of Irrigation Component of Cost of Sardar Sarovar Project between Gujarat and Rajasthan and Clause XIV as regards Machinery are all made subject to review at any time after a period of 45 years from the date of publication of the Decision of the Tribunal in the Official Gazette.

Chapter 3

Clearances and Institutional Aspects

Clearances

Ministry of Environment and Forests, Environmental Clearance, 1987

Source: Government of India, Ministry of Environment and Forests, Office Memorandum (Environmental Clearance), 24 June 1987, No. 3-87/80-IA.

Office Memorandum

Subject: Approval of Narmada Sagar Project, Madhya Pradesh and Sardar Sarovar Project, Gujarat from Environmental Angle.

1. The NSP, Madhya Pradesh and the SSP, Gujarat have been referred to this department for environment clearance.

2. On the basis of examination of details on these projects by the environmental appraisal committee for river valley projects and discussions with the central and state authorities the following details were sought from the project authorities:

i. Rehabilitation master plan;
ii. Phased catchment area treatment scheme;
iii. Compensatory afforestation plan;
iv. Command area development;
v. Survey of flora and fauna;
vi. Carrying capacity of surrounding area;
vii. Seismicity; and
viii. Health aspects.

3. Field Surveys are yet to be completed. The first set of information has been made available and complete details have been assured to be furnished by 1989.

4. The NCA has been expanded and its terms of reference have been amplified to ensure that environmental safeguard measures are planned and implemented to

depth and in its pace of implementation *pari passu* with the progress of work on the project.

5. After taking into account all relevant facts, the NSP, Madhya Pradesh and the SSP, Gujarat are hereby accorded environmental clearance subject to the following conditions:

i. The Narmada Control Authority (NCA) will ensure that environmental safeguard measures are planned and implemented *pari passu* with progress of work on projects.

ii. The detailed surveys/studies assured will be carried out as per the schedule proposed and details made available to the department for assessment.

iii. The catchment area treatment programme and the rehabilitation plans be so drawn as to be completed ahead of reservoir filling.

iv. The department should be kept informed of progress on various works periodically.

6. Approval under the Forest (Conservation) Act, 1980 for diversion of forest land will be obtained separately. No work would be initiated on forest area prior to this approval.

7. Approval from environmental and forestry angles for any other irrigation, power or development projects in the Narmada Basin should be obtained separately.

Ministry of Environment and Forests, Clearance for Diversion of Forest Land, 1987

Source: Government of India, Ministry of Environment and Forests, 8 September 1987, No. 8 37283 PC; letter from R.S. Bisht, Under Secretary to the Government of India addressed to The Secretary, Agriculture, Forest and Cooperative Department, Government of Gujarat; The Secretary, Forest Department, Government of Madhya Pradesh; The Secretary, Revenue & Forest Department, Government of Maharashtra.

Sub: Diversion of 13,388.48 ha (6,488.84 ha in Maharashtra, 4,165.91 ha in Gujarat and 2,731.00 ha in Madhya Pradesh) of forest land in Dhule, Bharuch and Khargone district respectively for Sardar Sarovar Project.

Sir,

1. I am directed to refer to your letter No. (1) FLD 1282.78159-V-1 dated 17 February 1983 (Gujarat), (2) 8/58/83/10/3 dated 31 August 1984 (Madhya Pradesh) and (3) CLD 1080/111531-II-F3 dated 8 September 1983 (Maharashtra) on the above mentioned subject seeking prior approval of the Central Government

under Section 2 of the Forest (Cons.) Act, 1980 and to say that the proposal has been considered by the Advisory Committee constituted by the Central Government under Section 3 of the Forest (Cons.) Act, 1980.

2. After careful consideration of the proposal, the Central Government hereby conveys its approval for diversion of 13,388.48 ha of forest land for Sardar Sarovar Project as per details given below:

	State	Forest Land to be diverted (ha)
i.	Gujarat	4,165.91
ii.	Madhya Pradesh	2,731.00
iii.	Maharashtra	6,488.84

3. This approval is strictly subject to the following conditions:

i) Legal status of land will remain unchanged.

ii) The full details of the non-forest lands for raising compensatory afforestation with complete details viz Khasara No., village etc. will be reported by the State governments before 30 September 1987.

iii) The non-forest areas available for rehabilitation of all the oustees will be reported by the State governments or a proposal to the satisfaction of the Government of India in this regard will be furnished by the State governments before 30 November 1987.

iv) No work on the project in forest area will be commenced until and unless conditions under (ii) and (iii) above are fulfilled.

v) Since the project involves violation and also most of the non-forest areas for compensatory afforestation are away from the project area, the State governments will raise compensatory afforestation in double the degraded forest lands also in the project impact areas in addition to the afforestation on equivalent non-forest land. A scheme for this will be submitted by 30 November 1987.

vi) The State governments will prepare by 30 November 87 a plan for the treatment of catchments areas failing which the Central Government will appoint a team for this purpose at the cost of the project.

vii) No forest land will be utilised for the rehabilitation of oustees.

viii) Tree felling will be permitted in submergence areas only up to 4 metres below FRL.

ix) Tree planting will be done on either side of the canals, roads, foreshore of the reservoir and in the wasteland/vacant lands under the control of the Irrigation Department.

x) Water will be supplied free of cost to the Forest Department for raising nursery and for irrigating forestry plantations in the command area.

xi) In order that the construction labour & staff while working on the project in the forest area may not cause destruction to the forest area for meeting their fuelwood needs, the user agency will establish fuelwood depots and will provide suitable alternative domestic fuel such as fuelwood, coal, kerosene oil etc. to them free of cost or at a cost deducted from their salary and wages.

Parliamentary Questions on Environmental Impact Assessment, 1993

Source: Annual Report of the MoEF for the Year 1992–93 (Presented to the Rajya Sabha on 9 December 1993).

Sardar Sarovar Project

62. The Committee made a reference to a recommendation made in an independent review of the Sardar Sarovar Project commissioned by the World Bank namely, that detailed assessment of social, human and environmental impacts need to be carried out for the project. The Committee enquired whether work on such a detailed assessment has commenced as yet or not.

63. The representative of the Ministry clarified that the project authorities were not ready with the data and field surveys/studies to carry out realistic environmental impact assessment in 1987 when a decision to go ahead with the project was taken at the highest level. It was decided at that time that the project authorities would be in a position to furnish environmental impact assessment reports and action plans by the end of 1989. However, it was informed that all the requisite environmental action plans had not yet been finalised.

64. The Committee fails to comprehend the actual factors responsible for the non-implementation of the specific recommendation made by the World Bank for carrying out realistic environmental impact assessment of the Sardar Sarovar Project. The question of environmental impact assessment is not merely a technical requirement. In fact, it involves a broader question of human welfare and affects society in a direct manner. Before work on any project commences, the project authorities should see to it that proper plans are chalked out for the rehabilitation of the oustees. The Committee is surprised to note that absence of these recommended studies did not deter the government from going ahead with the project. The Committee feels that the independent review of the Sardar Sarovar Project commissioned by the World Bank should have been accorded due importance. The Committee is pained to note that in spite of the existence of an environmental sub-group under the Chairmanship of Secretary (Environment &

Forests) for ensuring preparation of satisfactory environment action plan in a time bound manner, all the requisite plans have not yet been finalised.

Clearance of the Planning Commission, 1988

Source: Planning Commission, Government of India, 5 October 1988, Letter No. 2 (194)/88-I&CAD addressed to The Secretary, Planning Department, Government of Gujarat, Gandhinagar.

Subject: Sardar Sarovar Project in Gujarat, Acceptance of

Sir,

I am directed to convey that the Sardar Sarovar Project, estimated to cost Rs 6,406.04 crores (Rupees six thousand four hundred six crores and four lakhs) as per the salient features vide Annexure I enclosed herewith has been considered acceptable for investment subject to the conditions laid down below:

(i) The State shall comply with the conditions as laid down in the O.M. No. 3-87/80-IA dated 8 September 1987 issued by the MoEF while according the environmental clearance and the approval for diversion of forest lands for this project respectively.

(ii) Looking to the size and importance of this project, the State government will give sufficient priority to this project in the eighth Plan by ensuring adequate funding to match with the construction schedule as indicated in the concurrence of State Planning and Finance Department vide government of Gujarat in Narmada Development Department's letter No. NPP/1084/GOI-4/Pat.V/J dated 3 October 1988. The State will also complete other ongoing projects at advance stage in time to ensure that there is no difficulty in funding the peak requirements of the Sardar Sarovar Project.

(iii) A programme of drainage and groundwater balance studies has been completed for Mahi Narmada–Doab. Such a programme must be completed for the areas beyond the Mahi. The Bhal, Saurashtra, Kachchh, Sami-Harij and other areas require this as a precondition. The State should submit to Planning Commission a detailed programme of studies, with milestones of achievements, duly vetted through Central Water Commission for monitoring the same by Planning Commission.

(iv) The State should take suitable advance measures, as may be necessary, to ensure that annual revenue to be accrued from this project covers at least annual operation and maintenance charges including depreciation charges by setting the water rates suitably.

(v) The State should set up a special group of experts to study the siltation aspect in the main canals under all operating conditions since such siltation, if occurs, is likely to pose a serious problem during the actual operation of this project and may require a huge expenditure for desilting as well as result into serious operational difficulties.

(vi) The State should draw up a detailed time schedule for completion within five years of the investigation, detailed survey, planning and working out the detailed cost estimates for micro-level network system for the balance area of the total command of this project.

(vii) Past experience of irrigation projects has revealed that the main branch of canals are completed up to the end but, in the absence of micro-level networks to take irrigation water up to the outlet, corresponding irrigation benefits do not start accruing in spite of huge financial investments made. To avoid this, the State should draw up an implementation schedule, segmentwise, for completion of the canal network in such a way that a segment of the canal network, taken up from the head reaches, is completed in all respects so as to make the irrigation waters available, for the designed potential of that segment, up to the outlet in that particular segment.

2. The project may be executed as per the approved outlay from year to year.

Institutional Aspects

Constitution of the Narmada Control Authority, 1980

Source: Ministry of Irrigation Notification, Narmada Water Scheme, 10 September 1980 (as amended).

S.O. 770(E) Whereas the Central Government had constituted by a notification of the Government of India in the Ministry of Irrigation and Power S.O. No. 4054, dated 6 October 1969, issued under Section 4 of the Inter-State Water Dispute Act, 1956 (33 of 1956), the Narmada Water Disputes Tribunal to adjudicate upon the water dispute regarding the inter-State river Narmada and the river valley thereof;

And whereas the said Tribunal investigated the matters referred to it and forwarded to the Central Government under sub-section (2) of section 5 of the said Act, a report setting out the facts as found by it and giving its decision on the matters referred to it;

And whereas upon consideration of the said decision the Central Government and the Governments of the States of Gujarat, Madhya Pradesh, Maharashtra and

Rajasthan made references to the said Tribunal under sub-section (3) of section 5 of the said Act and the Tribunal, on such references, forwarded to the Central Government under that sub-section a further report;

And whereas the decision of the said Tribunal as modified by the explanation and guidance given in its further report was published in the Official Gazette by the Central Government as required by the section 6 of the said Act, vide notification of the Government of India in the Ministry of Agriculture & Irrigation (Department of Irrigation) No. S.O. 792(E) dated the 12 December 1979, whereupon the decision became final and binding on the partners to the dispute;

And whereas the decision of the said Tribunal provides for the setting up of the machinery, namely, Narmada Control Authority and a Review Committee for implementing its decisions and directions;

Now, therefore, in exercise of the powers conferred by Section 6-A of the said Act, the Central Government hereby frames the scheme, *inter alia* constituting the Narmada Control Authority (hereinafter referred as the Authority) and the Review Committee to give effect to the decision of the Narmada Water Dispute Tribunal, namely:

1(1) This scheme may be called the Narmada Water (Amendment) Scheme, 1990.

1(2) It shall come into force on the date of its publication in the Official Gazette.

I. Narmada Control Authority

2. Status and constitution of the Authority

(1) The Narmada Control Authority shall be a body corporate having perpetual succession and a common seal and shall sue and be sued.

(2)(a) The Authority shall consist of the following members namely:

- i) Secretary to the Government of India Ministry of Water Resources (Chairman)
- ii) Secretary to the Government of India, Ministry of Energy, Department of Power (Member)
- iii) Secretary to the Government of India, Ministry of Environment & Forests (Member)
- iv) Secretary to the Government of India, Ministry of Social Justice and Empowerment (Member)
- iv.a) Secretary to the Government of India, Ministry of Tribal Affairs (Member)
- v) Chief Secretary to the Government of Gujarat (Member)
- vi) Chief Secretary to the Government of Madhya Pradesh (Member)
- vii) Chief Secretary to the Government of Maharashtra (Member)
- viii) Chief Secretary to the Government of Rajasthan (Member)

ix) to xi) Three persons not below the rank of Chief Engineer to be appointed by the Central Government as independent members one of whom shall be designated as the executive member of the Authority (Member)

xi.a)One person not below the rank of a Joint Secretary to the Government of India or an Additional Secretary in a State government having experience in the fields of Environment & Rehabilitation of displaced persons, to be appointed by the Central Governments as independent member (Member)

xii) to xv) Four persons of the rank of engineer-in-chief, chief engineer in charge of the Irrigation Department or the Power Department or the State Electricity Board, one each to be appointed by the State governments of Gujarat, Madhya Pradesh, Maharashtra and Rajasthan (member).

(b) The executive member will be in charge of the administrative work of the Authority under the general supervision and control of the Chairman. The Central Government, or as the case may be, the State government shall have the power to remove or suspend any of the members appointed under clauses (ix) to (xv) if in its opinion he is not suitable to continue as a member.

(c) [omitted]

(d) The Secretaries to the Government of India or the Chief Secretaries to the State governments when unable to attend the meetings of the Authority may send their representative not below the rank of Joint Secretary in the Government of India or as the case may be a Secretary in the State.

(3) Each independent member shall be a full-time member and be appointed for a term not exceeding five years. The members appointed by the State governments shall be part-time members. The appointing authority for independent members or that for part-time members, as the case may be, shall determine the terms and conditions of appointment in each case.

(4) On any vacancy occurring in the offices of the three independent members, the Central Government shall appoint a person to such vacant office and on any vacancy occurring in the office of the four members other than the independent members, the State government by whom the member whose office falls vacant was appointed shall appoint a person to the vacant office.

Provided that in case of illness or absence for any cause whatever of a member, the Central Government or State government by whom he was appointed, as the case may be, may appoint a person as an acting member during such illness or absence and such acting member shall, while so acting, have all the powers and perform all the duties & be entitled to the indemnities of the member, in whose stead he so acts.

(5) [omitted]

3. Secretary of the Authority

The Authority shall employ a secretary, who shall be an engineer. He shall not be a member of the Authority.

4. Quorum and voting

The quorum to constitute a meeting of the Authority (other than routine business) shall be eight members. Of these at least one member each from any three of the four participating States shall be present. No decision affecting a State's interest shall, however, be taken without the presence of at least one member of that State government being present in the meeting. The quorum for transaction of routine business shall be five members. Every question, except a routine business, brought before any meeting of the Authority shall be decided by a majority of members present and voting at the meeting before which the matter is brought. The Authority shall not prescribe as routine any business in which decisions are to be taken on any matter affecting the interest of more than one State. In the absence of the chairman, the member elected at the meeting shall preside over the meeting. In the case of equality of votes the chairman or member presiding over the meeting shall have a second or casting vote. Save as aforesaid the members shall have equal powers.

5. Disposal of business by the Authority

(1) Subject to the provisions of sub-paragraph (2), the Authority may dispose of any matter before it either by circulation or by holding a meeting. However, it will be open to any member of the Authority to require that a matter shall not be disposed of by circulation but at a meeting.

(2) On the following matters, the Authority shall record its decision by a Resolution at a meeting in which the Chairman and all the members from the party States are present:

i) Framing of Rules of Business;

ii) Delegation of functions to a member or Secretary or any official of the Authority;

iii) Categorising any part of the business of the Authority as of a formal or routine nature;

iv) Any other matter which any of the four party states require that it shall be decided at a meeting where all the members from the party states are present.

Provided that, if any particular item cannot be disposed of at two successive meetings owing to the absence of one or more members from the party States, it shall be disposed of as provided in paragraph 4.

(3) Subject to the foregoing provisions, the Authority shall frame its own rules for the conduct of its business.

(4) The Authority shall cause proper minutes or records of all its proceedings to be kept as permanent record.

(5) Where the Secretary to the Government of India, Ministry of Environment & Forests, is of the opinion that any decision of the Authority is inconsistent with the needs of environmental protection, the matter shall be reserved for the decision of the Review Committee.

6. Indemnity of members

No member, officer or employee of the Authority shall be liable for loss, injury or damages resulting from (a) action taken by such member, officer or employee in good faith and without malice under the apparent authority of the orders, even though such action is later determined to be unauthorised or (b) the negligent or wrongful act of omission of any other person, employed by the Authority and serving under such member, officer or employee unless such member, officer or employee failed to exercise due care in the appointment of such other person or the supervision of his work.

7. Officers and servants of the Authority

The Authority may from time to time appoint or employ such and so many officers and employees as it thinks fit and remove or dismiss them, under the rules & regulations applicable to the appointment, removal and dismissal of the Central Government officers and employees. All such officers and employees shall be subject to the sole control of the Authority. The Authority may, with the previous approval of the Central Government, make regulations to regulate conditions of service of all such officers and employees in respect of residential accommodation, house rent allowance, travelling allowance, daily allowance, conveyance allowance and medical reimbursement. The scales of pay and other service conditions shall be as applicable to Central Government employees.

Persons employed in the services of the four states may be appointed or employed by the Authority in such proportions as the Authority may deem fit. The Authority shall arrange with the State governments to spare the services of the persons employed in State governments for whole-time employment with the Authority, or for the performance of any work or services for the Authority. The Authority may also make direct recruitment of any personnel or obtain the same from the Centre or other source as considered appropriate.

8. Administrative and field organisation costs

(1) All expenses of the Authority (including the salary and expenses of the independent members) shall be borne by the State governments of Madhya Pradesh, Gujarat, Maharashtra and Rajasthan in equal shares. The expenses pertaining to a member representing a State shall be borne by the State concerned.

The expenses pertaining to the members representing the Central Government shall be borne by the government. The cost of maintaining, operating and controlling the gauging and other hydrological stations in each State and the communication system for communicating the data shall be borne by the State concerned.

(2) The costs of construction of the storages, power installations, diversion works, headworks and canal networks shall be borne wholly by the State government in whose territory the work is located except for works whose cost has been ordered by the Tribunal to be shared between two or more party States. Where the capital cost is thus shared, the operation and maintenance cost shall also be shared in the same proportion.

9. Powers, functions, and duties of the Authority

(1) The role of the Authority will mainly comprise of overall coordination and direction of the implementation of all the projects including the engineering works, the environmental protection measures and the rehabilitation programme and to ensure the faithful compliance of the terms and conditions stipulated by the Central Government at the time of clearance of the aforesaid projects.

2. The Authority shall be charged with the power and shall be under a duty to do any or all things necessary, sufficient and expedient for the implementation of the Order of the Tribunal with respect to:

i) the storage, apportionment, regulation and control of the Narmada waters;

ii) sharing of power benefits from Sardar Sarovar Project;

iii) regulated releases by Madhya Pradesh;

iv) acquisition by the concerned States for Sardar Sarovar project of lands and properties likely to be submerged under Sardar Sarovar;

v) compensation and rehabilitation and settlement of oustees; and

vi) sharing of costs.

(2)(a) The Authority may constitute one or more sub-committees and assign to them such of its function and delegate such of its powers as it think fit.

(3) In particular and without prejudice to the generality of the foregoing functions, the Authority shall perform *inter alia*, the following functions:

i) Madhya Pradesh or Gujarat as the case may be shall submit to the Authority the SSP Report, the NSP Report, the Omkareshwar Project Report and the Maheshwar Project Report. The Authority shall point out to the States concerned, the Central Water Commission, the Central Electricity Authority and Planning Commission any features of these projects which may conflict with the implementation of the orders of the Tribunal. Any subsequent changes in the salient features or substantial increase in cost in respect of dams, power houses and canal headworks shall be reported to the Authority for taking appropriate action in the matter.

ii) The Authority shall decide the phasing and shall coordinate construction programmes of the NSP and Sardar Sarovar Unit II (canals) with a view to obtaining expeditiously optimum benefits during and after the completion of the construction of the projects having due regard to the availability of funds.

iii) The Authority shall obtain from the concerned States periodical progress reports both as to works and expenditure and shall on receipt of such reports review the progress of construction of different units of the projects and whenever necessary advise the State concerned on the steps to be taken to expedite the work, except in respect of Unit I (dam and appurtenant works) and Unit III (power complex of Sardar Sarovar Project). The States shall submit completion reports to the Authority in respect of projects referred to in sub-paragraph (3)(i).

iv) The Authority shall issue appropriate directions whenever necessary for timely and full compliance by the concerned States within the orders of the Tribunal in the matter of acquisition for and making available to Gujarat lands and properties likely to be submerged under the Sardar Sarovar Project and in the matter of compensation and rehabilitation of oustees thereunder.

v) The Authority shall cause to be established, maintained and operated by the State governments concerned or any one or more to them, such stream and other gauging stations, equipped with automatic recorders, where necessary, discharge, silt and evaporation observation stations and measuring devices as may be necessary from time to time for securing the records required for carrying out the provisions of the orders of the Tribunal. If deemed necessary, the Authority may require the installation, maintenance and operation by the State concerned of measuring devices of approved type at the head of main canals as also at the offtake of the canal for Rajasthan for measuring the amount of water diverted from the Narmada river system.

vi) Concurrent records shall be kept of the flow of the Narmada at all stations considered necessary by the Authority and the records correlated.

vii) The Authority shall frame rules of regulation and water accounting as per guidelines given in Clause IX of the order of the Tribunal. It shall determine the share of water of each State for every ten-day period for purposes of regulation and water accounting.

viii) The Authority shall ensure implementation of the orders of the Tribunal in respect of (a) quantum and pattern of regulated releases by Madhya Pradesh; (b) payment for such regulated releases and sharing of costs.

ix) The Authority shall collect from the State concerned data for the areas irrigated by Narmada waters in each season, of power generated at each hydroelectric power station at and downstream of Narmada Sagar, of withdrawals for domestic, municipal and industrial or any other purposes and of waters going down the river from Sardar Sarovar Project.

x) The Authority shall determine the volume of water flowing in the river Narmada and its tributaries in a water year (1 July of a year to 30 June of the next year).

xi) The Authority shall determine from time to time the volume of water stored by each State in reservoirs and other storages and may for that purpose adopt any device or method.

xii) The Authority shall determine at appropriate periodic intervals the use of Narmada Water made by the states, or such of them as necessary, at any place or in any areas at any time and for that purpose it may take note of all diversions or obstructions, whether natural or artificial or partly natural and partly artificial from the river Narmada and its tributaries and measure such use by any method as it deems fit.

xiii) The Authority or any of its duly authorised representative shall have power to enter upon any land and property upon which any project or development of any project, or any work of gauging, evaporation or other hydrological station or measuring device has been or is being constructed, operated or maintained by any state for the use of Narmada water. Each state through its appropriate departments shall render all cooperation and assistance to the Authority and its authorised representatives in this behalf.

xiv) The Authority shall meet as often as necessary and decide on a proper management of waters including in particular the manner and details of withdrawals of waters from the storages on the Narmada river system in accordance with the orders of the Tribunal. In particular, the Authority shall meet at the end of the filling season, and review the availability of waters in the storages on the Narmada river system and decide upon the pattern of their regulation for the next irrigation season, taking into account the carry over storage.

xv) The Authority shall give directions for a phased programme of construction for generation and transmission of power in fulfilment of the shares of power allocated to the three States of Madhya Pradesh, Maharashtra and Gujarat from Sardar Sarovar and for payments therefore in accordance with the orders of the Tribunal. The Authority shall also ensure that generation and transmission of power from Sardar Sarovar complex are in accordance with the orders.

xvi) The Authority shall issue appropriate directions for establishment, and maintenance and operation of an effective system of flood forecasting and flood control including reporting of heavy precipitation and telecommunication systems. The safety of a structure shall primarily be the responsibility of the Chief Engineer in charge of the structure and no decision or order shall be binding on him if in his opinion the safety of the structure will be endangered thereby. The Authority shall publish annually and make available to party States the data regarding the operation of reservoirs during floods.

(4) In the light of its experience, the Authority may modify or add to the functions enumerated in sub-paragraph (3)(i) to sub-paragraph 3(xvi) by a resolution.

(5) All the concerned States shall submit to the Authority all the relevant information called for by the Authority in connection with the Narmada Valley Development expeditiously.

10. Annual report of the Authority

The Authority shall prepare and transmit to each of the four party States as early as possible and in any case before the end of the current Water Year (1 July of the year to 30 June of the next year) an annual report covering the activities of the Authority for the preceding year and to make available to the Central Government and to the government of each of the party states, on its request, any information within its possession any time and always provide access to its records to the Central Government and to the government of each of the party states and their representatives. The Central Government shall cause the annual report to be laid before each house of the Parliament.

11. Records of the Authority and their location

The Authority shall keep a record of all meetings and proceedings, maintain regular accounts, and have a suitable office where documents, records, accounts and gauging data shall be kept open for inspection by the Central Government and the government of each of the party states or their representatives at such times and under such regulations as the Authority may determine.

The location of the central, regional and sub-regional offices of the Narmada Control Authority shall be determined by the Authority. The headquarters of the Authority shall be at New Delhi until such time at it decides on its permanent location.

12. Contracts and agreements

The Authority shall enter into such contracts and agreements as may be necessary and essential for the full and proper performance of the functions, and duties conferred or imposed on it.

13. Financial provision

(1) All the capital and revenue expenditure required to be incurred by the Authority shall be borne by the State governments of Madhya Pradesh, Gujarat, Maharashtra and Rajasthan equally. The governments of the said States shall provide the necessary funds to the Authority to meet all capital and revenue expenditure required to be incurred by the Authority for the discharge of its functions. For this a fund to be called 'The Narmada Control Authority Fund' shall be constituted to

which the sums paid by the States and other sums received by the Authority shall be credited.

(2) On the constitution of the Authority, the governments of the States of Madhya Pradesh, Gujarat, Maharashtra and Rajasthan shall contribute each a sum of Rs 5,00,000 to the fund of the Authority in the first instance.

(3) The Authority shall in the month of September of each year prepare detailed estimate of the amounts of money required during the 12 months from the first day of April of the ensuring year, showing the manner in which it is proposed to spend such money. The Authority shall on or before 15 October forward a copy of such detailed estimate to the concerned Chief Secretaries of the four States and indicate the amount required to be contributed by each State for the ensuing financial year. Each of the State governments shall pay to the Authority its contribution as indicated by the Authority on or before 30 April of the ensuing year.

(4) The Authority shall maintain detailed and accurate accounts of all receipts and disbursements and shall, after the close of each financial year, prepare an annual statement of accounts and send copies thereof to the Accountant General as well as the concerned Chief Secretaries of the four States. The form of the Annual Statements of Accounts shall be such as may be prescribed by rules framed by the Authority. The accounts maintained by the Authority shall be open for inspection at all reasonable times by the Central Government and the governments of the party states through their duly authorised representative or representatives.

(5) Disbursement shall be made from the funds of the Authority only in such manner as may be prescribed by the Authority. The Authority may incur such expenditure as it may think fit to meet any emergency in the discharge of its functions.

(6) The accounts maintained by the Authority shall be audited by the Comptroller and Auditor General of India or his nominee who shall certify subject to such observation as he may wish to make on the annual accounts of the Authority. The Authority shall forward to the Accountant General and the concerned Chief Secretaries of the four States copies of the Report of the Comptroller and Auditor General of India and shall include the same in its Annual Report.

14. Decisions of the Authority

The decisions of the Authority on all matters covered under paragraph 9 shall be final and binding on all the party States.

15. Construction outside jurisdiction of the Authority

Save and except to the extent otherwise prescribed in the order of the Tribunal, the planning and construction of the projects will be carried out by each State government through its own agencies.

16. Review Committee

(1) There shall be a Review Committee which may *suo moto* or on the application of any party State or Secretary to the Government of India, Ministry of Environment and Forests review any decision of the Authority. In urgent cases the Chairman of the Review Committee may on the application of the government of any party State, or Secretary to the Government of India, Ministry of Environment and Forests grant a stay of any order of the Authority pending final decision on review.

(2) The Review Committee shall consist of six members including a Chairman as under:

i) Union Minister in charge of Water Resources (Chairman)

i(a) Union Minister of Environment & Forests (Member)

ii) Chief Minister of Madhya Pradesh (Member)

iii) Chief Minister of Gujarat (Member)

iv) Chief Minister of Maharashtra (Member)

v) Chief Minister of Rajasthan (Member)

The Secretary to the Government of India, Ministry of Water Resources shall be the convenor of the Review Committee but shall not have any voting right. In case there is a President's Rule in any of the four party States, the governor of the State or his authorised representative will act as member of the Review Committee.

(3) The Chief Ministers of the party states may nominate their respective ministers in charge of irrigation either generally or specially as the alternate member with full powers of voting, taking decisions, etc.

(4) The Review Committee may review the decision of the Authority at a meeting at which the Chairman and all the members of the Review Committee are present. The decisions of the Review Committee will be by consensus. In cases, where no consensus is possible, the decision shall be by majority of votes of members including the chairman.

(5) Advance notice of the proposed meeting of the Review Committee, its agenda and agenda notes will be forwarded by the convenor to the governments of the party States.

(6) The decision of the Review Committee shall be recorded in writing and shall be final and binding on all the States.

17.

The Authority may with the previous approval of the Central Government, make regulations for giving effect to the purpose of the Scheme.

Annex III: Environment Sub-group – Composition and Functions

(a) Composition [omitted]

(b) Functions

i) To work out the environmental safeguard measures to be planned and implemented for the entire Narmada basin so that environmental safeguard measures are executed and remain fully in consonance with the clearance accorded to the Narmada Sagar and Sardar Sarovar Projects.

ii) To determine the terms of reference of required surveys and studies necessary for implementation of environmental safeguard measures inclusive of database required, the methods by which the database is to be prepared and also to identify the institutions/individuals to undertake the preparation of such documents.

iii) To get prepared for clearance by the Ministries and NCA the action plans with regard to all environmental safeguard measures and the assessment criteria thereof.

iv) To devise a suitable monitoring and evaluation mechanism so that the action plans are effectively implemented in consonance with stipulations at the time of clearance of the projects.

v) To assess the necessary organization with management capability being set up for adequate implementation of environmental safeguard measures.

vi) To undertake all measures necessary to assist Narmada Control Authority in the planning and implementation of environmental safeguard measures.

Annex IV: Resettlement and Rehabilitation Sub-group – Composition and Functions

(a) Composition [omitted]

(b) Functions

i) To monitor the progress of land acquisition in respect of submergence land of SSP.

ii) To monitor the progress of implementation of the action plan of rehabilitation of project affected families (PAFs) in the affected villages of SSP in concerned States.

iii) To review the resettlement and rehabilitation (R&R) action plan from time to time in the light of results of the implementation.

iv) To review the reports of the agencies entrusted by each of the States in respect of monitoring and evaluation of the progress in the matter of resettlement and rehabilitation.

v) To monitor and review the implementation of R&R programmes *pari passu* with the raising of the dam height, keeping in view the clearance granted to SSP from the environmental angle by the Government of India and the Ministry of Environment and Forests.

vi) To coordinate state/agencies involved in the R&R programmes of SSP.

viii) To undertake any or all activities in the matter of R&R pertaining to SSP.

Notification Setting up the Grievances Redressal Authority of Madhya Pradesh, 2000

Source: Government of Madhya Pradesh, Narmada Valley Development Department, Notification No.5/7/27/2/2000/474, Bhopal, 30 March 2000.

Preamble

1. That in accordance with stipulations in the NWDT Award and provisions in the R&R Policy of the State of Madhya Pradesh, the State of Madhya Pradesh has been taking all steps for R&R including providing civic amenities to displaced persons (oustees/PAPs) from submergence areas of Sardar Sarovar Dam in Madhya Pradesh.

2. Complaints and grievances, if any, relating to R&R are being attended to and necessary action is being taken through existing machinery. In pursuance of Appeal mechanism as provided in paragraph 13 of the R&R Policy of Madhya Pradesh 1992, the Government of Madhya Pradesh in addition is desirous of setting up a high-level autonomous authority before whom the oustees (PAPs) already resettled and to be resettled hereafter in Madhya Pradesh can ventilate grievances for redressal after their resettlement until the process of R&R is completed. Such authority will be empowered to ensure that all oustees (PAPs) receive all benefits and amenities due to them in accordance with stipulations of the NWDT Award, provisions in the R&R policy of MP and various orders made from time to time by the Government of Madhya Pradesh. Accordingly, the Government of Madhya Pradesh is pleased to resolve as follows:

Resolution

1. The Government of Madhya Pradesh shall, by notification, establish an authority to be known as the Grievances Redressal Authority for Sardar Sarovar Project Affected Persons (PAPs) resettled in Madhya Pradesh.

2. The Authority shall consist of a chairman and two members to be appointed by the Government of Madhya Pradesh. The appointment of the members shall be made in consultation with the chairman.

3. The chairman or the member may, by a letter addressed to the Governor, resign from office. The chairman shall not be removed from office except by an order made by the Government with the concurrence of the Hon'ble Chief

Justice of India. A member shall not be removed from office except on the recommendations of the chairman. Such removal shall only be on the ground of proved misbehaviour and incapacity after holding due enquiry.

4. The chairman shall be a retired Supreme Court judge or retired Chief Justice or a judge of the High Court. The members shall be from amongst men of eminence in social public life and/or administrative service who have distinguished themselves in their field of activity.

5. The term of office, salaries and allowances and other terms and conditions of the chairman and the members shall be determined separately in consultation with the first chairman.

6. The Authority shall be provided with establishment, officers, employees as well as other amenities as necessary. The officers and employees of the Authority shall discharge their functions and duties under the general superintendence and directions of the chairman.

7. The chairman shall exercise such financial and administrative powers as may be vested in him by the Government of Madhya Pradesh after consultation with him. Such financial and administrative powers may be delegated to a member or an officer of the Authority by the chairman.

8. The appointment of secretary shall be done in consultation with the chairman.

9. The Authority shall have the power and authority to decide its own method to deal with complaints or grievances from a resettled oustee in Madhya Pradesh ensuring that such persons receive all benefits and amenities due to them in accordance with stipulation in the N.W.D.T. Award and provisions in the policy of Madhya Pradesh, and the various orders/resolutions made from time to time on that behalf by the Government of Madhya Pradesh.

10. The Authority, without prejudice to the generality of overall supervision power, shall exercise the following powers and authority:

    i. To set up an appropriate mechanism before which oustees (PAPs) would be enabled to lodge a complaint/grievance in relation to any matter arising out of R&R.

    ii. To devise a procedure by which the complaint/grievance may reach as expeditiously as possible to the secretary of the Authority.

    iii. To ensure that the complaint/grievance received by the secretary or referred to him by the Authority is examined and communicated to the persons and agencies concerned working under him as well as to the Authority.

iv. To take such steps as are necessary to satisfy itself that the grievances raised from time to time by the oustees (PAPs) have been suitably redressed at the appropriate levels.

11. The Authority may make rules, regulations or may frame guidelines for its practice and procedure.

12. The Authority or any member or any officer nominated by the chairman may tour the different R&R sites and such other areas to enquire into the complaints/grievances and for redressal thereof.

13. The Authority will have power to take assistance of any expert or instrumentality having expertise in the discharge of its functions.

14. The Authority shall have the power to disqualify any body or person from representing before itself the cause of oustees (PAPs) resettled in Madhya Pradesh, if it is duly satisfied that it is necessary so to do in the interest of oustees (PAPs) or the public interest.

15. The decision and directions of the Authority in all matters relating to the redressal of all complaints/grievances of oustees (PAPs) shall be final and binding on the State of Madhya Pradesh and all concerned including the oustees (PAPs) and their associations, if any.

16. The Government of Madhya Pradesh shall:

    Take adequate steps to ensure that the oustees (PAPs) resettled in Madhya Pradesh are informed that if they have a complaint/grievance in the matter of their R&R and/or if they have any surviving grievance in connection therewith, it would be open to them to move the Authority and seek relief.

Chapter 4

Drinking Water and Benefits to Drought-Prone Areas

Sardar Sarovar Project Benefits to Saurashtra and Kachchh Areas in Gujarat, 1992

Source: Narmada Control Authority, Sardar Sarovar Project Benefits to Saurashtra and Kachchh Areas in Gujarat, NCA Publication No. 1/92, 1992.

1. In Gujarat State, 17.92 lakh ha of land, spread over 12 districts, 62 talukas and 3,393 villages would come under irrigation from the Narmada canals. About 75 per cent of this area is drought-prone. The project will make available to Gujarat 11.101 billion cu.m (9 MAF) of water out of which 9.794 billion cu.m (7.94 MAF) will be for irrigation and 1.307 billion cu.m (1.06 MAF) for domestic and industrial use. Of this, about 1.052 billion cu.m (0.853 MAF) will be used for domestic purposes while the remaining will be used for industrial purposes. The drinking water facilities will be provided to about 8,215 villages and 135 urban centres.

2. The surface water resources of Gujarat are very meagre. The utilisable resources have been assessed to be 26.47 billion cu.m (21.46 MAF) including 11.101 billion cu.m (9 MAF) allocated from the Narmada river. Of this, Saurashtra with its 71 river basins has only 2.08 billion cu.m (1.686 MAF) while Kachchh with 97 river basins possesses only 0.402 billion cu.m (0.326 MAF). Thus, Saurashtra and Kachchh are very adversely placed as far as the surface water resources are concerned. The picture of groundwater resources is also not bright. The central plateau of Saurashtra overlies basalt formations, which do not contain any significant groundwater aquifers. The coastal areas are saline. Cavernous milliolite limestone in this region contains some groundwater but it is getting significantly replaced by the ingress of saline sea water on account of over exploitation of the groundwater due to the lack of adequate sweet surface water. The northern areas bordering the Rann of Kachchh are also saline and the groundwater there is obviously brackish. Similarly, large parts of the Kachchh region are occupied by deserts and the coastal areas are saline. The central plateau of Kachchh possesses

some groundwater but this storage is dwindling fast because of its extraction without adequate corresponding recharge from the surface. The average annual rainfall in Saurashtra ranges between only 40 and 60 cm and its variability is very high. Kachchh has even less average annual rainfall in the range of 20 to 40 cm and its variability is extremely high. The topography of Saurashtra is like an inverted saucer and its rivers flow radially from the central high land. The river slopes are steep and they dry out after the monsoon. Similar is the condition in Kachchh, which is made more adverse by extremely low rainfall. Frequent drought conditions in Saurashtra and Kachchh are a common phenomenon and the scarcity conditions occur almost every second or third year. Sometimes the drought conditions extend over a period of 3 to 4 consecutive years as experienced during the period 1985–86 to 1987–88. There was an extreme shortage of even drinking water during this period and thousands of people and lakhs of heads of cattle had to migrate to South Gujarat in search of food, water and fodder. During these years, the State government had to spend nearly Rs 1,500 crores on drinking water and relief measures. Water had to be transported by rail to Rajkot city over a distance of 250 km and the cost of supplying water was about Rs 250 per cu.m (more than three times the cost of desalination of sea water), of all the 4,877 villages of Saurashtra, 1,619 have only saline water. Several others have water with a high fluoride content. Other villages also have no dependable sources of water. In these circumstances, the State government is forced to go to the help of the distressed people by taking up relief works and thus the investment, which could otherwise have been made in the construction of permanent development works is required to be spent on emergency works.

3. Of the 64 lakh ha gross areas of Saurashtra, 42 lakh ha are cultivable. Of these, only 2.3 lakh have received the benefit of surface irrigation through major, medium and minor irrigation works. The projects in progress at present will add another area of about 1 lakh ha, while new schemes in the pipeline will add about 25,000 ha. Thus with all the local surface irrigation works, the irrigated area is not likely to exceed about 3.55 lakh ha. As mentioned earlier, the groundwater resources are very limited and, therefore, unless additional surface water is imported from surplus basins outside, the picture about irrigated agriculture cannot be improved. In the case of Kachchh also, out of 45.65 lakh ha of geographical area, only 23.68 lakh ha are cultivable and of this area only 70,000 ha receives the benefit of surface irrigation. With all the additional surface irrigation schemes in progress and contemplated, the total irrigation potential in Kachchh will not exceed about 1.2 lakh ha.

4. It is against the background of this dismal irrigation and drinking water scenario that the Saurashtra and Kachchh branches of the Narmada canal system have been planned. As compared to about 3.5 lakh ha of irrigation through other surface schemes, the Narmada canal will provide irrigation to 3.86 lakh ha in Saurashtra. Of this, 3.04 lakh ha will be in the Surendranagar district while the Bhavnagar and Rajkot districts will receive irrigation in 0.48 and 0.34 lakh has respectively. The

Kachchh branch will provide irrigation facilities to about 57,300 ha in Anjar, Rapar, Bhachau and Mundra talukas. All the 4,877 villages and 90 urban centres of Saurashtra as well as 948 villages and 10 urban centres of Kachchh, irrespective of their locations with reference to the command area, will receive drinking water from Narmada canals.

5. Drinking water supply obviously gets the first priority according to the National Water Policy. All villages and many towns of Saurashtra and Kachchh have either no source of drinking water or the available sources are highly inadequate and unreliable. All these 5,825 villages and 100 urban centres will, therefore, be supplied with Narmada waters irrespective of their locations with reference to the irrigation command. Urban centres with drainage facilities will be supplied 140 litres per capita per day (lpcd), while those without drainage facility will be provided 100 lpcd. Rural areas will be supplied with 70 lpcd. For Saurashtra villages and towns, major offtakes will be at the tail end of the Saurashtra branch, namely Bhogavo reservoir and the tail ends of Botad, Vallabhipur and Morbi branch canals. For serving Kachchh areas, offtakes have been proposed near Bhachau, Anjar and Mundra on the Kachchh branch canal.

6. The Saurashtra branch canal offtaking at 263.2 km from the Narmada main canal near the town of Kadi will cover 5,43,320 ha under its command in 18 talukas of five drought-prone districts viz. Mehsana, Ahmedabad, Surendranagar, Rajkot and Bhavnagar. Of this, the area to be irrigated in Saurashtra will be about 3.86 lakh ha (an additional area of 0.65 lakh ha in Surendranagar district will be irrigated through other branches of Narmada system). The Saurashtra branch will also carry the drinking water needs of the peninsula. Thus, this branch canal will be the lifeline of Saurashtra. There are no alternative competing sources to fulfil the water requirements of Saurashtra at comparable cost. It is, therefore, imperative to provide Narmada waters to Saurashtra even if this involves some small lifts in stages. The Saurashtra branch will be 104.56 km long and will tail into the existing Wadhwan-Bhogavo-II (Dholi Dhaja) reservoir. At its head, this largest branch canal of the system has a 38 m bed width and a 5 m full supply depth to serve the peak discharge requirement of 319.10 cu.m./s.

7. The Saurashtra branch canal crosses a natural valley in its initial reach of 68.40 km. The total drop is 42 m while the total rise beyond is about 63.3 m. If a conventional layout is adopted to cross this valley, the canal will require unusually high banks. Such high banks will not only be very costly but also unsafe. The topography and the submergence and drainage aspect would not permit a conventional level crossing also. It has, therefore, been planned to provide falls and lifts along the canal to negotiate the valley. This aspect had also been examined by the Narmada Water Dispute Tribunal and formally accepted. In the falling limb, three vertical falls have been planned while in the rising limb five pumping stations have been proposed. After the first lift, two branches take off from the Saurashtra branch canal. Another two canals branch off after the third lift and the last two

branches bifurcate after the last lift. The Saurashtra branch canal and these six sub-branches will together irrigate 3.86 lakh ha in the Surendranagar, Rajkot and Bhavnagar districts. It has been planned to generate electricity at the falls, by installing hydel stations. This energy, supplemented by the balance from the power grid system, will be utilised in lifting the water through five pumping stations along the alignment to feed various sub-branches. Studies for identifying the most appropriate generating and pumping installations are being made at present in consultation with various Indian and foreign suppliers and professional experts.

8. In the Kachchh branch canal there will be three falls where electricity will be generated and the water will be lifted at two pumping stations. The canal will irrigate about 57,300 ha in the coastal areas as per the present planning. However, extending this canal further to cover an area of about 1 lakh ha for irrigation, is under consideration.

9. The present planning of the irrigation in Saurashtra and Kachchh through the Narmada canals has been made keeping in view the water allocation allowed to Gujarat, the requirements en route, and the topographical as well as agroclimatic conditions in Saurashtra and Kachchh. If, however, additional waters become available, there is definite scope to provide additional irrigation to these regions via the Narmada canals. Additional waters from South Gujarat can be made available in two ways. It is obvious that the Sardar Sarovar will be spilling quite often in the initial decades of development. When the planned utilisation in the upstream area has not reached its peak, this water, after fulfilling all the committed uses will be available to Gujarat for irrigation in surplus areas. It can be conveyed to Saurashtra and Kachchh through the main canal and the concerned branches, which have already been sized keeping this possibility in view. Another source of water will be the storages constructed across the west flowing rivers south of the Tapi, which will be connected through a link channel to the Narmada main canal via Ukai reservoir. As these surplus supplies will be available during the monsoon, and particularly during the flood periods, it will be necessary to construct some on-line intermediate balancing storages to conserve these waters and ultimately convey them to suitable points in the Saurashtra and Kachchh branches. Several such locations have been identified in the proximity of the canal alignments in North Gujarat, Saurashtra and Kachchh. Some of the already existing storage reservoirs in these regions will also be integrated into the system.

10. Preliminary studies have shown that the above-mentioned surplus supplies can be utilised to augment the existing, ongoing and proposed projects of the Saurashtra region. These reservoirs can be serviced by the proposed terminal storage to be located near Chotila at an elevation of 180 m (FSL) with two link supply canals capacities 53.8 cu.m./s (1,900 cusecs) and 42.5 cu.m./s (1,500 cusecs). One link canal towards the west from this reservoir may replenish the reservoirs of the Rajkot and Jamnagar districts and the other canal (towards the south) may replenish the reservoirs of the Bhavnagar and Surendranagar districts.

There are some two major, 57 medium and 158 minor irrigation schemes in Saurashtra which are either existing or under construction or planned in the near future. It has been found that, in general, the existing storage in the districts of Saurashtra (except parts of Junagadh) get filled up only to the extent of 25 to 30 per cent of their live capacity. These tanks will be linked up with the aforesaid link canals and their irrigation dependability, which is 50 per cent to 60 per cent, will be raised to 80 per cent or even more.

11. From the surplus water available, 0.370 billion cu.m to 0.616 billion cu.m (0.3 to 0.5 MAF) could perhaps be utilised for Kachchh to extend irrigation to about 1 lakh ha including the areas originally planned under the Kachchh branch system. The irrigation dependability of the existing schemes, which is at present 35 per cent to 40 per cent, will be raised to 80 per cent or even more.

12. All the nearby tanks in the command will be deepened for borrowing earth for the canal system. This will help in harnessing and conserving more water in these arid areas.

13. The work on the Saurashtra branch canal from 0 to 46 km has been started and the tenders for its seven branches are under invitation. It is proposed to construct branch canals and sub branch canals by 1998 and the distribution system by 1999. The tenders for the work of the Kachchh branch for the initial 33 km are yet to be invited. The work of the Kachchh branch canal is likely to start in 1993–94 and get completed in 1999 or so.

14. It has also been planned to develop, as a part of the project, the following wildlife sanctuaries.

Saurashtra: Black Buck Sanctuary at Velavadar.

Kachchh: 1. Wild Ass Sanctuary in the Rann of Kachchh; 2. Great Indian Bustard Sanctuary in about 600 km^2 near Jakhau.

15. It has also been proposed to develop compensatory afforestation in an area of 5,738 ha in 16 villages in the Kachchh district. This will provide protection against advancing little Rann of Kachchh and Rajasthan Desert.

16. It will thus be seen that the Narmada canals will be immensely beneficial to the chronically drought-affected and water-deprived regions of Saurashtra and Kachchh.

Narmada Control Authority, Drinking Water from Sardar Sarovar Project, 1991

Source: Narmada Control Authority, Drinking Water from Sardar Sarovar Project, NCA Publication No. 3/91, 1991.

Drinking Water from Sardar Sarovar Project

1.0 Gujarat's population of 4.12 crores stays in over 18,000 villages and 205 urban centres. Providing potable water to these people is one of the prime responsibilities of the State. Unfortunately, the utilisable surface water resources of the State are very meagre (only 2.65 MHM as compared to 69 MHM in the country). Again, these are very unevenly distributed. South Gujarat has been blessed with abundant water while North Gujarat, Saurashtra and Kachchh are semi-arid and suffer from frequent droughts. The groundwater scenario is also disappointing. The eastern hilly and tribal belts are rocky terrains with insignificant groundwater. The groundwater bowl of North Gujarat is being continuously emptied due to over extraction in the absence of adequate surface waters. As a result, the water table has gone down to 200 to 300 m depth and excessive fluorides are now found in the groundwater of vast areas. Reliable water sources have disappeared. The areas bordering the desert and the coast are heavily saline. Large parts of Saurashtra are underlain by traps, which do not bear any significant groundwater. The coastal areas are saline and the salinity is ingressing into areas further inland due to the over extraction of sweet groundwater. Waters in substantial areas in Amreli and Junagadh districts are fluoride affected. Large parts of Kachchh are covered by desert and the water in coastal strips is saline. There is some groundwater in the central plateau but it has been dwindling fast due to lack of adequate replenishment.

2.0 In these circumstances, most of the existing irrigation surface water storages in Saurashtra, Kachchh and North Gujarat are slowly getting converted into, predominantly, water supply storages. Even then, the domestic water supply problem has been getting worse year after year. Of the over 18,000 villages in the state, more then 7,000 have no source of water supply. This figure rises to 12,000 or more during drought years, which occur very frequently. There was a severe drought period of a consecutive three years from 1985–86 to 1987–88 in the recent past. Some 13,700 villages and 24 urban centres experienced acute water scarcity and in some cases special trains had to be run to carry drinking water over a distance of 250 km. During the last drought year alone, a sum of Rs 115 crores had to be spent on providing a water supply to the affected villages and towns.

3.0 This, then, is the water supply scenario in Gujarat and the only dependable source from where the thirst of all these areas can be quenched is the Narmada. Appreciating this, the Narmada Water Dispute Tribunal (NWDT) has set apart special allocation for water supply from the total allocated share of Gujarat. Out of 34.537 billion cu.m (28 MAF) utilisable water resources of Narmada at 75 per cent reliability, the NWDT has allocated 11.101 billion cu.m (9 MAF) to Gujarat. Out of this quantity, 1.307 billion cu.m (1.06 MAF) have been set aside for domestic, municipal and industrial uses. About 1.052 billion cu.m (0.853 MAF) of water will be utilised for drinking purposes and the balance for industrial purposes. The

NWDT in their further report allowed the State government to use the allocated water according to its own will.

4.0 The GWSSB and the State Industries Department were requested by the SSNNL to formulate a detailed plan of utilising 1.307 billion cu.m (1.06 MAF) of Narmada water. They were also requested to indicate their offtake points along the canal system. A number of meetings were held between the senior officials of SSNNL, the GWSSB and the Industries Department. As a result of in-depth discussion, it was agreed that planning for use of Narmada waters should be done for 30 years, i.e., 2021 considering 1991 as the base year. It was also agreed that out of the 1.307 billion cu.m earmarked for municipal and industrial uses, planning for domestic and municipal water supply should be done for 1.05 billion cu.m and as for industries that should be done for 0.257 billion cu.m. It was further decided to adopt the following norms for working out the water requirement for domestic purposes (these compare with national norms recently adopted for the assessment of water requirements for municipal and industrial uses).

	Litres per capita per day (LPCD)
Urban centres having drainage facilities	140
Urban centres without drainage facilities	100
Rural areas	70

It was also agreed that the no-source or inadequate-source villages of the whole of Saurashtra, Kachchh and North Gujarat, as well as the Panchmahals and Sabarkantha districts should be considered for water requirement of rural areas. Another decision was that the water requirement for domestic consumption in rural and urban areas within the command area up to Mahi will be met with from the groundwater and local sources including recharge. Also, it was agreed that villages within the command will receive water from the groundwater recharge or village tank recharge. However, it was agreed that villages having groundwater with high salinity and fluoride content should be considered for water supply from Narmada. It was further decided that existing sources of water supply should be fully tapped and allowed for before projecting water supply from Narmada.

5.0 On the basis of these principles, the GWSSB has prepared a paper on domestic water demand from the Narmada canal system. As per this latest study, 135 urban centres and 8,215 villages are considered for providing water supply from the Narmada canal system. The broad break up of location of urban centres and villages is given in the table that follows:

Sr. No.	Area	Urban Centres	Village
1.	Saurashtra (all the 6 districts)	90	4,877
2.	Kachchh	10	948
3.	Banaskantha	3	490
4.	Mehsana	13	542
5.	Sabarkantha	4	568
6.	Ahmedabad	12	377
7.	Panchmahal	3	413
	Total	135	8,215

The GWSSB has tentatively fixed major offtake points along the main canal and branch canals for serving the areas of the Sabarkantha, Mehsana and Banaskantha districts, and Kachchh and Saurashtra. For serving the Sabarkantha area, the offtake point is indicated near Dehgam on the main canal. For serving the Mehsana district, offtakes are proposed at the heads of the Kharaghoda and Dholera branch canals. For serving Kachchh area, offtakes have been proposed near Bhachau, Anjar and Mundra on the Kachchh branch canal. For serving areas in Saurashtra, major offtakes are proposed at the tail end of the Saurashtra branch canal i.e. Bhogavo reservoir, tail end of Botad branch canal, tail end of the Vallabhipur branch canal and tail end of the Morbi branch canal. Areas of the Bhavnagar district will be served mostly from the offtake on the Botad branch canal. Offtake from the Bhogavo reservoir will serve areas of the Rajkot and Junagadh districts and the offtake from the Morbi branch canal will serve Jamnagar district.

6.0 Detailed demand statements and tapping points will be prepared by the GWSSB so that a project report is ready within about a year. This project, whose cost would run into several thousand crores, will have to be sanctioned and implemented so that by the time the Narmada canals reach the needy areas, the water supply scheme is implemented and kept ready to supply the required quantity of drinking and industrial water to various regions in North Gujarat, Saurashtra and Kachchh.

7.0 As regards industrial water supply, the Industries Department is finalising its plan for use of Narmada waters based on the following principles:

1) Narmada waters are precious and limited and should be used very economically and efficiently. The industries to which water will be supplied should treat the effluent to the acceptable standard and make it usable for

irrigation purposes. Recycling of water within the industry should be encouraged.

2) Water available for industrial use should be allocated to different regions so as to disperse the industries over a larger area and avoid concentration.

3) As far as possible, there should be common offtake and conveyance system for water supply and industrial uses of water.

4) Narmada waters should be supplied for the use of industries located in the areas north of the Shedhi river as areas south of the Shedhi river can get their water requirements from natural sources including fair weather flows in the Mahi and the Narmada.

5) The canals will have to be shut down for a month or so for maintenance and hence, industries should provide for at least one month's storage.

6) Water will be supplied on a volumetric basis and the water rate will be decided by the Nigam in due course and revised from time to time which will comprise a fixed charge plus a variable charge.

7) Narmada canals up to 8.5 cumecs are to run on the controlled volume concept and hence offtakes for M&I water supply should be located either on the main canal or branch canals having capacity of 8.5 cumecs and above.

8) Industries may be given a rebate for the treated water released by them and made available for irrigation purposes.

8.0 With the growth in population, the water requirements for various uses will increase. Efforts have, therefore, to continue on two fronts; namely development of water resources both surface and underground, so that more and more water becomes available for beneficial uses. At the same time, the efficiency of water use in various sectors, notably irrigation and drinking water supply is very low and it requires to be improved by a major effort through the modernisation of the existing schemes, education of the people, involvement of beneficiaries in water distribution and use, charging appropriate water rates so that water is considered a valuable commodity, and such other strategies. We also have to guard against pollution, so that the established uses are not fouled up.

Kachchh Jal Sankat Nivaran Samiti v. Gujarat, 2005

Source: *Kachchh Jal Sankat Nivaran Samiti v. State of Gujarat and 8 Ors*, Special Civil Application No. 3358 of 1999, Civil Application No. 1947 of 2005 in Special Civil Application No. 3358 of 1999, Civil Application No. 2374 of 2005 in Special Civil Application No. 3358 of 1999 and Civil Application No. 2241 of 2005 in Civil Application – for Injunction No. 1947 of 2005, High Court of Gujarat at Ahmedabad (Kshitij R. Vyas and Akshay H. Mehta, JJ.), Decision of 4 October 2005.

Judgment

Kshitij R. Vyas, J.

1. The petitioner, Kachchh Jal Sankat Nivaran Samiti, through its Convener and other Citizens of Kachchh, by way of this Public Interest Litigation voiced grievance as regards the meagre allocation of water from the SSP by the government to the District of Kachchh which constitutes one-fourth of the total area of the State which is a 100 per cent drought-prone district. The locus standi of the petitioners, as indicated in paragraph 3.1 of the petition, makes the petition of a representative character on behalf of the people of Kachchh voicing their grievances as well as the aspirations of the people. To appreciate the grievances of the petitioners, it is necessary to give certain background of the factual aspects.

2. In the year 1969, the Government of India constituted a Tribunal named the Narmada Water Dispute Tribunal (hereinafter referred to as NWDT/Tribunal) headed by Justice V Ramaswamy, a retired Judge of the Supreme Court to decide the inter-State dispute of sharing water of the river Narmada. On 16 August 1978, the Tribunal declared its award under Section 5(2) read with Section 5(4) of the Inter-State Water Disputes Act, 1956 (for short, 'the Act'). Thereafter, references No. 1, 2, 3, 4 and 5 of 1978 were filed by the Union of India and the states of Gujarat, Madhya Pradesh, Maharashtra and Rajasthan respectively under section 5(3) of the said Act which were heard by the Tribunal and on 7 December 1979, gave its final order. The same was published in the Extraordinary Gazette by the Government of India on 12 December 1979. In arriving at its final decision, issues regarding the allocation of water, height of the dam, hydrology and other related issues came to be subjected to a comprehensive and thorough examination by the Tribunal. Extensive studies were done by the Irrigation Commission and the Drought Research Unit of India, the Meteorological Department in matters of catchment area of the Narmada basin, climate, rainfall, variability of rainfall, and the semi-arid zones and scarcity area of Gujarat. The report also suggests that the Tribunal has taken into consideration various technical literature before giving its award.

3. The issue of the allocation of Narmada waters at Sardar Sarovar Dam site was determined on the basis of 75 per cent dependability at 28 MAF. It further ordered that out of the utilisable quantum of Narmada water, the allocation between the States should be as under: Madhya Pradesh 18.25 MAF, Gujarat 9.00 MAF, Rajasthan 0.50 MAF and Maharashtra 0.25 MAF.

As a result of the award of the Tribunal from the Sardar Sarovar Dam and related constructions, the main canal system known as the Narmada main canal, 458 km long which is to carry away water meant for irrigation and drinking purposes to the canal systems of Gujarat and Rajasthan is constructed. It may be stated at this stage that the State of Gujarat demanded 20.73 MAF of water out of the total demand of 22.02 MAF of water before NWDT for a total 71.38 lakhs acres area. The demand included 6.57 MAF of water for reclaiming and/or irrigating 12.17 lakhs acres of land of the district of Kachchh for area under zone XI-C, Banni and Ranns. However, the Tribunal did not consider the demand of the State of Gujarat for Banni and Ranns of Kachchh on the ground that these areas are barren and sparsely populated and the soil is highly saline having, very low permeability and a vertical permeability of nearly nil, a high groundwater table and an impervious layer near the groundwater surface, high evaporation and low rainfall. Thus, the Tribunal rejected the claim of the State for irrigating 11 lakhs acres in Ranns and Banni areas and granted a limited allocation of 9.00 MAF of water in favour of the State of Gujarat. The Tribunal has, however, given the concerned states choice to utilise the quantity of water falling to their share in their own way.

4. The State government, out of 9 MAF water, provided 1.94 MAF water diverted or for irrigation and 1.06 MAF allocated for domestic and industrial use. It appears that because of the limited water allocation, the proportionate water requirement for Kachchh region was worked out as 0.15 MAF.

5. The petitioners in this petition, have challenged the decisions of the meagre allocation of water to the Kachchh areas on various grounds and are also seeking directions for the reconsideration of the decisions taken by the State in relation to the allocation of water to Kachchh.

6. We have heard Mr Mihir Joshi, learned Sr. Counsel appearing with Mr N.V. Anjaria for the petitioners, Mr S.N. Shelat, learned Advocate General appearing for respondent No. 1, State of Gujarat and Mr Kamal Trivedi, learned Addl. Advocate General appearing with Ms Sangeeta Vishen for the SSNNL, respondent No. 2 at length. We have also gone through the entire pleadings and various reports and decisions cited before us.

7. Learned Counsel for the petitioner submitted that the decisions of the State government regarding the allocation of water for irrigation to the Kachchh district ignores the constitutional imperative of distributing material resources of the community so as best to sub-serve the common good. According to the learned counsel, water of the Narmada river apportioned to the State of Gujarat by the NWDT to be allocated for such use and in such areas within its territory as determined by the State, constitutes a material resource of the community. The State is obliged under Article 38(2) of the Constitution to strive to minimise inequalities of income, to eliminate inequalities in status, facilities and opportunities not only amongst individuals but also amongst group of people residing in public areas. By inviting our attention to Article 39(a) of the Constitution, it was submitted that the State was obliged to direct its policy towards securing the right to an adequate means of livelihood and under (b) that the ownership and control of the material resources of the community are so distributed as best to sub-serve the common good. According to the learned counsel, 'common good' envisaged in the Constitution is not merely the greatest good of the greatest number but is governed by the key word 'distribute' which is

the genus of the article and must be given full play as it fulfils the basic purpose of restructuring the economic order. Each word in the article has a strategic role and the whole article has a mission. By inviting our attention to various authorities on the point as under:

(i) *State of Karnataka v. Ranganath Reddy*, AIR 1978 SC 80,

(i) *Jacob v. Kerala Water Authority* (1991) 1 SCC 28,

(iii) *Madan Mohan Pathak v. Union of India*, (1978) 2 SCC 50,

(iv) *Video Electronic v. State of Punjab*, (1990) 3 SCC 87,

it was submitted that the directive principles enshrined in Articles 38 and 39 must be harmonised with economic unity as well as economic development of developed and under developed areas. Elaborating his submission, it was stated that the special disadvantages of Kachchh in relation to availability of water, availability of an alternative source of livelihood other than agriculture and dependence of a large number of people on such availability of water, would oblige the State to undertake an affirmative action by unequal (higher) allocation of water to the region to bring about real equality. The learned counsel highlighted the circumstances by showing that Kachchh suffers special disadvantages regarding availability of water and its need for water would take precedence over the necessity and/or requirement of water to other regions by inviting our attention to various grounds narrated given in the petition and details of the factual aspects of the case.

8. On minutely examining the aforesaid authorities cited before us wherein, it has been laid down that although Articles 38 and 39 of the Constitution and the directive principles laid down therein are fundamental in the governance of the country, the State is under the obligation to implement them in making laws. The Court can also give vitality and effect by using them as criteria of reasonableness while deciding the question of validity of legislation and executive action. There cannot be any dispute with respect to the principles laid down in the aforesaid decisions. The question which requires to be considered is whether allocation of the water is consistent with the provisions of Articles 39(b) and (c) of the Constitution of India. In other words, whether distribution of Narmada water made by the State to various districts is to sub-serve the common good. It is required to be borne in mind that the directive principles of State policy and Article 39 of the Constitution cannot in the very nature of things be enforced by a court of law and therefore, the said decision taken by the State in the present case, cannot be challenged on the ground of violation of the said Article. However, in the instant case, the parties have argued extensively on this question and, therefore, we have to decide the legality and validity of the decision of the State government regarding the distribution of the Narmada Water for the Kachchh region.

9. Learned counsel for the petitioner submitted that the decisions of the State government are not based on the relevant criteria/parameters but on irrelevant considerations. The learned counsel has given the relevant parameters for allocation of water as under:

i. That requirement of irrigation should have priority over power;

ii. That irrigation should be extended to the maximum area within physical limits of command subject to availability of water and in particular to arid border areas to encourage settlement;

iii. Culturable area;

iv. Population dependent on water;

v. Drought areas;

vi. Economic needs including requirements for irrigation;

vii. Geography, hydrology, climate;

viii. Economic and social needs;

ix. Comparative costs of alternative means of satisfying the economic and social needs of the region;

x. Availability of other resources;

xi. To cover as large a part of the State to be commanded by the project as possible;

xii. To cover special areas where water is relatively scarce and rainfall is undependable even though such areas would be relatively difficult from the point of irrigation;

xiii. To spread water thinly over the area in order to maximise total returns from water;

10. In the submission of the learned counsel for the petitioners, the decision taken by the State in relation to the allocation of water to Kachchh did not take into account the aforesaid factors but are based on and/or supported by the NWDT award and the reports of the Narmada Planning Group (in short 'NPG') and the D.T. Buch Committee, which are irrelevant, inconsequential and in any case do not even support such decisions on facts. He invited our attention to the aforesaid award/reports and submitted that the decision to allocate water of 0.15 MAF for irrigating 42,800 hectares of Kachchh is stated to be based only on the NWDT award. He further submitted that NWDT has not assessed or adjudicated that Kachchh is entitled to irrigation in respect of 1,14,000 acres (42,800 hectares) as this was not even the issue before the Tribunal, which was only concerned with the equitable apportionment of water amongst party states. He also submitted that in the absence of any determination by the NWDT regarding the area to be irrigated

in Kachchh and the decision to adopt the purported determination is wholly misconceived. In any case, in absence of any independent justification for seeking to irrigate only 42,800 hectares (1,14,000 acres) out of 23,63,000 hectares of Kachchh (excluding the Ranns and Banni), by inviting our attention to the award of NWDT wherein it was specifically stipulated that for a variety of reasons, it may become necessary to provide water for irrigation to areas not covered by proposals for use of Narmada waters at present and it is, therefore, reasonable to give the party states the freedom to vary within their share of water the pattern of water use and the areas to be served by such use within their respective State boundaries as it may consider necessary, the State could not have considered to allocate the water on account of NWDT award.

11. Learned counsel for the petitioner also challenged the said decision of the revision in the allocation of water from 0.203 MAF to 0.256 MAF which is stated to be only on the basis of the report submitted by the Narmada Project Group. By contending that the NPG report does not justify or even incorporate a decision/finding regarding enhancement of area or water allocation. Similarly, the revision by the State in the area from 42,800 hectares to 57,300 hectares and of allocation of water from 0.203 MAF to 0.256 MAF which is stated to be only on the basis of this assessment was challenged on the ground that the assertion by the State is vague purported to have been placed on record. He has submitted that the report of D.T. Buch committee whereby a decision was taken to the effect that the revision in the area from 57,300 hectares to 1,12,700 and of the allocation of water from 0.256 MAF to 0.496 MAF is stated to be on the basis of the said report. The learned counsel has challenged the decision of the State government on the basis of D.T. Buch committee's report by contending that this is stated to be a technical study undertaken with the objectives of studying how best irrigation water allocation for the Kachchh region from the SSP can be further augmented within the constraints of the NWDT Award. The study recommended the option of augmenting water supply by storage of surplus water. This was calculated to be 0.223 MAF. Therefore, it was submitted that this study was not concerned with the enhancement of the allocation of water from 9 MAF allocated to the State of Gujarat. It was further submitted that enhancement of area from 57,300 hectares to 1,12,700 hectares is the mere mathematical extrapolation on the basis that 0.256 MAF would serve 55,700 hectares and therefore, additional 0.223 MAF would pro rata serve 55,700 (sic) hectares. This indicates that if more water were to be allocated to the region, further additional areas could be brought under irrigation. If this is accepted as a basis, the decision of the State is wholly irrational since determination of area to be irrigated based on the relevant parameters must precede the consequential allocation of water and it cannot be the other way round, more so when the study itself recommends that it is not possible to allocate more water to Kachchh since farmers in Central Gujarat, North Gujarat and Saurashtra want more water. Learned counsel criticised the report of the Buch Committee by contending that there is no study, finding or averment that land other than 42,800 hectares of

Kachchh is unsuitable for irrigation for any reason whatsoever. It is submitted that the Buch Committee report in fact finds that there is scope of putting a considerable part of the lands to productive use with spread out irrigation. The fact that the study has recommended pro rata revision of the culturable command area to 1,12,700 hectares underlines the potential of irrigating additional land. However, the study does not assess or determine why land more than 1,12,700 hectares should not be considered for irrigation.

12. On the basis of the Water and Power Consultancy Services (WAPCOS) report in relation to the allocation of surplus water, recommending irrigation of 3,12,000 hectares, it was submitted that the said fact once again highlights the fact that the restriction of area to 42,800 hectares at the first stage was wholly unjustified and by correctly identifying land for irrigation a substantially higher allocation of water ought to have been made by the State at the first stage itself. It was further submitted that the claim for 42,800 hectares appears to be based on redundant planning. Initially a limited area of Kachchh was proposed to be irrigated from the Saurashtra branch canal by lift system. This is why Kachchh is placed under zone XI relating to Saurashtra, though it is an independent district. The other part of Kachchh district was proposed to be served through the Banni branch canal. Presently the Kachchh branch canal takes off from the main canal itself and carries water by flow, and the Banni branch was done away with after the NWDT award, though it was to carry water for irrigating 1.60 lakhs acres of Kachchh mainland and reinstituted only in part thereafter, despite which the same area of 42,800 hectares has been proposed to be irrigated.

13. Learned counsel also challenged the decisions of the State government by contending that the same are arbitrary, discriminatory and unconstitutional since far from undertaking affirmative action by positive discrimination in water allocation for Kachchh on account of its special facts and disadvantages as contended above, the State has allocated a lower share of water and adopted a lesser area of Kachchh, than would otherwise come to its share even on proportionate/equal allocation by producing the following relevant comparative data regarding allocation of water:

District	Cultivable area (ha)	Average rainfall (mm)	Alternative sources (local irrigation projects-groundwater)	SSP irrigation (ha)	Water allocation (MAF)
Bharuch	5,07,800	877	Yes	1,59,559	0.668
Vadodara	5,93,000	917	Yes	2,62,310	1.170
Panchmahals	5,47,400	1,026	Yes	6,201	0.028
Kheda	5,36,100	769	Yes	1,01,876	0.454
Gandhinagar	51,900	533	Yes	9,608	0.043

Ahmedabad	6,76,200	782	Yes	3,67,379	1.484
Mehsana	7,53,200	613	Yes	1,93,194	0.740
Banaskantha	9,25,600	751	Yes	2,55,071	1.138
Surendranagar	7,82,500	487	Yes	3,47,711	1.494
Bhavnagar	7,03,100	620	Yes	64,475	0.267
Rajkot	8,10,800	594	Yes	44,716	0.199
Kachchh	23,63,100	340	No	1,12,700	0.496

It was submitted by the learned counsel that the above table indicates the discriminatory allocation of water and area for Kachchh, which would amount to discrimination even otherwise. Finally, he submitted that the decision-making process suffers from lack of transparency, undue haste and premeditation. To elaborate the same, it was submitted that the decision/s regarding the allocation of water to Kachchh, despite their public importance, have not been formally expressed at all at any stage and that the State has not even produced the decisions regarding allocation of water to Kachchh. It was submitted that it is not known as to who has taken the decision or the procedure adopted for the same. There is no contemporaneous record establishing considerations/factors which have weighed with the government in taking the decisions. In support of this, our attention was invited to a decision rendered by the apex court in the case of *Gulabrao v. State of Gujarat*, (1996) 2 SCC 26 wherein the apex court observed that the cabinet is the driving and steering body responsible for governance and its primary function is to formulate policies in conformity with the directive principles of the Constitution and carry on the executive function of the state as per the Constitution and the laws.

13.1. One more decision was cited by the learned counsel in the case of *R.K. Jain v. Union of India*, (1993) 4 SCC 119. In paragraph 25 it was observed that the cabinet is a constitutional mechanism to ensure that before important decisions are reached, many sides of the questions are weighed and considered which would mean that much work must be done beforehand in inter-departmental discussion and preparation.

14. Applying the aforesaid principles to the present case, it was submitted that the decision to allocate water for 42,800 hectares of Kachchh area for irrigation was taken within 24 hours of the award of the NWDT on 16 May 1978, this would go to show that the pre-meditation is apparent since even after the NWDT clarified the issue in its amended final award on 7 December 1979 that within its territory each State had the freedom to vary the area and use of water as found necessary, there was no reconsideration of the matter. It was further submitted that there is no decision on record adopting either wholly or in part any recommendations of the

NWDT, NPG report or the D.T. Buch Committee report, particularly in the context of the fact that all the reports contemplate further studies.

15. Mr Shelat has submitted that this petition itself is misconceived. Since there is no enforceable right in favour of the petitioners to receive water for irrigation purposes, only out of a limited 7.94 MAF allocated by the Tribunal to the State. He has submitted that the state has taken decisions from time to time to increase the quantity of water to Kachchh district and at present total availability of water for irrigation purposes would be 1.496 MAF. It is his submission that while determining the quantity of water for allocation to the Kachchh district, the State has kept in view the interest of all concerned and also the factors relevant for this purpose. He has placed reliance on several decisions of the apex court which will be discussed at appropriate places in the judgment.

15.1. Mr Trivedi submitted that the initial allocation of water for irrigation in case of the Kachchh district was based on adjudication of the water dispute by the NWDT which was made after an extensive study of all relevant factors which came to be approved even by the apex court. He has further submitted that against the total demand of 22.02 MAF water by the State of Gujarat before the NWDT, including the demand of 6.57 MAF for Kachchh, only 9 MAF water, i.e. 7.94 MAF for irrigation and 1.06 MAF for domestic purposes was allocated to the State. He has further submitted that even thereafter the demand of the Kachchh district has been considered several times and a gradual increase in the quota for the said district has been made. He has brought to our notice the fact that initially as per the award of the Tribunal 0.15 MAF was allocated for the Kachchh district which was increased from time to time and ultimately 1.583 MAF water has been earmarked for the said district. He has also stated that in terms of percentage, the increase is from 6.25 per cent to 13.7 per cent. According to Mr Trivedi, out of the total irrigable mainland i.e. 10.25 lakh ha, 5.14 ha, will be irrigated with the help of Narmada water which forms a substantial portion of the mainland. Lastly, he has submitted that so far as the requirement of water is concerned, the allocation of water to the Kachchh district will fully meet the requirement. He has also placed reliance on certain decisions of the apex court and we will refer to the same during the course of discussion.

16. In order to appreciate the aforesaid contentions, it is necessary to examine the stand taken by the respondents in the replies filed by them. In the affidavit in reply filed by Mr S.N. Khajanchi, Deputy Secretary, Narmada Water Resources and Water Supply Department, Government of Gujarat, Sachivalaya, it was *inter alia* pointed out that the allocation of the share of water for a region within the State of Gujarat is a policy decision arrived at by the State government having regard to the award of the NWDT, detailed multi-disciplinary planning studies undertaken and completed in the period 1981–84 in consultation with the World Bank to formulate a comprehensive Sardar Sarovar Project (SSP) and technical report of augmentation of water allocation for Kachchh region under the Sardar Sarovar

Project by Shri Buch. At this stage, it may be stated that there is no dispute to the fact that part of the statement of claim of the State before the NWDT, report of the Narmada Planning Group, report of WAPCOS, correspondence between the officers of the State/SSP and others, minutes of meetings etc. have been produced only at the time of hearing in the course of arguments of the respondents, and that the petitioners have been given copies of such documents at the time of their production. Reverting back to the reply affidavit of the State government wherein it has been pointed out that the State has made attempts to maximise social and economic benefits from the SSP within the framework of the decision of the NWDT for 13 agro-climatic regions for agriculture water planning and water allocations arrived at for all the regions of the State. It is also stated that any attempts to enhance water allocation for one region is likely to affect the need of the other regions. The policy decision involved consideration of competing claims on scarce water resources to be made available i.e. 9 MAF from SSP. In paragraph 4 of the said affidavit-in-reply, it is pointed out that the State of Gujarat demanded 22.02 MAF of water for 71.38 lakh acres before the NWDT. The demand included 6.57 MAF of water for the district of Kachchh for 12.17 lakh acres including the areas of Rann and Banni.

17. The NWDT, however, did not concede the demand of the State of Gujarat for Banni and Great and Little Rann of Kachchh area on the ground that the area is barren and sparsely populated, the soil is highly saline, very low permeability, a vertical permeability of nearly nil, a high groundwater table and an impervious layer near the groundwater surface, high evaporation and low rainfall. It was concluded with the quantity of water indicated by the State, the project would be highly uneconomic. The Tribunal, therefore, rejected the claim of the State for irrigating 11 lakhs areas in Rann and Banni regions. We may reproduce the observations of the Tribunal in that behalf as under (NWDT volume I, page No. 75):

> Culturable Command Area (CCA) in Gujarat
>
> 6.5.2. In the first place, we see no reason why the area under Mahi command (6.33 lakh acres) should be included under Narmada command. This area is already irrigated or intended to be irrigated by Mahi waters under the sanctioned Mahi right bank canal project, Stage I (Ex.G/342/iv/i). Stage I has already been completed by Gujarat which comprises a diversion weir at Wanakbori and Mahi right bank canal works. Gujarat made no proposal for including this area in Narmada command originally before the Khosla Committee but Dr Khosla on his own initiative, suggested that Mahi area should be brought under the Narmada Command so that 1.58 MAF of water may be released for the use of border areas of Rajasthan. As regards the Great and Little Rann of Kachchh and Banni area also, we see no justification for Gujarat's claim to irrigate these areas from Narmada. Gujarat has claimed to 6.36 MAF of water for this area on the basis of CCA of 11.03 lakh acres and delta of 5.8 ft (at canal head). Gujarat made no claims for the Great Rann of Kachchh and Banni area before the Khosla Committee. So far as the Little Rann is concerned, the Dutch team was of the opinion that desalination was a great problem and the soil studies made by Gujarat

did not furnish sufficient basis to show that desalination was possible. In any case, these areas are admittedly barren and sparsely populated. The soil conditions in this area are characterised by high salinity, a very low horizontal permeability, a vertical permeability of nearly nil, a high groundwater table and an impervious layer near the groundwater surface. The whole area is also subject to high evaporation and low rainfall. There is no adequate evidence produced by Gujarat that these areas are capable of being reclaimed at reasonable cost. Neither the past experiments conducted at the Soil Research Institute, Baroda nor the experiments conducted at Umreth on 36 acres of land could be extrapolated to this area. The pilot plot in Banni area on the light soils has no doubt shown the possibility of growing crops but Gujarat has not investigated or furnished data from which design parameters for effective reclamation of the area could be derived. Even if it is assumed that the area could be reclaimed and developed with the quantity of water indicated by Gujarat, the project would be highly uneconomic. A delta of 3.8 ft at field head has been proposed for the area. Taking into consideration 50 per cent towards transit loss, the delta at canal head will be 5.8 ft. We, therefore, accept the argument of Maharashtra and Madhya Pradesh that the claim of Gujarat for 6.36 MAF of water for irrigating 11 lakh acres in Ranns and Banni should be rejected. Our assessor, Dr Ambika Singh has expressed the same view in his report, ex. C-5. For these reasons, we are of the opinion that the Mahi Command area, the Little and Great Ranns of Kachchh and Banni area should be excluded from the computation of the equitable share of Gujarat.

18. The Tribunal also did not accede to the request of the state for Mahi Command Area and considered 5,402 lakh acres as CCA. The Tribunal, however, considered the entitlement of water for 49,796 lakhs acres for the State of Gujarat in respect of zones 1 to 11 indicated in the statement of claim. The district of Kachchh was covered under zone XI-C in the said statement of claim, which was in addition to the claim for the Banni area and the Rann area of Kachchh. The claim in respect of the Kachchh region was 1.14 lakhs acres.

19. The claim of the State government in respect of the share of water was 22.02 MAF of water. The Tribunal did not accept the demand and concluded that the requirement of the State of Gujarat was 11,656 MAF of water. Even then, the Tribunal, having regard to the availability of water, limited the allocation to 9 MAF of water in favour of the State.

20. The deponent has further deposed in paragraph 4.6 of the affidavit that the claim having been reduced with the limited water allocation, proportionate water requirement for the Kachchh region was worked out as 0.15 MAF. Annexure D to the reply is a statement showing the water requirement to be considered in favour of the Kachchh region with reduced water allocation and reduced CCA.

21. We may, at this stage state that this was the first decision taken wherein the proportionate water requirement for the Kachchh region was considered and worked out. It is further stated by the deponent that after the award of the Tribunal, multi-disciplinary planning studies were undertaken in consultation with the World Bank to formulate a comprehensive SSP. The NPG prepared a project plan for SSP and forwarded it to the Government of India and the World Bank. These

documents form the basis for investment clearance of the Planning Commission and the Government of India in October 1988. The project plan carved out 13 agro-climate regions for agriculture water planning and water allocation. It was based upon the consideration of the mean annual rainfall, land irrigability class, groundwater quality and quantity and drainage conditions. After the Narmada Command was divided into 13 agro-climate zones in which Kachchh district falls under zone No. 13 for 42,800 ha water of 0.203 MAF was allocated for Kachchh instead of 0.15 MAF. Subsequently, the area was revised to 57,300 ha after detailed assessment with a water allowance of 0.256 MAF. Subsequent to whatever has been stated in paragraph 5 of the affidavit, it has further been pointed out that the region of Kachchh has also been allocated water to serve 948 villages and 10 urban centres by allocation of 0.049 MAF and allocation of 0.037 MAF for industries. The following table would show the operative water allocation for the region of Kachchh:

Item	Allocation
a) Allocation for irrigation to serve 57,300 ha of CCA	0.256 MAF
b) Allocation for water supply to 948 villages and 10 urban centres	0.049 MAF
c) Allocation for industries	0.037 MAF
Total:	0.0342 MAF

It was therefore, submitted that the allocation of water supply is not 2 per cent as indicated by the petitioners but water allocation is 0.342 MAF, which constitutes 3.8 per cent out of the allocated water (9 MAF). Out of water allocation of 9 MAF, 7.94 MAF is for irrigation purposes. Considering this, the water allocated to Kachchh for irrigation is 0.256 MAF which is 3.22 per cent of the water share to the state for irrigation. According to the deponent, no injustice has been done to the residents of the Kachchh region. The deponent has also produced extracts from the technical report examined as paragraph 7 and 7.2 of the affidavit in reply. Paragraph 8 of the affidavit in reply, it has been pointed out the total area in the district for land utilisation purpose during 1993–94 is 19.58 lakh ha. By producing the statistics of land utilisation in Kachchh during the year 1993–94 and the details given therein, it was submitted that the cultivable area of the district is 11.66 lakh hectares after deducting area of forests, barren and uncultivable land and land put to non-agricultural use from a reporting area of 19.58 lakh hectares. According to the deponent, this clearly shows the submission made by the petitioner regarding 23.63 lakhs hectares of cultivable area of the Kachchh district is not correct. Thus, considering the worked out figure of 1.6 per cent of cultivable area, the water allocation to Kachchh for irrigation is 3.22 per cent of water allocation to the State for irrigation, the deponent has thus disputed the figure of 2.11 per cent shown by the petitioners as incorrect.

22. In paragraph 9 of the affidavit in reply by producing a statement showing district wise area as per the original plan and subsequently revised by the State government, it has been pointed out that under the revised plan, the region of Kachchh has been considered for 57,307 hectares as against 42,817 hectares originally planned which is more than 33 per cent from the original plan. Similarly, in paragraph 10, by producing a statement showing the population figure of different districts and water allocation for the Narmada command, vide annexure 'H', it has been pointed out that the population of the Kachchh region is 10.501 lakhs which is 4.89 per cent of the total population of the State, to which the government has allocated 3.8 per cent of water from allocation of 9 MAF to the State. The water allocation to Kachchh for irrigation use works out to 3.22 per cent out of allocation of 7.94 MAF and for domestic and industrial use, it works out to 8.2 per cent of the allocation of 1.06 MAF to the State. Likewise, in paragraph 11, by taking a sympathetic view looking to the need of the Kachchh region, a study has been carried out for augmentation of water to Kachchh by Shri D.T. Buch, Consultant to the Narmada Planning Group and the recommendations of Shri D.T. Buch are under active consideration and if it is feasible, the respondents will not hesitate to implement them.

23. On behalf of the State government, Mr I.I. Momin, Under Secretary, Narmada Water Resources and Water Supply Department, has filed a further affidavit in reply to the rejoinder to the affidavit filed by Keshavji Dedhiya, petitioner No. 2 dated 16 October 2000. In the rejoinder affidavit, in paragraph 3(b) he has stated after referring to recommendations of the technical committee of Shri D.T. Buch as under:

(b) The utilisable water of the Kachchh district using surplus monsoon water of Narmada under SSP through the balancing storage and lifting of water thereafter, so as to augment the culturable command to 1,12,700 ha from 57,300 ha. He has in paragraph 3.3 highlighted various aspects considered by him. In paragraph 5, he has stated that apart from the petitioners, other leading citizens of the Kachchh district who had made representations to the State government on the same line as being made by the petitioners, who were informed by communication dated 22 November 1993 regarding the discussion that took place at the Nigam as well as before the Chief Minister on 13 June 1993.

24. At this stage, it is also necessary to see the further affidavit in reply filed on behalf of respondent No. 1 by the Under Secretary to the Government, Narmada Water Resources and Water Supply Department which is at page No. 316 of the record of the petition. In paragraph 4 of the said reply, it has been pointed out that for the sizing of the canal system only technical considerations have weighed with the government. The command of SSP has been divided into 13 different agro-climatic zones based on mean annual rainfall, land irrigability class, groundwater quality and quantity and drainage condition. It is stated that zone-wise requirements, region-wise delta in mm/ha are worked out and capacity of branch

canals has been worked out for peak water requirement. As regards governing criteria for utilising, allocating, distributing and conveying water, he has submitted that after the award of the Tribunal, a detailed project report was prepared considering the area pleaded before the Tribunal and admitted by the Tribunal. For utilising allocated water for irrigation, the command has been divided into 13 agro-climatic zones and zone wise delta was finalised. It is stated that the allocation of water, therefore, is done having regard to the provisions made in the award. The area posed before the Tribunal is finally adopted for coverage excluding the Rann/Banni area which has not been accepted by the Tribunal.

24.1. In paragraph 5, it has been pointed out that the Khosla Committee report was not accepted by the states and, therefore, the matter was required to be referred to the Tribunal. The State government has, in fact, placed its demand of water for the Rann and Banni area. However, the Tribunal did not consider the requirement of irrigation/reclamation of Banni and Rann areas. In that view of the matter, the Khosla Committee report may not be relevant for the determination of the allocation of water available pursuant to the award of the NWDT. It was further pointed out that the area posed before the Tribunal in Zone I to XI was considered by the Tribunal and the same is followed for implementation.

24.2. We have also gone through the affidavit filed by Mr S.J. Desai, Secretary (Narmada), Narmada Water Resources, Water Supply and Kalpsar Department on behalf of respondent No. 2 wherein he has practically taken the stand taken by respondent No. 1.

25. Having considered the issues involved in the petition and the submissions advanced before us and having considered the various authorities cited before us, it really appears to us that the NWDT made the award as a result of ten long years of adjudication wherein the State of Gujarat provided various details based on various data and studies, well-established norms of engineering and planning and the principles laid down in IS: 5968-1970 through as many as 1,288 exhibits and has demanded 22.02 MAF of water for irrigation including drinking and industrial water requirements. This includes demand of 6.57 MAF of water for reclaiming and/or irrigating 12.17 lakh acres of land of the Kachchh district. Therefore, it cannot be said that the demand of the State for allocation of water as a whole and for the Kachchh district in particular, as placed before the NWDT, is without any study or data. The aforesaid set of facts would further go to suggest that the State has, in fact, while keeping in mind the interest of the Kachchh district, demanded before the NWDT, 6.57 MAF of water for reclaiming and/or irrigating 12.17 lakh acres of land for the Kachchh district. However, while allocating in all 9 MAF for irrigation and 1.06 MAF (for domestic purposes) for the Gujarat State, the NWDT declined to consider 11.03 lakh acres (4.46 lakh hectares) of land of the Rann and Banni and granted only 0.15 MAF of water for irrigating 1.05 lakh acres (42,800 hectares) of land. We have already reproduced the reasoning of the Tribunal in the foregoing paragraphs and, therefore, it is not necessary to reproduce the same.

Thus, the NWDT has, in fact, reduced the share available to Gujarat, i.e., 7.94 MAF for irrigation as against the demand of 20.73 MAF. Under the circumstances, the State has no option but to allocate the water in favour of the regions within, keeping in mind the limited availability of water, full supply levels of the main canals, topography of the area, height of the dam and the FSL of the Narmada main canal at its offtake point having been fixed by the Tribunal. The full supply level of the Kachchh branch canal at its offtake point is 49.65 and compliance with the parameters laid down in Indian Standard Guide for planning and lay out of the canal system for irrigation. In this circumstance, we do not find fault with the State in accepting the award of the NWDT in toto without resorting to the options available to it in refixing the distribution of the water to regions within the State. It is quite understandable that the State did not want to waste further time by depriving the people of the benefits arising out of the award of the Tribunal, else no public interest would have been served by the present allocation. From the subsequent exercises undertaken by the State, it clearly appears to us that the State has in fact looked after the welfare of the people of Kachchh by setting up the NPG by Resolution dated 7 April 1981 to function at the policy making level and to report directly to the Narmada High Power Committee with the objective of formulating medium- and long-term plans including distribution system. Chapter VI of the NPG provides discussion on the command. The command area is fixed on the basis of the areas included by NWDT for the purpose of considering requirements of Narmada water for irrigation in Gujarat. Three-fourths of the command is drought-prone. The command area has essential need for agriculture development as well as financial and economical capacity to absorb irrigation water. It is to be noted that the CCA was revised in the light of the discussion with the World Bank Mission in May 1983 on 21 July 1983 to revise the CCA to 57,300 hectares (0.256 MAF). For the Kachchh region it was revised to 57,300 hectares with 0.256 MAF. Not only that but at the request of the State Government Technical Report on augmentation of water allocation for the Kachchh region was presented by Shri D.T. Buch. There were deliberations about the said report with the petitioners. The State government approved the same on 24 February 2000. The petitioners have virtually challenged the legality and propriety of the award and the reports on various grounds stated in the petition. In view of the above, it is not possible for us to accept the same. Frankly speaking, the various details given by the petitioners challenging those decisions have made the entire petition involving disputed questions of facts and, therefore, it is just not possible for this court to deal with the same in this petition under Article 226 of the Constitution of India.

26. Having heard the learned advocates for the petitioners and the respondents, the main question that may arise for our consideration is as to the relief prayed for in the petition to direct the State government to allocate the water of the river Narmada from the SSP to the Kachchh district for irrigation of minimum 15 lakh acres of land, out of 7.94 MAF water made available for irrigation purposes by the NWDT, is justified or not.

27. As already stated above, the learned Advocate General appearing for respondent No. 1 has submitted that there is no enforceable right to receive water for irrigation purpose only out of a limited 7.94 MAF allocated pursuant to the award of the Tribunal. He submitted that drinking water is the basic need of every individual and every village of Kachchh is to be provided with drinking water. Therefore, no grievance can be made by the petitioners if, for irrigation purpose, in view of the limited resources a particular quantity of water is made available. He has submitted that at the instance of the petitioner, water allocated to other regions cannot be reduced resulting into disequilibria effects in the State. To substantiate his submissions, he has invited our attention to certain authorities. In the case of *Olga Tellis v. Bombay Municipal Corporation*, AIR 1986 SC 180 194, while dealing with the right to work and adequate means of livelihood, the Supreme Court has observed as under:

> The State may not, by affirmative action, be compellable to provide adequate means of livelihood or work to the citizens. But, any person who is deprived of his right to livelihood except according to just and fair procedure established by law, can challenge the deprivation as offending the right to life conferred by Article 21.

28. In the case of *Punjab v. Ram Lubhaya Bagga*, AIR 1998 SC 1703 the Supreme Court has made following observations:

> It is not normally within the domain of any court to weigh the pros and cons of the policy or scrutinise it and test the degree of its beneficial or equitable disposition for the purpose of varying, modifying or annulling it, based on howsoever sound and good reasoning, except where it is arbitrary or violative of any constitutional, statutory or any other provision of law. When the Government forms its policy, it is based on a number of circumstances on facts, law including constraints based on resources. It is also based on expert opinion. It would be dangerous if the court is asked to test the utility, beneficial effect of the policy or its appraisal based on facts set on affidavits. The Court would dissuade itself from entering into this realm which belongs to the executive.

In our opinion, the above observations would answer the submissions advanced by the learned counsel of the petitioners. We are not here to weigh the pros and cons of the policy or scrutinise it and test the degree of its beneficial or equitable disposition for the purpose of varying, modifying or annulling it, unless it is arbitrary or violative of any constitutional, statutory or any other provision of law. Needless to say that the petitioners have not challenged these decisions on the ground that they are arbitrary nor have they pointed out that they are unconstitutional or violative of statutory or any other provisions of law. The government, in the instant case, decided to accept the award of the NWDT which is based on expert opinion and now we are asked to test the utility, beneficial effect etc. of the policy on the basis of the affidavit filed before us. Besides this, the issues raised in the petition are not justiciable in view of the following:

(i) There are no judicially manageable standards for adjudication for allocation of water in favour of any region within the State. How much water is to be released from the canal, how much water is to be left for other regions are issues which are not within the purview of judicial review. There are no manageable standards to appreciate the evidence. It requires expertise in irrigation management system and water resources management in respect of which matters, there is always more than one opinion. The government is the best judge as it gathers information, material from various sources including the administrators assigned with the duty to manage the irrigation system.

(ii) The decisions involve a balance of competing claims of the public resources. The issues to be determined are polycentric. The State government examined the matter from the overall interest of the State. The enhancement of allocation for one region will result in a reduction of the irrigation water to the other region. The interest of one region would be pitted against the other region. All decisions which require delicate balancing and consideration of complex, social and economical considerations cannot be brought under judicial scrutiny.

(iii) The State is divided into several districts and the inter se distribution of the water is not to be considered in isolation. If the contention of the petitioners were to be accepted, the court has to come to the conclusion that no public interest is served by allocating water to the other regions.

(iv) The allocation of water is, therefore, as per the allocation preferred before the Tribunal. Out of the total water available only 12 districts are given water. Out of 12 districts some villages are left out. The constraints in the allocation of water are technical, topographical and geographical.

29. Apart from that, determining the choice of priorities and formulating a perspective thereof is a matter of policy and it is not within our domain to interfere with the sole question of efficacy or otherwise of such policy unless the same is vitiated or in violation of any provisions of the statute or the Constitution of India.

29.1. In the case of *Narmada Bachao Andolan v. Union of India*, reported in (2000) 10 SCC 664, it has been observed by the apex court that:

> in case of projects of national importance where the Union of India and/or more than one state(s) are involved and the project would benefit a large section of the society and there is evidence to show that the said project had been contemplated and considered over a period of time at the highest level of the states and the Union of India and more so when the project is evaluated and approval granted by the Planning Commission, then there should be no occasion for any court carrying out any review of the same or directing its review by any outside or 'independent' agency or body. In a democratic set up, it is for the elected Government to decide what project should be undertaken for the benefit of the people. Once such a decision had been taken then unless and until it can be proved or shown that there is a blatant illegality in the undertaking of the project or in its execution, the Court ought not to interfere with the execution of the project.

29.2. In the case of *BALCO Employees Union v. Union of India*, AIR 2002 SC 350, it was observed that public interest litigation was not meant to be a weapon to challenge the financial or economic decisions which are taken by the government in the exercise of their administrative power. No doubt, a person personally aggrieved by any such decision, which he regards as illegal, can impugn the same in a court of law, but a public interest litigation at the behest of a stranger ought not to be entertained.

30. Mr Kamal Trivedi, learned Additional Advocate General appearing for respondent No. 2 submitted that this court may not intervene in the matter of judicial review of administrative action and/or decision and/or policy relating to any matter (including allocation of water), for balancing the relevant consideration by deciding as to what quantity of water should be available to each region since such a direction may result in an unmanageable situation leading to unimaginable consequences. In support, he has drawn our attention to the following decisions:

(i) *Shri Sachidanand Pandey v. WB*, AIR 1987 SC 1109;

(ii) *Tata Cellular v. Union of India*, (1994) 6 SCC 651;

(iii) *Pennar Delta Auyacutdars Association v. Government of Andhra Pradesh*, AIR 2000 AP 317;

(iv) *Union of India v. S.B. Vohra*, (2004) 2 SCC 150.

As observed earlier, the challenge in this petition is against the administrative action taken after due deliberations, in the absence of any arbitrariness and/or malafide, we see no reason to interfere in the matter. Even at the cost of repetition, we may say that this court cannot go into the question as to how much water is to be released from the canal for Kachchh and how much water is to be left for other regions which are issues not falling under the purview of judicial review. In our opinion, the government is the best judge, as said earlier. The various criteria/parameters for allocation of water suggested by the learned counsel for the petitioners have been appreciated and considered by the NWDT and subsequent reports by the experts in great detail. Maybe they were not appreciated in the line and the way in which the petitioners wanted them to be appreciated. That would not render the decisions bad by calling as 'non-application' of mind which warrants our interference. Thus, in our view, the issues raised in the petition are not justiciable.

31. At this juncture, we may consider the claim of respondent No. 2 which is in the form of submissions made by Mr Trivedi to the effect that the petitioners have got more than they actually demanded. In support of this submission, Mr Trivedi has drawn our attention to the relief clause and in particular to clause (A) of paragraph 29 of the petition which is:

> (A) Your Lordship will be pleased to issue a writ of or in the nature of mandamus or any other appropriate writ, order or direction directing the respondents to reconsider their decision of allocating 2 per cent share of water to the district of Kachchh in view of what is pleaded in this petition.

The aforesaid clause would clearly show that the allocation of water that has been demanded for the district of Kachchh by the petitioner is 2 per cent. Mr Trivedi has thereafter drawn our attention to the representation made to the concerned Minister dated 5 September 1995 wherein it has been demanded that through three branch canals of Kachchh Gross Command Area of total 6 lakhs hectares is required to be covered. Further, our attention has also been drawn to the annexure to the said representation which is in the form of a statement which shows that demand for the 2,94,297 hectares as a cultivable command area is made against the cultivable command area of 5,88,594 hectares (6 lakh acres). On the basis of this demand, a detailed discussion has been made in the meeting which was convened for this purpose wherein the representatives of the petitioners were present. Thus, it is clear that during that time, the demand of the petitioners for a cultivable common area was to the extent of 2,94,297 hectares. Mr Trivedi has thereafter brought to our notice that how from time to time, in accordance with the various recommendations and increase in the allocation of Narmada water to the Kachchh district has been made. He has drawn our attention to the details which have been arranged in a tabular form which show that between the period 1968 and 1978, demand before the NWDT of the Kachchh district (excluding the Banni and Rann areas) was 0.21 MAF covering about 46,150 hectares of irrigated land. Later on as per the award of the NWDT dated 16 August 1978, it was reduced to 0.15 MAF covering about 42,800 hectares. The award is dated 16 August 1978 and in the month of August 1978 itself the respondent increased the quantity of water from 0.15 MAF to 0.203 MAF and thereafter in the month of July 1983, it was increased to 0.256 covering an area of 57,300 hectares. These details also include the demand made by the petitioners vide its representation dated 5 September 1995 wherein it was requested to cover CCA to the extent of 2,94,297 hectares. It further shows that besides this, on 24 March 2000, the quantity of water was increased from 0.256 to 0.496 (0.58 inclusive of domestic and irrigation) hectares which was sufficient to cover 12,700 hectares. This increase was made on the basis of the recommendation made in the report of Shri D.T. Buch. These details further show that in the month of October 2002, the respondent decided to allocate 1.00 MAF water from the monsoon spill over on the basis of the WAPCOS report which should cover 2,00,000 hectares of lands. Over and above this, a provision was also made for domestic consumption which increased the quantity of water from the main supply to 0.583 MAF. In view of the same, the respondents say that at present, the total quantity of water that has to be allocated to the Kachchh district is 1.583 MAF covering an area of 3,12,700 hectares against the claim of 2,94,297 hectares. The aforesaid details would clearly show that as compared to the original claim made by NWDT for the Kachchh district is 0.21 MAF, as per the present decision, the quantity of water to the extent of 1.583 MAF is decided to be allocated. Consequently, the area likely to receive the benefit is increased from 42,150 hectares to 3,12,700 hectares, which in our opinion is a substantial increase even considering the demand that was made by the petitioners in September 1995, i.e. 2,94,297 hectares.

Thus, in our opinion, the demand of the petitioners is adequately met and there is no reason now for the petitioners to make any grievance on the ground that the Kachchh district has been meted out with discriminatory treatment.

32. A faint attempt has been made by the learned counsel Mr Joshi that a quantity of 1.583 MAF water includes 1.00 MAF from monsoon spill over which cannot be taken into account at all since it is not certain that every year the State supply would be available. He has, therefore, submitted that the supply of water to Kachchh district should be to the extent of 2 per cent of the total Narmada water allocation. We do not agree to the submission of Mr Joshi since it has been stated by the respondents that so far as the Narmada allocation is concerned, certainty is to the extent of 75 per cent whereas it is to the extent of 65 per cent in the allocation of monsoon spill over. There is hardly any difference between the two. If there is failure in monsoon, there would be hardly any water supply from the Narmada allocation also, leave apart the monsoon spill over. However, when there is 65 per cent of certainty of the supply from the spill over, the same cannot be ignored and it is required to be taken into account.

32.1. Further, it is also evident from the record that the total irrigable main land in the Kachchh district comes to 10.95 lakh hectares as per the official figure. Out of this, 3,12,700 hectares of land in the district is to be irrigated with the help of 1.538 MAF water. Added to this there is 1,40,000 hectares of land which at present is being irrigated with the help of groundwater and also 67,000 hectares of land which is being irrigated by other schemes, total 5.40 lakh hectares to be irrigated out of 10.95 lakh hectares of irrigable main land. These figures would make the picture very clear and they show that the petitioners are to receive substantial benefit than what they have actually demanded. We, therefore, do not find any discriminatory treatment having been meted out to the petitioners.

33. In view of what has been discussed above, we are clearly of the view that the petitioners have no enforceable right to receive water for irrigation as per their claim from the allotted share of 7.92 MAF to the State of Gujarat, determined by the NWDT. This is particularly in view of the fact that each and every part of Kachchh district has been provided with drinking water which is the basic requirement.

In view of what is discussed above, we answer the questions posed to us:

(i) There are no judicially manageable standards for the adjudication for allocation of water in favour of any region within the State. How much water is to be released from the canal, how much water is to be left for other regions is the subject matter of irrigation management system and water resources management in respect of which there is always more than one opinion and the government is the best judge to allocate resources and to consider the overall interest of the public of the region.

(ii) All decisions which require delicate balancing and consideration of complex, social and economic considerations cannot be brought under judicial scrutiny.

(iii) If the petitioners' contention is accepted, this court will have to record its conclusions, namely:

(a) Public interest is not served by the present allocation.

(b) It would be better served if other regions are not entitled to allocation to the extent determined by the State so as to make water available to the Kachchh region.

(c) Water is made available to only 12 districts. Out of those 12 districts, some villages are left out, and therefore, it is necessary that the State should reconsider its policy of allocation of water.

We are afraid, it is not possible for us to record these conclusions as any directions that may be given may result in an unmanageable situation leading to unimaginable consequences.

(iv) When the State government has accepted the decisions of the tribunals regarding allocation of water, it cannot be termed as 'arbitrary'. The fact that there is no opposition from any other region except the Kachchh district, it is reasonable to conclude that the other regions are satisfied with the allocation of water.

The State was justified to accept the recommendations of the NWDT without any modification. Any decision on the part of the State to change the decision of the Tribunal regarding the allocation of water to its districts would have further delayed the benefits already accrued from the award of the Tribunal.

(v) The State shall in particular direct its policy towards ownership and control of the material resources of the community so that they are distributed as best to subserve the common good within the meaning of Article 39(b) of the Constitution of India which cannot be attracted in this case in view of the fact that the word common good used in the said Article is not for the Kachchh region only since it applies to other regions of the State. Therefore, the decision of the State government in the instant case is consistent with the provisions of Article 39 of the Constitution of India.

34. In view of these findings, we see no merit in this petition and the same is rejected. Rule discharged. The interim relief , if any, stands vacated.

In view of the order passed in the main matter, Civil Applications No. 194, 2374 and 2241 of 2005 stand disposed of.

Chapter 5

Documents Pertaining to the 1994 Writ Petition of the Narmada Bachao Andolan

Supreme Court Order Restraining Further Construction of the Dam, May 1995

Source: *Narmada Bachao Andolan v. Union of India and Others*, Writ Petition (Civil) No. (2) 319 of 1994, Supreme Court of India, Order of 5 May 1995 (interim measures).

Order

Learned Solicitor General on instructions made the following statement which is placed on record:

(i) For the time being the effective height of the dam would be maintained at 80.30 metres which is the existing height between blocks 30 to 46, and

(ii) According to the report of the Dam Safety Panel dated 23/27 March 1995 adopted by the Sardar Sarovar Construction Advisory Committee (SSCAC) at its meeting held on 3 May 1993, the height of the dam in blocks 29 and 47 to 50 should be raised uniformly to the EL 105 metres, a streamlined hump should be constructed along the down-stream edge of the spillway crest, and in blocks 32 to 37 and 42 to 46 the hump should be of 3 metres height and in blocks 30, 31 and 38 to 41, the maximum height of the hump should be not more than 1.2 metres.

The particulars of these recommendations are contained in a copy of the minutes and are shown in a sketch produced before us for this purpose. The same are kept on record as Annexures A and B.

Learned Solicitor General prayed that permission be granted for making the above constructions only in the meantime, retaining the effective height of the dam at 80.30 metres in the manner indicated above. He submits that this is essential to ensure safety of the dam and the stilling basin, according to the opinion given in the above report.

Learned counsel for the concerned State governments submitted that even though they may not fully agree with these recommendations, yet at this stage, as an interim measure, to ensure the safety of the dam, they would not object to grant of permission by the court to this extent only. Learned counsel for the governments of Madhya Pradesh and Maharashtra added that in this event their governments would take the necessary measures to ensure at least the temporary rehabilitation of all families that may be affected by the larger submergence as a result of this further construction on the dam. We place this statement on record. Learned counsel for the petitioner, however, submitted that it is not necessary to either construct the humps or make the other construction as indicated above and that certain alternative measures could be adopted to achieve the same or better results.

Having considered these submissions, we are of the opinion that the statement made by the learned Solicitor General which has been placed on record should be accepted as an interim measure at this stage, and the concerned authorities be permitted to make the construction limited to the extent indicated for which permission is sought by the learned Solicitor General on behalf of the concerned authorities. The concerned State governments having given their assurance to take all the necessary measures to provide the requisite relief and rehabilitation needed by the affected families as a result of likely submergence of a larger area on account of this extra construction, no direction in this behalf by us is necessary. It is, however, made clear that the concerned governments would take at least temporary measures wherever it is not possible for them due to paucity of time to take permanent measures. We make it clear that the merits of the submission made on behalf of the petitioner by Shri Prashant Bhushan are not to be treated as rejected by this court and this order is being made only on the ground of expediency at this stage.

Supreme Court Order Allowing Construction up to 85 Metres, 1999

Source: *Narmada Bachao Andolan v. Union of India and Others*, Writ Petition (Civil) No. 319 of 1994, Supreme Court of India, Order of 18 February 1999.

Order

At an interim stage of these proceedings, on 5 May 1995 Mr Dipankar P. Gupta, the learned Solicitor General appearing for the Union of India, made a statement in this court to the effect that the effective height of the dam would be maintained at RL 80.30 metres, which was the existing height between blocks 30 to 46, until further orders. This is the height at which the dam remains to date. However, certain humps were permitted to be constructed to maintain the safety of the dam subsequently.

Learned counsel for the State of Gujarat states that the State of Gujarat has now taken all R&R measures required by Narmada Water Disputes Tribunal Award (NWDT Award) for resettlement of the oustees up to RL 90 metres. He is supported in this statement by learned counsel for the Union of India. Learned counsels have further submitted that the Narmada Control Authority (NCA) has allowed the level of the dam to be increased to RL 90 metres. Mr Shanti Bhushan, learned senior counsel appearing for the petitioner, however, joins issue and asserts that the assertion made by learned counsel for the State of Gujarat and Union of India regarding making provisions for resettlement and rehabilitation is not correct.

Mr H.N. Salve, learned senior counsel appearing for the State of Gujarat, has placed before us copies of two resolutions dated 16 February 1999 and 17 February 1999 and the corrigendum dated 18 February 1999. Vide Resolution dated 17 February 1999, the Government of Gujarat has constituted an Independent Grievances Redressal Authority (hereinafter the Authority) for redressing the grievances in the work of resettlement and rehabilitation under the Sardar Sarovar Project. The Authority has Mr Justice P.D. Desai, retired Chief Justice, as the Chairman. The powers, terms and conditions of the Authority have also been detailed in the resolutions.

Since the main concern of this court in this writ petition has been with regard to provision of resettlement and rehabilitation of the oustees (PAFs and PAPs), we are of the view that before we proceed further to hear the writ petition, the Authority be requested to survey the rehabilitation sites which have been established by the State of Gujarat for the oustees from the States of Madhya Pradesh and Maharashtra as also of the State of Gujarat itself, to find out whether the resettlement and rehabilitation measures as mandated by the NWDT Award (including provisions for civic amenities) have been substantially complied with so as to accommodate the oustees from the above-mentioned states up to height of RL 90 metres. We, therefore, request the Authority to ascertain the ground realities with regard to provision for resettlement and rehabilitation and in the first instance make an interim report to this court with regards to its findings concerning resettlement and rehabilitation measures. This may be done on or before 9 April 1999.

The request of Mr H.N. Salve, learned senior counsel, for the State of Gujarat, which is supported by the Union of India and the State of Maharashtra also, to permit the State of Gujarat to raise the height of the dam to RL 90 metres shall be considered after 9 April 1999 after examining the interim report of the Authority. At this stage, however, we permit the State of Gujarat to raise the level of the dam to RL 85 metres, excluding the humps necessary for maintenance of safety of the dam.

We would like to emphasise that the oustees affected by the raising of the dam height up to RL 85 metres should be rehabilitated by the State of Gujarat as required by the NWDT Award, if not already done. The Government of Gujarat shall file an affidavit on or before 9 April 1999, disclosing the status of resettlement and rehabilitation work concerning the PAFs affected by raising the height up to RL 85 metres. The State shall disclose what further resettlement and rehabilitation measures are being undertaken for the prospective oustees after RL 85 metres.

Affidavit of the State of Madhya Pradesh, 2000

Source: Affidavit on behalf of the State of Madhya Pradesh in response to the Order of 9 May 2000 of the Honourable Supreme Court, Writ Petition (Civil) No. 319 of 1994, Supreme Court of India, 6 July 2000.

I H.N. Tiwari, aged 61 years, son of late Ramadhin Tiwari, Director (TW), Narmada Valley Development Authority, Government of Madhya Pradesh, Bhopal (MP) do hereby solemnly affirm and state as under:

1. That, I am the Director (TW), Narmada Valley Development Authority, Government of Madhya Pradesh and have been appointed officer-in-charge in the case and as such I am competent to swear this affidavit.

2. That, this Hon'ble Court vide Order dated 9 May 2000 was pleased to pass the following order:

> All the three State governments (Gujarat, Madhya Pradesh and Maharashtra) are also directed to file affidavits on or before 1 July 2000 disclosing the latest status of resettlement and rehabilitation work for the existing as well as prospective oustees likely to be affected by raising the height of the dam.

3. Pursuant to the aforesaid order of this Hon'ble Court, the State of Madhya Pradesh submits that the present effective height of the dam is at EL 85 m and the next rising height at the interval of 5 m would be 90 m. Since the construction of a hump of 3 m over the present height of 85 m has been allowed vide order dated 7 May 1999 of this Hon'ble Court, bringing the effective height of the dam to EL 88 m, the State of Madhya Pradesh, in response to the aforesaid order dated 9 May 2000 of this Hon'ble Court, is setting out the following further progress in respect of R&R of the PAFs to be affected at EL 85 m, EL 90 m, and at EL 95 m.

4. With a view to arrange the resettlement of the PAFs to be affected at aforestated levels, detailed instructions to the field officers of the submergence area including Commissioner Indore Division, Indore and Collectors and Director (Field) were issued vide letter No. DIR(TW)/WP(C)/154/VC/NVDA/MP dated 20 May 2000 in respect of all the following aspects of the resettlement of the PAFs:

(i) The land area identified by the collectors, by reduction of the grazing land area of the villages, for allotment to those PAFs who are losing more than 25 per cent of their holding of the SSP be examined by a team of officials comprising officers of the Agriculture Department, Revenue Department, Irrigation Department, PHE Department to find out what development process is required and to prepare an estimate therefor. The development process is to be undertaken by the collector through the agency under them so that the work may not be held up for want of procedural delay and lack of co-ordination as it is a time-bound programme. Besides, collectors were asked to initiate removal of encroachment as per rules from those cultivators other than scheduled caste, scheduled tribes and small & marginal farmers for making the land available to the PAFs of SSP out of the land reported to have been encroached. Apart from this, in regard to certain land areas identified from forest land, collectors were directed to send the proposal for obtaining clearance from the Government of India, Ministry of Environment & Forest as per rules. Similarly, the Collectors were also instructed to expedite fixation of price, as per provisions in the R&R Policy of MP State and as per rules in vogue, in respect of the land proposed to be acquired by purchase from prospective sellers.

(ii) The processing of all pending land acquisition cases and the passing of awards, specially in the case of six villages which have already been processed and which have reached the stage of passing of awards.

(iii) All the PAFs to be affected at EL 90 m and the next rising height of 95 m and those allocated for resettlement in Gujarat be taken to Gujarat for selection of land. Efforts in this regard should be made to get the land selected by the PAFs as per their choice where their brethren PAFs have been allotted land and house plots.

(iv) In regard to those PAFs who are affected at EL 90 m and 95 m and have consented for resettlement in MP, they, as per entitlement, should be allotted house plots at the relocation sites as well as payments of compensation, grant-in-aid, transportation charges etc. taking the principle of resettlement of PAFs as a village unit as far as practicable. Those PAFs who are entitled to land for land packages should be shown land out of the lands which have been identified for allotment to the PAFs for selection by them. The development processes like tractorisation, bunding, ripping, development of drainage system etc. may be carried out only after PAFs have selected the land. It was also instructed (to the field officers) that the PAFs be very clearly intimated that the land being shown to them would be allotted to them only after due development process is carried out and the land is made irrigable. Notices containing offer of the land with such specific provisions be issued to the PAFs who are entitled for land to land packages. Notices may also indicate the option by the PAFs either for allotment of land or payment of compensation in cash.

(v) All the relocation sites being developed and/or established for the PAFs to be affected at EL 90 m and the next rising height at EL 95 m be fully developed at the earliest. All civic amenities as mandated in NWDT Award and R&R Policy of the State of MP should invariably be created at the relocation sites. Here it may be mentioned that for the PAFs to be resettled in MP, there exists a requirement of establishing 92 sites. And out of these, 18 sites have been fully developed, development at 23 sites is under progress, 18 sites are such where location has been determined and land identified but development work has not started, and 33 sites are such where location and land for the development are to be decided by the task force constituted for this purpose. The details showing civic amenities as mandated by the NWDT Award and these amenities provided at each site are given in the statements enclosed and marked as Annexure MP-01.

5. The aforesaid aspects of the resettlement of the balance of the PAFs to be affected at aforesaid levels are under process. Villagewise status of the R&R of the PAFs to be affected at aforesaid levels, and also resettlement status of those PAFs who are required to be resettled in MP, are given in the enclosed statements and marked as Annexures MP-02 and MP-03. Its summary is given below:

Villagewise status of the resettlement and rehabilitation of the project affected families

Elevation level (EL)	Total No. of PAFs affected	Allocated for R&R		Present status of R&R			Balance PAFs to be resettled		
		In Gujarat	In MP	Allotted land & House plots in Gujarat	Paid compensation in lieu of land or otherwise, dead etc. in MP	Total	In Gujarat	In MP	Total
85 m	1,077	974	103	796**	91	887	178	10	188
		(976)	(101)	(798)					
90 m	2,776	1962*	814*	1363**	379	1742	599	435	1034
		(2053)	(723)	(1365)	(377)**				
95 m	5,357	3318*	2038*	1575*	483	2058	1743	1556	3299
		(3409)	(1948)	(1592)	(466)**				

Note:

* 91 PAFs who have changed their option from R&R in Gujarat to Madhya Pradesh have been deleted from the number of PAFs allocated for R&R in Gujarat and added to the figures of Madhya Pradesh.

** 2 PAFs to be affected at EL 85 m and EL 90 m and similarly 15 PAFs to be affected at EL 95 m thus totalling 17 PAFs who were earlier shown to have been allotted land and house plots in Gujarat have actually been allotted house plots and paid compensation as per their entitlement in Madhya Pradesh.

( ) Figures in parentheses were reported earlier.

(i) R&R Status of PAFs to be affected at EL 85 m

Out of 188 balance PAFs to be resettled, 178 PAFs have consented for resettlement in Gujarat and 10 PAFs who have changed their option for resettlement from Gujarat to MP are required to be resettled in Gujarat and Madhya Pradesh respectively. Out of 178 PAFs to be resettled in Gujarat, 133 PAFs are resisting their resettlement under the influence of anti-dam activists. They have already been given an offer of land by the Government of Gujarat. 19 PAFs out of 178 PAFs have selected land in Gujarat; their pattas are awaited from the Government of Gujarat. And the balance 26 PAFs of villages Sakarja and Jhandana have consented to resettlement in Gujarat and accordingly they are being taken for selection of land in Gujarat. As regards the 10 PAFs of village Dharamrai who have changed their earlier option for settlement from Gujarat to MP, they are required to be allotted land in Madhya Pradesh in lieu of land acquired from them. They are being shown the land area in MP.

(ii) R&R Status of PAFs to be affected at EL 90 m

(a) Out of the 2,776 PAFs to be affected at EL 90 m 2,053 PAFs had earlier consented for resettlement in Gujarat and 723 PAFs in MP, but 91 PAFs have changed their option for resettlement from Gujarat to MP. Thus, the final allocation for resettlement in Gujarat and MP comes to 1,962 PAFs and 814 PAFs respectively.

(b) Out of the balance 435 PAFs to be resettled in MP, 228 PAFs are required to be allotted land in lieu of land acquired from them, 20 PAFs are adult sons of the PAFs (losing more than 25 per cent of the holding) and have no land recorded in their names in the revenue record; they, as per stipulations in NWDT Award and R&R Policy of MP, are required to be allotted house plots and payment of grant-in-aid and financial assistance etc. Similarly, 132 PAFs are such who are not losing their homesteads though losing land less than 25 per cent are entitled for payment of compensation only. Remaining 30 PAFs are entitled to house plots and compensation whereas 18 PAFs are entitled to house plots only. These PAFs belong to only those six villages, namely Dehar, Jangarwa, Kothada, Sondul, Bhavati, Bijasen, where awards in land acquisition cases have not been passed for which all out efforts are being made; these land acquisition cases have reached the final stage of passing of awards. Once the awards in these villages are passed they will be allotted house plots, and compensation will be paid. Similarly, payment of rehabilitation grants and other financial assistance for purchase of productive assets (so that the PAFs may be established on some self-employment avocation/non agricultural pursuits as provided in paragraph 9.1 and 9.2 of the R&R Policy of MP will also be released as soon as the PAFs shift to the site and start construction of their houses. It is feared that if aforestated grants are paid to the oustees before they move, they may possibly squander the grant and the State government may be required to pay again to establish them in some self-employment venture. However, it is now being thought to make payment of these grants to the PAFs and to keep it in a joint account in the bank in the name of the PAFs and the concerned rehabilitation officer so as to release the grants as soon as the PAFs start construction of houses at the relocation sites.

(c) For the PAFs to be resettled in MP there exists a requirement of the development of ten relocation sites, namely Dharamrai, Kikarwas, Chandankhedi, Bhavati-I & Bhavati-II, Bijasen, Nisarpur, Sondul, Dehar and Jangarwa. Out of these, the first five have been fully developed but, except at Chandankhedi where a few PAFs have started construction of their houses, PAFs of other relocation sites are yet to shift and start construction of their houses. Here it may be mentioned that as per provision in the R&R Policy of MP, the PAFs are paid the replacement cost of the house as compensation and those PAFs receiving less than Rs 20,000 as compensation are paid the difference as subsidy, and PAFs are required to construct their houses from the compensation so paid to them.

The other three sites, namely Bijasen, Nisarpur (for the village Kothada along with other villages) and Sondul are under development; only a few civic amenities are to be provided. As regards, the remaining two sites, namely Dehar and Jangarwa, the construction of civic amenities at the former will start soon after the formalities in respect of invited tenders are complete, and the land for the latter is procured.

Details of civic amenities provided against, as mandated in the NWDT Award at these sites alongwith other sites, and the site plans as approved by the Town & Country Planning are enclosed and marked as Annexure MP-01 & MP-04 respectively.

(d) As regards the balance of 599 PAFs to be resettled in Gujarat, 114 PAFs have selected land in Gujarat though their 'pattas' are awaited from the Government of Gujarat, and 163 PAFs are resisting their shifting under the influence of anti-dam activists though they have been given notices containing the offer of the land and house plots by the Government of Gujarat. The remaining 322 PAFs who were earlier resisting have been persuaded, and arrangement for the selection of land in Gujarat has already been initiated.

(iii) R&R status of PAFs to be affected at EL 95 m

(a) Taking into account those 91 PAFs who have changed their option for R&R from Gujarat to MP, out of the 5,357 PAFs to be affected at EL 95 m, 3,318 PAFs are to be resettled in Gujarat and 2,039 PAFs in Madhya Pradesh.

(b) Out of the balance 1,556 PAFs to be resettled in Madhya Pradesh, the entitlement of the various categories of the PAFs in accordance with stipulations in the NWDT Award and R&R Policy of Madhya Pradesh, is as below:

Sl. No.	Entitlement	No. of PAFs
1	Entitled for culturable land, compensation and house plot.	755
2	Entitled for house plot only being adult sons and neither losing land & house plots nor recorded in their name in revenue record.	399
3	Entitled for compensation for land only as losing less than 25 per cent of holding and homestead not affected.	268
4	Entitled for compensation for homestead and house plots as belonging to landless agricultural labourers and other landless labourers.	134
	Total	**1,556**

(c) The PAFs that are losing more than 25 per cent of their holding and are required to be allotted culturable land were given notices containing offer of the land from some of the parcels of the land identified by the collectors by reduction of grazing land area of the villages from 5 to 2 per cent and also other lands identified and as mentioned in the affidavit submitted in the Supreme Court on 11 April 2000. This notice also contained specific intimation to the PAFs that the land in question will be brought under due development process and will be made irrigable and if selected by the PAFs only then will the development process be taken up and thereafter the land would be allotted. It also contained a request to the PAFs to indicate their preference either for allotment of land or payment of compensation in cash in lieu of land to land package in accordance with the provisions in paragraph 5.1 of R&R Policy of Madhya Pradesh. But the PAFs after seeing the land have indicated that the land shown to them is hilly, stony and some parcels of the land have been encroached upon and as such they have rejected all the land shown to them. A copy of the notices given to the PAFs and proceedings regarding rejection of the land are enclosed herewith as Annexures MP-05 and MP-06. Under the situation, the PAFs are being shown other parcels of land out of the already identified land.

(d) As regards payment of compensation to the PAFs for the land acquired from them and the submerging homesteads, the same has not been paid to only those PAFs belonging to these villages where Awards in land acquisition cases has not been passed or the cases have not been processed. Out of 70 villages to be affected at EL 95 m, one village is such where only government land is being submerged and no land acquisition case is required to be processed. Awards have been passed in 44 villages and the process of land acquisition in 20 villages is at various stages.

In the remaining five villages, the NBA, the petitioner, is not allowing a survey to be conducted for the demarcation of the submergence area and the identification of the PAFs to be affected at EL 132.86 m (436 ft). And therefore, the process of land acquisition cases in such villages is still to be initiated. Here it may be mentioned that the NBA is not allowing surveys to be conducted in such 30 villages and therefore the processing of land acquisition cases in these villages is at a standstill. The issue pertaining to such survey in village Kadmal was raised by the petitioner during a tour of the Chairman, GRA, MP on 14 June 2000, but despite the instructions of the Chairman, GRA MP the survey work was not allowed to be conducted. Reports of the Director (Field) and other field officers in this regard are enclosed as Annexures MP-07 and MP-08.

(e) For the PAFs to be affected at EL 95 m, the State of MP is required to develop 25 relocation sites. Out of these 25 relocation sites, six sites are fully developed and the creation of civic amenities is under progress in seven sites. Where there are formalities regarding the acceptance of tenders, possession of acquired land is in progress in five sites and identification of land for another six sites by the task force has been initiated. In the remaining one site, namely Kadmal, the anti-dam activists are not allowing the development of this site. As and when the house sites are developed and established the PAFs are being allotted house plots.

(f) As regards 1,743 PAFs remaining to be resettled in Gujarat, 137 PAFs have selected land in Gujarat though the 'patta' is awaited. 163 PAFs are under the influence of anti-dam activists and have not agreed to shift from the submergence area. Processes to select land in Gujarat by the remaining 1,443 PAFs has already been initiated. Here it may be mentioned that 249 PAFs of villages of Alirajpur tehsil, namely Kukdia, Roligaon, Mahalgaon and Kakrana, to be affected at various levels, were taken to Gujarat for the selection of land. And, similarly, the 18 PAFs of villages of Badwani tehsil to be affected at EL 90 and 95 m and 25 PAFs of villages of Kukshi tehsil, namely Rekti, Retwa and Kadmal, to be affected at 95 m, were taken to Gujarat for the selection of land, but the officials of the government of Gujarat refused to show land as well as to allot land to these PAFs.

(g) The PAFs are not being affected at the present height of the dam, i.e. 85 m, and with humps at 90 m. Here it may be mentioned that the PAFs have desired to be resettled along with their brethren PAFs who have already been allotted house plots and land in Gujarat. Thus, such refusal is not in consonance with the decision taken by the NCA and its R&R sub-group at its various meetings that 'PAFs will be resettled as a community/village unit as far as possible taking physical constraints into consideration'. The government of Gujarat has confirmed before the Hon'ble Supreme Court that adequate culturable land is available for allotment to PAFs to be affected at EL 455 ft and 138.68 m and, that (ii) the relocation site established for, have additional capacity to absorb additional PAFs to be affected at higher elevation beyond 90 m. However, despite this congenial atmosphere, such a refusal by the government of Gujarat is against their own stand and assurances.

Here it may also be mentioned that in the submergence area there exist strong forces of anti-dam activists, and that persuasion efforts to get the PAFs to agree to resettlement either in Gujarat or in MP is a difficult task. Despite this atmosphere, the State of MP is getting many PAFs to agree to the resettlement, but the aforesaid process of R&R of Gujarat is not only thwarting the persuasion efforts but consequently giving a chance to anti-dam activists to say that the efforts of MP are a farce and ridiculous. This state of affairs has been brought to the notice of NCA and a copy of the correspondence in this regard is enclosed herewith and marked as Annexure MP-09.

Narmada Bachao Andolan v. Union of India, Majority Judgment, October 2000

Source: *Narmada Bachao Andolan v. Union of India*, Writ Petition (Civil) No. 319 of 1994, Supreme Court of India, Judgment of 18 October 2000.

32. Kirpal, J. (for himself and on behalf of Dr A.S. Anand, Chief Justice of India) (Majority view). Narmada is the fifth largest river in India and largest west-flowing river of the Indian Peninsula. Its annual flow approximates to the combined flow of the rivers Sutlej, Beas and Ravi. Originating from the Maikala ranges at Amarkantak in Madhya Pradesh, it flows westwards over a length of about 1,312 km, before draining into the Gulf of Cambay, 50 km west of Bharuch City. The first 1,077 km stretch is in Madhya Pradesh and the next 35 km stretch forms the boundary between the States of Madhya Pradesh and Maharashtra. Again, the next 39 km forms the boundary between Maharashtra and Gujarat and the last stretch of 161 km lies in Gujarat.

33. The basin area of this river is about 1 lakh km^2. The utilisation of this river basin, however, is hardly about 4 per cent. Most of the water of this peninsula river goes into the sea. Inspire of the huge potential, there was hardly any development of the Narmada water resources prior to Independence.

34. In 1946, the then Government of Central Provinces and Berar and the then Government of Bombay requested the Central Waterways, Irrigation and Navigation Commission (CWINC) to take up investigations on the Narmada river system for basin-wise development of the river with flood control, irrigation, power and extension of navigation as the objectives in view. The study commenced in 1947 and most of the sites were inspected by engineers and geologists who recommended detailed investigation for seven projects. Thereafter in 1948 the Central Ministry of Works, Mines & Power appointed an ad hoc committee headed by Shri A.N. Khosla, Chairman, CWINC to study the projects and to recommend priorities. This ad hoc committee recommended as an initial

step detailed investigations for the following projects keeping in view the availability of men, materials and resources:

1) Bargi Project,

2) Tawa Projects near Hoshangabad,

3) Punasa Project, and

4) Broach Project.

35. Based on the recommendations of the aforesaid ad hoc committee, estimates for investigations of the Bargi, Tawa, Punasa (Narmada sagar) and Broach Projects were sanctioned by the Government of India in March 1949.

36. The Central Water & Power Commission carried out a study of the hydroelectric potential of the Narmada basin in 1955. After the investigations were carried out by the Central Water & Power Commission, the Navagam site was finally decided upon in consultation with the erstwhile Government of Bombay for the construction of the dam. The Central Water & Power Commission forwarded its recommendations to the then Government of Bombay. At that time the implementation was contemplated in two stages. In Stage I, the Full Reservoir Level (FRL) was restricted to 160 ft with provision for wider foundations to enable raising the dam to 300 ft in Stage II. A high level canal was envisaged in Stage-II. The erstwhile Bombay government suggested two modifications, first the FRL of the dam be raised from 300 to 320 ft in Stage II, and second the provision of a power house in the river bed and a power house at the head of the low level canal be also made. This project was then reviewed by a panel of consultants appointed by the Ministry of Irrigation and Power, who in a report in 1960 suggested that the two stages of the Navagam dam as proposed should be combined into one and the dam be constructed to its final FRL of 320 ft in one stage only. The consultants also stated that there was scope for extending irrigation from the high-level canal towards the Rann of Kachchh.

37. With the formation of the State of Gujarat on 1 May 1960, the Narmada Project stood transferred to that state. Accordingly, the Government of Gujarat gave an administrative approval to Stage I of the Narmada Project in February 1961. The project was then inaugurated by the late Pandit Jawaharlal Nehru on 5 April 1961. The preliminary works such as approach roads and bridges, colonies, staff buildings and remaining investigations for dam foundations were soon taken up.

38. The Gujarat Government undertook surveys for the high level canal in 1961. The submergence area survey of the reservoir enabled assessment of the storage capability of the Navagam reservoir, if its height should be raised beyond FRL 320 ft. The studies indicated that a reservoir with FRL +460 ft would enable the realisation of optimum benefits from the river by utilising the untapped flow below Punasa dam and would make it possible to extend irrigation to a further area of over 20 lakh acres. Accordingly, explorations for locating a more suitable site in

the narrower gorge portion were taken in hand and finally in November 1963, site No. 3 was found to be the most suitable on the basis of the recommendations of the Geological Survey of India and also on the basis of exploration and investigations with regard to the foundation as well as construction materials available in the vicinity of the dam site.

39. In November 1963 the Union Minister of Irrigation and Power held a meeting with the Chief Ministers of Gujarat and Madhya Pradesh at Bhopal. As a result of the discussions and exchange of views, an agreement (Bhopal Agreement) was arrived at. The salient features of the said Agreement were:

a) That the Navagam Dam should be built to FRL 425 ft by the Government of Gujarat and its entire benefits were to be enjoyed by the State of Gujarat.

b) Punasa Dam (Madhya Pradesh) should be built to FRL 850 ft. The costs and benefits of Punasa Power Project shall be shared in the ratio 1:2 between the Governments of Gujarat and Madhya Pradesh. Out of the power available to Madhya Pradesh half of the quantum was to be given to the State of Maharashtra for a period of 25 years for which the State of Maharashtra was to provide a loan to the extent of one-third the cost of Punasa Dam. The loan to be given by the State of Maharashtra was to be returned within a period of 25 years.

c) Bargi Project was to be implemented by the State of Madhya Pradesh, Bargi Dam was to be built to FRL 1365 ft in Stage I and FRL 1390 ft in stage II, and the Government of Gujarat was to give a total loan assistance of Rs 10 crores for the same.

40. In pursuance of the Bhopal Agreement, the Government of Gujarat prepared a brief project report envisaging the Navagam Dam at FRL 425 ft and submitted the same to the Central Water and Power Commission under the Gujarat Government's letter dated 14 February 1964. Madhya Pradesh, however, did not ratify the Bhopal Agreement. In order to overcome the stalemate following the rejection of the Bhopal Agreement by Madhya Pradesh, a high level committee of eminent engineers headed by Dr A.N. Khosla, the then Governor of Orissa, was constituted on 5 September 1964 by the Government of India. The terms of reference of this committee were decided by the Government of India in consultation with the States of Madhya Pradesh, Maharashtra and Gujarat. The same read as under:

i. The drawing up of a master plan for the optimum and integrated development of the Narmada water resources;

ii. The phasing of its implementation for maximum development of the resources and other benefits;

iii. The examination, in particular of Navagam and alternative projects, if any, and determining the optimum reservoir level or levels.

iv. Making recommendations of any other ancillary matters.

41. The Khosla Committee submitted the unanimous report to the Government of India in September 1965 and recommended a master plan for the Narmada Water Development. In Chapter XI of the said report, the Khosla Committee outlined its approach to the plan of the Narmada development. An extract from this chapter is reproduced below:

> 11.1 In the meeting from 14 to 18 December 1964, at which the State representatives were also present, the Committee laid down the following basic guidelines in drawing up the master plan for the optimum and integrated development of the Narmada water resources:
>
> 1. National interest should have over-riding priority. The plan should, therefore, provide for maximum benefits in respect of irrigation, power generation, flood control, navigation etc. irrespective of State boundaries;
> 2. Rights and interests of the State concerned should be fully safeguarded subject to (1) above;
> 3. Requirements of irrigation should have priority over those of power;
>
> Subject to the provision that the suitable apportionment of water between irrigation and power may have to be considered, should it be found that with the full development of irrigation, power production is unduly affected;
>
> 4. Irrigation should be extended to the maximum area within physical limits of command, irrespective of State boundaries, subject to the availability of water; and in particular, to the arid areas along the international border with Pakistan both in Gujarat and Rajasthan, to encourage sturdy peasants to settle in these border areas (later events have confirmed the imperative need for this); and
> 5. All available water should be utilised to the maximum extent possible for irrigation and power generation and, when no irrigation is possible, for power generation. The quantity going to waste to the sea without being used for irrigation or generating power should be kept to the unavoidable minimum.

42. The master plan recommended by the Khosla Committee envisaged 12 major projects to be taken up in Madhya Pradesh and one, viz. Navagam, in Gujarat. As far as Navagam Dam is concerned, the Committee recommended as follows:

1. The terminal dam should be located at Navagam,
2. The optimum FRL of the Navagam worked out to RL 500 ft,
3. The FSL of the Navagam canal at offtake should be RL 300 ft,
4. The installed capacity at the river bed power station and canal power station should be 1000 MW and 240 MW respectively with one stand-by unit in each

power station (in other words the total installed capacity at Navagam would be 1,400 MW).

The benefits of the Navagam Dam as assessed by the Khosla Committee were as follows:

(i) Irrigation of 15.80 lakh hectares (39.4 lakh acres) in Gujarat and 0.4 lakh hectares (1.00 lakh acres) in Rajasthan. In addition, the Narmada waters when fed into the existing Mahi canal system would release Mahi water to be diverted on higher contours enabling the additional irrigation of 1.6 to 2.0 lakh hectares (4 to 5 lakh acres) approximately in Gujarat and 3.04 lakh hectares (7.5 lakh acres) in Rajasthan.

(ii) Hydro-power generation of 951 MW at 60 per cent LF in the mean year of development and 511 MW on ultimate development of irrigation in Gujarat, Madhya Pradesh, Maharashtra and Rajasthan.

43. The Khosla Committee stressed an important point in favour of a high Navagam Dam, namely, additional storage. They emphasised that this additional storage will permit a greater carryover capacity, increased power production and assured optimum irrigation and flood control and would minimise the wastage of water to the sea. The Khosla Committee also observed that instead of a higher Navagam Dam as proposed, if Harinphal or Jalsindhi dams were raised to the same FRL as at Navagam, the submergence would continue to remain about the same because the cultivated and inhabited areas lie mostly above Harinphal while in the intervening 113 km (70 mile) gorge between Harinphal and Navagam there was very little habitation or cultivated areas.

44. The Khosla Committee report could not be implemented on account of disagreement among the States. On 6 July 1968 the State of Gujarat made a complaint to the Government of India under Section 3 of the Inter-State Water Disputes Act, 1956 stating that a water dispute had arisen between the State of Gujarat and the respondent States of Madhya Pradesh and Maharashtra over the use, distribution and control of the waters of the inter-state river Narmada. The substance of the allegation was that executive action had been taken by Maharashtra and Madhya Pradesh which had prejudicially affected the State of Gujarat and its inhabitants. The State of Gujarat objected to the proposal of the State of Madhya Pradesh to construct Maheshwar and Harinphal Dams over the river Narmada in its lower reach, and also to the agreement reached between the States of Madhya Pradesh and Maharashtra to jointly construct the Jalsindhi Dam over Narmada in its course between the two States. The main reason for the objection was that if these projects were implemented, the same would prejudicially affect the rights and interests of Gujarat State by compelling it to restrict the height of the dam at Navagam to FRL 210 ft or less. Reducing the height of the dam would mean the permanent detriment of irrigation and power benefits that would be available to the inhabitants of Gujarat and this would also

make it impossible for Gujarat to reclaim the desert area in the Ranns of Kachchh. According to the State of Gujarat, the principal matters in disputes were as under:

i. The right of the State of Gujarat to control and use the waters of the Narmada river on well-accepted principles applicable to the use of waters of inter-State rivers;

ii. The right of the State of Gujarat to object to the arrangement between the State of Madhya Pradesh and the State of Maharashtra for the development of Jalsindhi Dam;

iii. The right of the State of Gujarat to raise the Navagam dam to an optimum height commensurate with the efficient use of Narmada waters including its control for providing requisite cushion for flood control; and

iv. The consequential right of submergence of areas in the States of Madhya Pradesh and Maharashtra and areas in the Gujarat State.

45. Acting under Section 4 of the Inter-State Water Disputes Act, 1956, the Government of India constituted a tribunal headed by the Hon'ble Mr Justice V Ramaswamy, a retired Judge of this court. On the same day, the Government made a reference of the water dispute to the Tribunal. The reference being in the following terms:

> In exercise of the powers conferred by sub-section (1) of Section 5 of the Inter-State Water Disputes Act, 1956 (33 of 1956), the Central Government hereby refers to the Narmada Water Disputes Tribunal for adjudication of the water dispute regarding the inter-state river Narmada and the river-valley thereof, emerging from letter No. MIP-5565/C-10527-K dated 6 July 1968, from Gujarat.

46. On 16 October 1969 the Government of India made another reference of certain issues raised by the State of Rajasthan to the said Tribunal.

47. The State of Madhya Pradesh filed a demurrer before the Tribunal stating that the constitution of the Tribunal and reference to it were *ultra vires* of the Act. The Tribunal framed 24 issues, which included the issues relating to Gujarat having a right to construct a high dam with FRL 530 ft and a canal with FSL 300 ft or thereabouts. Issues 1(a), 1(b), 1(A), 2, 3 and 19 were tried as preliminary issues of law and, by its decision dated 23 February 1972, the said issues were decided against the respondents herein. It was held that the Notification of the Central Government dated 16 October 1969, referring the matters raised by the State of Rajasthan by its complaint, was *ultra vires* of the Act but constitution of Tribunal and, making a reference of the water dispute regarding the Inter-State river Narmada, was not *ultra vires* of the Act, and the Tribunal had jurisdiction to decide the dispute referred to it at the instance of the State of Gujarat. It further held that the proposed construction of the Navagam project involving consequent submergence of portions of territories of Maharashtra and Madhya Pradesh, could form the subject matter of a water dispute within the meaning of Section 2(c) of the

1956 Act. It also held that it had the jurisdiction to give appropriate direction to Madhya Pradesh and Maharashtra to take steps by way of acquisition or otherwise for making submerged land available to Gujarat in order to enable it to execute the Navagam project, and the Tribunal had the jurisdiction to give consequent directions to Gujarat and other party States regarding the payment of compensation to Maharashtra and Madhya Pradesh, for giving them a share in the beneficial use of Navagam Dam, and for rehabilitation of displaced persons.

48. Against the aforesaid judgment of the Tribunal on the preliminary issues, the States of Madhya Pradesh and Rajasthan filed appeals by special leave to this court and obtained a stay of the proceedings before the Tribunal to a limited extent. This Court directed that the proceedings before the Tribunal should be stayed but discovery, inspection and other miscellaneous proceedings before the Tribunal may go on. The State of Rajasthan was directed to participate in these interlocutory proceedings.

49. It appears that on 31 July 1972, the Chief Ministers of Madhya Pradesh, Maharashtra, Gujarat and Rajasthan had entered into an agreement to compromise on the matters in dispute with the assistance of the Prime Minister of India. This led to a formal agreement dated 12 July 1974 being arrived at between the Chief Ministers of Madhya Pradesh, Maharashtra and Rajasthan and the Advisor of the Governor of Gujarat on a number of issues which the Tribunal otherwise would have had to go into. The main features of the Agreement, as far as this case is concerned, were that the quantity of water in Narmada available for 75 per cent of the year was to be assessed at 28 MAF, and the Tribunal, in determining the disputes referred to it, was to proceed on the basis of this assessment. The net available quantity of water for use in Madhya Pradesh and Gujarat was to be regarded as 27.25 MAF, which was to be allocated between the States. The height of the Navagam Dam was to be fixed by the Tribunal after taking into consideration various contentions and submissions of the parties, and it was agreed that the appeals filed in this court by the States of Madhya Pradesh and Rajasthan would be withdrawn. It was also noted in this agreement that the 'development of Narmada should no longer be delayed in the best regional and national interests'.

50. After the withdrawal of the appeals by the States of Madhya Pradesh and Rajasthan, the Tribunal proceeded to decide the remaining issues between the parties.

51. On 16 August 1978 the Tribunal declared its Award under Section 5(2) read with Section 5(4) of the Inter-State Water Disputes Act, 1956. Thereafter, reference numbers 1, 2, 3, 4 and 5 of 1978 were filed by the Union of India and the States of Gujarat, Madhya Pradesh, Maharashtra and Rajasthan respectively under Section 5(3) of the Inter-State Water Disputes Act, 1956. These references were heard by the Tribunal, which on 7 December 1979 gave its final order. The same was published in the extraordinary Gazette by the Government of India on 12 December 1979. In arriving at its final decision, the issues regarding allocation,

height of dam, hydrology and other related issues came to be subjected to comprehensive and thorough examination by the Tribunal. Extensive studies were conducted by the Irrigation Commission and Drought Research Unit of India and Meteorological Department in matters of the catchment area of the Narmada basin, the major tributaries of the Narmada basin, drainage area of the Narmada basin, climate, rainfall, variability of rainfall, arid and semi-arid zones and scarcity area of Gujarat. Perusal of the report shows that the Tribunal also took into consideration various pieces of technical literature before giving its Award.

Award of the Tribunal

The main parameters of the decision of the Tribunal were as under:

a) Determination of the height of the Sardar Sarovar Dam:

The height of the Sardar Sarovar Dam was determined at FRL 455 ft. The Tribunal was of the view that the FRL +436 ft was required for irrigation use alone. In order to generate power throughout the year, it would be necessary to provide all the live storage above minimum drawn down level (MDDL) for which an FRL of +453 ft with MDDL +362 ft would obtain gross capacity of 7.44 MAF. Therefore, the Tribunal was of the view that the FRL of the Sardar Sarovar Dam should be +455 ft providing gross storage of 7.70 MAF. It directed the State of Gujarat to take up and complete construction of the dam.

b) Geological and seismological aspects of the dam site:

The Tribunal accepted the recommendations of the Standing Committee under the Central Water and Power Commission that there should be a seismic coefficient of 0.10 g for the dam.

c) Relief and rehabilitation:

The final Award contained directions regarding submergence, land acquisition and rehabilitation of the displaced persons. The Award defined the meaning of the land, oustee and family. The Gujarat Government was to pay to Madhya Pradesh and Maharashtra all costs including compensation, charges, expenses incurred by them for and in respect of the compulsory acquisition of land. Further, the Tribunal had provided for the rehabilitation of oustees and for civic amenities to be provided to the oustees. The Award also provided that if the State of Gujarat was unable to resettle the oustees or the oustees were unwilling to occupy the area offered by the States, then the oustees will be resettled by the home state and all expenses for this were to be borne by Gujarat. An important mandatory provision regarding rehabilitation was the one contained in Clause XI Sub-clause IV(6)(ii), which stated that no submergence of any area would take place unless the oustees were rehabilitated.

d) Allocation of the Narmada waters:

The Tribunal determined the utilisable quantum of water of the Narmada at Sardar Sarovar Dam site on the basis of 75 per cent dependability at 28 MAF. It further ordered that out of the utilisable quantum of Narmada water, the allocation between the States should be: Madhya Pradesh: 18.25 MAF, Gujarat: 9.00 MAF, Rajasthan: 0.50 MAF and Maharashtra: 0.25 MAF.

e) Period of non reviewability of certain award terms:

The Award provided for the period of operation of certain clauses of the final order and decision of the Tribunal as being subject to review only after a period of 45 years from the date of the publication of the decision of the Tribunal in the official gazette. What is important to note however is that the Tribunal's decision contained in clause II relating to the determination of 75 per cent dependable flow as 28 MAF was non-reviewable. The Tribunal's decision of the determination of the utilisable quantum of Narmada water at Sardar Sarovar Dam site on the basis of 75 per cent dependability at 28 MAF is not a clause which is included as a clause whose terms can be reviewed after a period of 45 years.

52. The Tribunal in its Award directed for the constitution of the inter-state administrative authority, i.e. Narmada Control Authority, for the purpose of securing compliance with and implementation of the decision and directions of the Tribunal. The Tribunal also directed for constitution of a review committee consisting of the Union Minister for Irrigation (now substituted by Union Minister for Water Resources) as its Chairperson and the Chief Ministers of Madhya Pradesh, Maharashtra, Gujarat and Rajasthan as its members. The review committee might review the decisions of the Narmada Control Authority and the Sardar Sarovar Construction Advisory Committee. The Sardar Sarovar Construction Advisory Committee headed by the Secretary, Ministry of Water Resources as its Chairperson was directed to be constituted for ensuring efficient, economical and early execution of the project.

53. Narmada Control Authority is a high powered committee having the Secretary, Ministry of Water Resources, Government of India as its Chairperson, Secretaries in the Ministry of Power, Ministry of Environment and Forests, Ministry of Welfare, Chief Secretaries of the concerned four States as members. In addition, thereto, there are a number of technical persons like chief engineers as members.

54. Narmada Control Authority was empowered to constitute one or more sub-committees and assign to them such of the functions and delegate such of its powers as it thought fit. Accordingly, the Narmada Control Authority constituted the following discipline-based sub-groups:

i. Resettlement and Rehabilitation Sub-group under the Chairmanship of Secretary, Ministry of Welfare;

ii. Rehabilitation Committee under Secretary, Ministry of Welfare to supervise the rehabilitation process by undertaking visits to R&R sites and submergence villages;

iii. Environment Sub-group under the Chairmanship of Secretary, Ministry of Environment and Forests;

iv. Hydromet Sub-group under the Chairmanship of Member (Civil), Narmada Control Authority;

v. Power Sub-group under the Chairmanship of Member (Power), Narmada Control Authority;

vi. Narmada Main Canal Sub-committee under the chairmanship of Executive Member, Narmada Control Authority.

55. The Award allocated the available water resources of the Narmada river between the four States. Based on this allocation, an overall plan for their utilisation and development had been made by the States. Madhya Pradesh was the major sharer of the water. As per the water resources development plan for the basin it envisaged, in all, 30 major dams, 135 medium dam projects and more than 3,000 minor dams. The major terminal dam at Sardar Sarovar was in Gujarat, the remaining 29 being in Madhya Pradesh. Down the main course of the river, the four major dams were the Narmada Sagar (now renamed Indira Sagar), Omkareshwar and Maheshwar, all in Madhya Pradesh, and Sardar Sarovar in Gujarat. Rajasthan was to construct a canal in its territory to utilise its share of 0.5 MAF.

Relevant details of the Sardar Sarovar Dam

56. As a result of the Award of the Tribunal, the Sardar Sarovar Dam and related constructions, broadly speaking, are to comprise the following:

a) Main dam across the flow of the river with gates above the crest level to regulate the flow of water into the Narmada main canal.

b) An underground river bed power house through which a portion of the water is diverted to generate power (1200 MW). This water joins the main channel of the Narmada river downstream of the dam.

c) A saddle dam located by the side of the main reservoir through which water to the main canal system flows.

d) A canal head power house located at the toe of the saddle dam, through which the water flowing to the main canal system is to be used to generate power (250 MW).

e) The main canal system, known as Narmada main canal, 458 km long, which is to carry away the water meant for irrigation and drinking purposes to the canal systems of Gujarat and Rajasthan.

Expected benefits from the project

57. The benefits expected to flow from the implementation of the Sardar Sarovar Project had been estimated as follows:

Irrigation: 17.92 lakh hectare of land spread over 12 districts, 62 talukas and 3,393 villages (75 per cent of which is drought-prone areas) in Gujarat and 73,000 hectares in the arid areas of Barmer and Jalore districts in Rajasthan.

Drinking Water facilities to 8,215 villages and 135 urban centres in Gujarat both within and outside command. These include 5,825 villages and 100 urban centres of Saurashtra and Kachchh which are outside the command. In addition, 881 villages affected due to high levels of fluoride will get potable water.

Power Generation: 1,450 MW.

Annual Employment Potential:

7 lakh man-years during construction

6 lakh man-years in post construction.

Protection against advancement of little Rann of Kachchh and Rajasthan desert.

Flood Protection to riverine reaches measuring 30,000 ha, 210 villages including Bharuch city and 7.5 lakh population.

Benefits to:

a) Dhumkhal Sloth Bear Sanctuary,

b) Wild Ass Sanctuary in Little Rann of Kachchh,

c) Black Buck Sanctuary at Velavadar,

d) Great Indian Bustard Sanctuary in Kachchh,

e) Nal Sarovar Bird Sanctuary.

Development of fisheries: Deepening of all village tanks of command which will increase their capacities, conserve water, will recharge groundwater, save acquisition of costly land for acquiring earth needed for constructing canal banks and will reduce health hazard.

Facilities of sophisticated communication system in the entire command.

Increase in additional annual production on account of

	(Rs in crores)
Agricultural production	900
Domestic water supply	100
Power generation	440

Total	1,400

Post Award clearances

58. In order to meet the financial obligations, consultations had started in 1978 with the World Bank for obtaining a loan. The World Bank sent its reconnaissance mission to visit the project site and carried out the necessary inspection. In May 1985 the Narmada Dam and Power Project and Narmada Water Delivery and Drainage Project were sanctioned by the World Bank under International Development Agency, credit No. 1552. Agreement in this respect was signed with the Bank on 10 May 1985 and credit was to be made available from 6 January 1986.

59. With regard to giving environmental clearance, a lot of discussion took place at different levels between the Ministry of Water Resources and the Ministry of Environment. Ultimately on 24 June 1987 the Ministry of Environment and Forests, Government of India accorded clearance subject to certain conditions. The said Office Memorandum containing the environmental clearance reads as follows:

(...)*

60. In November 1987, for monitoring and implementation of various environmental activities effectively, an independent machinery, the Environment Sub-group, was created by Narmada Control Authority. This Sub-group was appointed with a view to ensure that the environmental safeguards were properly planned and implemented. This Sub-group is headed by the Secretary, Ministry of Environment and Forests, Government of India, as its Chairperson and various other independent experts in various fields relating to environment as its members.

61. After clearance was given by the Ministry of Environment and Forests, the Planning Commission, on 5 October 1988, approved investment for an estimated cost of Rs 6,406 crores with the direction to comply with the conditions laid down in the environment clearance accorded on 24 June 1987.

62. According to the State of Gujarat and Union of India, the studies as required to be done by the Office Memorandum dated 24 June 1987, whereby environmental clearance was accorded, have been undertaken and the requisite work carried out. The construction of the dam had commenced in 1987.

63. In November 1990, Dr B.D. Sharma wrote a letter to this court for setting up of a National Commission for Scheduled Castes and Scheduled Tribes including proper rehabilitation of oustees of the Sardar Sarovar Dam. This letter was entertained and treated as a writ petition under Article 32 of the Constitution being Writ Petition No. 1201 of 1990.

---

* The text of the Office Memorandum is reproduced at page 77.

64. On 20 September 1991 this Court in the said Writ Petition bearing No. 1201 of 1990 gave a direction to constitute the Committee headed by the Secretary (Welfare) to monitor the rehabilitation aspects of SSP.

65. The NBA, the petitioner herein, had been in the forefront of agitation against the construction of the Sardar Sarovar Dam. Apparently because of this, the Government of India, Ministry of Water Resources vide Office Memorandum dated 3 August 1993 constituted a Five Member Group (FMG), to be headed by Dr Jayant Patil, Member, Planning Commission and Dr Vasant Gowarikar, Mr Ramaswamy R. Iyer, Mr L.C. Jain and Dr V.C. Kulandaiswamy as its members, to continue discussions with the Narmada Bachao Andolan on issues relating to the SSP. Three months was given to this group to submit its report.

66. During this time, the construction of the dam continued and on 22 February 1994 the Ministry of Water Resources conveyed its decision regarding closure of the construction sluices. This decision was given effect, and on 23 February 1994 the closure of ten construction sluices was effected.

67. In April 1994 the petitioner filed the present writ petition *inter alia* praying that the Union of India and other respondents should be restrained from proceeding with the construction of the dam and they should be ordered to open the aforesaid sluices. It appears that the Gujarat High Court had passed an order staying the publication of the report of the FMG established by the Ministry of Water Resources. On 15 November 1994 this court called for the report of the FMG, and the Government of India was also directed to give its response to the said report.

68. By order dated 13 December 1994 this court directed that the report of the FMG be made public and responses to the same were required to be filed by the States, and the report was to be considered by the Narmada Control Authority. This report was discussed by the Narmada Control Authority on 2 January 1995 wherein disagreement was expressed by the State of Madhya Pradesh on the issues of height and hydrology. Separate responses were filed in this court to the said FMG report by the Government of India and the Governments of Gujarat and Madhya Pradesh.

69. On 24 January 1995, orders were issued by this court to the FMG for submitting a detailed further report on the issues of:

a) Height

b) Hydrology

c) Resettlement and rehabilitation and environmental matters.

Dr Patil who had headed the FMG expressed his unwillingness to continue on the ground of ill-health and on 9 February 1985 this court directed the remaining four members to submit their report on the aforesaid issues.

70. On 17 April 1985 the Four Member Group submitted its report. The said report was not unanimous, unlike the previous one, and the members were equally divided. With regard to hydrology, Professor V.C. Kulandaiswamy and Dr Vasant Gowariker were for the adoption of 75 per cent dependable flow of 27 MAF for the design purpose, on the basis of which the Tribunal's Award had proceeded. On the other hand, Shri Ramaswamy Iyer and Shri L.C. Jain were of the opinion that for planning purposes, it would be appropriate to opt for the estimate of 23 MAF. With regard to the height of the dam, the views of Dr Gowariker were that the Tribunal had decided FRL 455 ft after going into exhaustive details, including social, financial and technical aspects of the project, and that it was not practicable at this stage, when an expenditure or Rs 4,000 crores had been incurred and an additional contract amounting to Rs 2,000 crores entered into, and the various parameters and features of the project having been designed with respect to FRL 455 ft, that there should be a reduction of the height of the dam. The other three members proceeded to answer this question by first observing as follows:

> We must now draw conclusions from the foregoing analysis, but a preliminary point needs to be made. The SSP is now in an advanced stage of construction, with the central portion of the dam already raised to 80 m; the canal constructed up to a length of 140 km; and most of the equipment for various components of the project ordered and some of it already wholly or partly manufactured. An expenditure of over Rs 3,800 crores is said to have already been incurred on the project; significant social costs have also been incurred in terms of displacement and rehabilitation. The benefits for which these costs have been and are being incurred have not materialised yet. In that situation, anyone with a concern for keeping project costs under check and for ensuring the early commencement of benefits would generally like to accelerate rather than retard the completion of the project as planned. If any suggestion for major changes in the features of the projcct at this juncture is to be entertained at all, there will have to be the most compelling reasons for doing so.

71. It then addressed itself to the question of whether there were any compelling reasons. The answer, they felt, depended upon the view they took on the displacement and rehabilitation problem. The two views, which it examined, were, firstly whether the problem of displacement and rehabilitation was manageable and, if it was, then there would be no case of reduction in the height. On the other hand, if relief and rehabilitation was beset with serious and persistent problems then they might be led to the conclusion that there should be an examination of the possibility of reducing submergence and displacement to a more manageable size. These three members then considered the question of the magnitude of the relief and rehabilitation problem. After taking into consideration the views of the States of Madhya Pradesh and Gujarat, the three members observed as follows:

> We find that the Government of India's idea of phased construction outlined earlier offers a practical solution; it does not prevent the FRL from being raised to 455 ft in due course if the necessary conditions are satisfied; and it enables the Government of Madhya Pradesh to take stock of the position at 436 ft and call a halt if necessary. We would, however, reiterate the presumption expressed in

> paragraph 3.9.2 above, namely that no delinking of construction from R&R is intended and that by 'phased construction' the Government of India does not mean merely tiered construction which facilitates controlled submergence in phases. We recommend phased construction in a literal sense, that is to say, that at each phase it must be ensured that the condition of advance completion of R&R has been fulfilled before proceeding to the next phase (i.e. the installation of the next tier of gates). This would apply even to the installation of the first tier. 'Judicious operation of the gates' (while necessary) cannot be a substitute for the aforesaid condition.

The possibility of further construction when the FRL 436 ft was reached or a stoppage at that stage was left open by the members. With regard to the environment, it observed that this subject had been by and large covered in the first FMG report.

Rival contentions

72. On behalf of the petitioners, the arguments of Shri Shanti Bhushan, learned senior counsel, were divided into four different heads, namely general issues, issues regarding environment, issues regarding relief and rehabilitation and issues regarding review of the Tribunal's Award. The petitioners have sought to contend that it is necessary for some independent judicial authority to review the entire project, examine the current best estimates of all costs (social, environmental, financial), benefits and alternatives in order to determine whether the project is required in its present form in the national interest or whether it needs to be restructured/modified. It is further the case of the petitioners that no work should proceed until an environment impact assessment has been fully conducted and its implications for the projects' viability being assessed in a transparent and participatory manner. This can best be done, it is submitted, as a part of the comprehensive review of the project.

73. While strongly championing the cause of the environment and of the tribals who are to be ousted as a result of submergence, it was submitted that the environmental clearance which was granted in 1987 was without any or proper application of mind as complete studies in that behalf were not available and until this is done the project should not be allowed to proceed further. With regard to relief and rehabilitation, a number of contentions were raised with a view to persuade this court that further submergence should not take place and the height of the dam, if at all it is to be allowed to be constructed, should be considerably reduced as it is not possible to have satisfactory relief and rehabilitation of the oustees as per the Tribunal's Award as a result of which their fundamental rights under Article 21 would be violated.

74. While the State of Madhya Pradesh has partly supported the petitioners in as much as it has also pleaded for a reduction in the height of the dam, so as to reduce the extent of submergence and the consequent displacement, the other States and the Union of India have refuted the contentions of the petitioners and of the State of Madhya Pradesh. While accepting that initially the relief and rehabilitation

measures had lagged behind, now adequate steps have been taken to ensure proper implementation of relief and rehabilitation at least as per the Award. The respondents have, while refuting other allegations, also questioned the bona fides of the petitioners in filing this petition. It is contended that the cause of the tribals and the environment is being taken up by the petitioners not with a view to benefit the tribals, but the real reason for filing this petition is to see that a high dam is not erected per se. It was also submitted that at this stage this court should not adjudicate on the various issues raised, specially those which have been decided by the Tribunal's Award.

75. We first propose to deal with some legal issues before considering the various submissions made by Shri Shanti Bhushan regarding environment, relief and rehabilitation, alleged violation of rights of the tribals and the need for review of the project.

Laches

76. As far as the petitioner is concerned, it is an anti-dam organisation and it is opposed to the construction of the high dam. It has been in existence since 1986 but has chosen to challenge the clearance given in 1987 by filing a writ petition in 1994. It has sought to contend that there was a lack of study available regarding the environmental aspects and also because of the seismicity the clearance should not have been granted. The rehabilitation packages are dissimilar and there has been no independent study or survey done before the decision to undertake the project was taken and construction started.

77. The project, in principle, was cleared more than 25 years ago when the foundation stone was laid by the late Pandit Jawahar Lal Nehru. Thereafter, there was an agreement of the four Chief Ministers in 1974, namely the Chief Ministers of Madhya Pradesh, Gujarat, Maharashtra and Rajasthan, for the project to be undertaken. Then dispute arose with regard to the height of the dam which was settled with the award of the Tribunal being given in 1978. For a number of years thereafter, final clearance was still not given. In the meantime some environmental studies were conducted. The final clearance was not given because of the environmental concern which is quite evident. Even though complete data with regard to the environment was not available, the Government did in 1987 finally given environmental clearance. It is thereafter that the construction of the dam was undertaken and hundreds of crores have been invested before the petitioner chose to file a writ petition in 1994 challenging the decision to construct the dam and the clearance as was given. In our opinion, the petitioner, which had been agitating against the dam since 1986, is guilty of laches in not approaching the Court at an earlier point of time.

78. When such projects are undertaken and hundreds of crores of public money are spent, individuals or organisations in the garb of PIL cannot be permitted to challenge the policy decision taken after a lapse of time. It is against the national

interest and contrary to the established principles of law that decisions to undertake developmental projects are permitted to be challenged after a number of years during which period public money has been spent in the execution of the project.

79. The petitioner has been agitating against the construction of the dam since 1986, before environmental clearance was given and construction started. It has, over the years, chosen different paths to oppose the dam. At its insistence the FMG was constituted, but its report could not result in the stoppage of construction *pari passu* with relief and rehabilitation measures. Having failed in its attempt to stall the project the petitioner has resorted to court proceedings by filing this writ petition long after the environmental clearance was given and construction started. The pleas relating to height of the dam and the extent of submergence, environment studies and clearance, hydrology, seismicity and other issues, except implementation of relief and rehabilitation, cannot be permitted to be raised at this belated stage.

80. This Court has entertained this petition with a view to satisfy itself that there is proper implementation of the relief and rehabilitation measures at least to the extent they have been ordered by the Tribunal's Award. In short it was only the concern of this court for the protection of the fundamental rights of the oustees under Article 21 of the Constitution of India which led to the entertaining of this petition. It is the R&R measures that this court is really concerned with and the petition in regard to the other issues raised is highly belated. Though it is, therefore, not necessary to do so, we, however, presently propose to deal with some of the other issues raised.

Award binding on the States

81. It has been the effort on the part of the petitioners to persuade this court to decide that, in view of the difficulties in effectively implementing the Award with regard to relief and rehabilitation, and because of the alleged adverse impact the construction of the dam will have on the environment, further construction of the dam should not be permitted. The petitioners support the contention on behalf of the State of Madhya Pradesh to the effect that the height of the dam should be reduced in order to decrease the number of oustees. In this case, the petitioners also submit that with regard to hydrology, the adoption of the figure of 27 MAF is not correct and the correct figure is 23 MAF and in view thereof the height of the dam need not be 455 ft.

82. The Tribunal in this Award has decided a number of issues which have been summarised hereinabove. The question which arises is as to whether it is open to the petitioners to directly or indirectly challenge the correctness of the said decision. Briefly stated, the Tribunal had in no uncertain terms come to the conclusion that the height of the dam should be 455 ft. It had rejected the contention of the State of Madhya Pradesh for fixing the height at a lower level. At the same time in arriving at this figure, it had considered the relief and

rehabilitation problems and had issued directions in respect thereof. Any issue which has been decided by the Tribunal would, in law, be binding on the respective States. That this is so has been recently decided by a Constitution Bench of the Court in *The State of Karnataka v. State of Andhra Pradesh and others*, 2000 (3) SCALE 505. That was a case relating to a water dispute regarding inter-state river Krishna between the three riparian States and in respect of which the Tribunal constituted under the Inter-State Water Disputes Act, 1956, had given an Award. Dealing with the Article 262 and the scheme of the Inter-State Water Disputes Act, this court at page 572 observed as follows:

> The Inter-State Water Disputes Act having been framed by the Parliament under Article 262 of the Constitution is a complete Act by itself and the nature and character of a decision made thereunder has to be understood in the light of the provisions of the very Act itself. A dispute or difference between two or more State governments having arisen which is a water dispute under Section 2(C) of the Act and complaint to that effect being made to the Union Government under Section 3 of the said Act, the Central Government constitutes a water disputes tribunal for the adjudication of the dispute in question, once it forms the opinion that the dispute cannot be settled by negotiations. The Tribunal thus constituted, is required to investigate the matters referred to it and then forward to the Central Government a report setting out the facts found and giving its decision on it as provided under sub-Section (2) of Section 5 of the Act. On consideration of such decision of the Tribunal, if the Central Government or any State government is of the opinion that the decision in question requires explanation, or that guidance is needed upon any point not originally referred to the Tribunal, then within three months from the date of the decision, reference can be made to the Tribunal for further consideration and the said Tribunal then forwards to the Central Government a further report giving such explanation or guidance as it deems fit. Thereby the original decision of the Tribunal is modified to the extent indicated in the further decision as provided under Section 5(3) of the Act. Under Section 6 of the Act the Central Government is duty bound to publish the decision of the Tribunal in the Official Gazette whereafter the said decision becomes final and binding on the parties to the dispute and has to be given effect to, by them. The language of the provisions of Section 6 is clear and unambiguous and unequivocally indicates that it is only the decision of the Tribunal which is required to be published in the Official Gazette and on such publication that decision becomes final and binding on the parties.

Once the Award is binding on the States, it will not be open to a third party like the petitioners to challenge the correctness thereof. In terms of the Award, the State of Gujarat has a right to construct a dam up to the height of 455 ft and, at the same time, the oustees have a right to demand relief and resettlement as directed in the Award. We, therefore, do not propose to deal with any contention which, in fact, seems to challenge the correctness of an issue decided by the Tribunal.

General issues relating to the displacement of tribals and the alleged violation of the rights under Article 21 of the Constitution

The submission of Shri Shanti Bhushan, learned senior counsel for the petitioners was that the forcible displacement of tribals and other marginal farmers from their land and other sources of livelihood for a project which was not in the national or public interest was a violation of their fundamental rights under Article 21 of the Constitution of India read with ILO Convention 107 to which India is a signatory. Elaborating this contention, it was submitted that this court had held in a large number of cases that international treaties and covenants could be read into the domestic law of the country and could be used by the courts to elucidate the interpretation of fundamental rights guaranteed by the Constitution. Reliance in support of this contention was placed on *Gramophone Co. of India Ltd. v. B.B. Pandey*, 1984 (2) SCC 534, *PUCL v. Union of India*, 1997 (3) SCC 433 and *CERC v. Union of India*, 1995 (3) SCC 42. In this connection, our attention was drawn to the ILO Convention 107, which stipulated that tribal populations shall not be removed from their lands without their free consent from their habitual territories except in accordance with national laws and regulations or reasons relating to national security or in the interest of national economic development. It was further stated that the said Convention provided that in such cases where removal of this population is necessary as an exceptional measure, they shall be provided with lands of quality at least equal to that of lands previously occupied by them, suitable to provide for their present needs and future development. Shri Shanti Bhushan further contended that while the SSP will displace and have an impact on thousands of tribal families, it had not been proven that this displacement was required as an exceptional measure. He further submitted that given the seriously flawed assumptions of the project and the serious problems with the rehabilitation and environmental mitigation, it could not be said that the project was in the national interest. It was also submitted that the question arose whether the Sardar Sarovar Project could be said to be in the national and public interest in view of its current best estimates of cost, benefits and evaluation of alternatives and specifically in view of the large displacement of tribals and other marginal farmers involved in the project. Elaborating this contention, it was contended that serious doubts had been raised about the benefits of the project – the very rationale which was sought to justify the huge displacement and the massive environmental impacts etc. It was contended on behalf of the petitioners that a project which was sought to be justified on the grounds of providing a permanent solution to water problems of the drought-prone areas of Gujarat would touch only the fringes of these areas, namely Saurashtra and Kachchh and even this water, which was allocated on paper, would not really accrue due to a host of reasons. It was contended that in spite of concentrating on small-scale decentralised measures which were undertaken on a large scale could address the water problem of these drought-prone areas. Huge portions of the State resources were being diverted to the Sardar Sarovar Project and as a result the small projects were ignored and the water problem in these areas persists. It was submitted that the Sardar Sarovar Project could be restructured to minimise displacement.

83. Refuting the aforesaid arguments, it has been submitted on behalf of the Union of India and the State of Gujarat that the petitioners have given a highly

exaggerated picture of the submergence and other impacts of this project. It was also submitted that the petitioner's assertion that there was large-scale relocation and uprooting of tribals was not factually correct. According to respondents, the project would affect only 245 villages in Gujarat, Maharashtra and Madhya Pradesh due to pondage and backwater effect corresponding to 1 in 100 year flood. The state-wise break up of affected villages and the number of project affected families (PAFs) shows that only four villages would be fully affected (three in Gujarat and one in Madhya Pradesh) and 241 would be partially affected (16 in Gujarat, 33 in Maharashtra and 192 in Madhya Pradesh). The total project affected families were 40,827. The extent of the submergence was minimum in the State of Madhya Pradesh. The picture of this submergence as per the Government of Madhya Pradesh Action Plan of 1993 is as follows:

> Abadi will be fully submerged in 39 villages and partially in 116 villages, agricultural land will be affected up to 10 per cent in 82 villages, up to 26 per cent in 32 villages, 26 to 50 per cent in 30 villages, 51 to 75 per cent in 14 villages, 76 to 90 per cent in 4 villages and 100 per cent in only 1 village. In 21 villages, only abadi will be affected and Government land only in 9 villages. Thus, in most of the villages, submergence is only partial.

The submergence area of the SSP can be divided into two areas:

i) Fully tribal, hilly area covering the initial reach of about 105 villages with mainly subsistence economy. It includes 33 villages of Maharashtra, 19 of Gujarat and about 53 of Madhya Pradesh.

(ii) Mixed population area in the plains of Nimad, with a well developed economy and connected to the mainstream. This area includes about 140 villages in Madhya Pradesh.

These two areas have quite different topographic and habitation features which result in totally different types of submergence impacts. The state of the hilly area to be affected by its submergence and where most of the tribal population exists is described by the Government of Madhya Pradesh Action Plan, 1993 as follows:

> The Narmada flows in hilly gorges from its origin to the Arabian Sea. The undulating hilly terrain in the lower submergence area of the SSP exhibit naked hills and depleted forests. Even small forest animals are rarely seen because of a lack of forest cover and water. The oft quoted symbiotic living with forests is a misnomer in this area because the depleted forests have nothing to offer except fuel wood. Soil is very poor, mostly disintegrated, granite and irrigation is almost nil due to the undulating and hilly land. Anybody visiting this area finds the people desperately sowing even in hills with a steep gradient. Only one rain fed crop of mostly maize is sown and so there is no surplus economy.
>
> PAPs inhabiting these interior areas find generous rehabilitation and resettlement packages as a means to assimilate in the mainstream in the valley.

84. In 193 villages of Madhya Pradesh to be affected by the project, a very high proportion of the houses would be affected whereas the land submergence was

only 14.1 per cent. The reason for this is that the river bed is a deep gorge for about 116 km upstream of the dam and as a result the reservoir will be long (214 km), narrow (average width of 1.77 km) and deep. The result of this is that as one goes further upstream, the houses on the river banks are largely affected while agricultural land which is at a distance from the river banks is spared. A majority of the 33,014 families of Madhya Pradesh (which would include 15,018 major sons) that would lose only their houses and not agricultural lands would be required to be resettled in Madhya Pradesh by constructing new houses in the new abadi. According to the Award, agricultural land was to be allotted only if the project affected families lost 25 per cent or more of agricultural land and on this basis as per the Government of Madhya Pradesh, only 830 project affected families of Madhya Pradesh were required to be allotted agricultural land in Madhya Pradesh.

85. According to the Government of Gujarat the tribals constituted the bulk of the PAFs who would be affected by the dam in Gujarat and Maharashtra, namely 97 per cent and 100 per cent respectively. Out of the oustees of PAFs of Madhya Pradesh, tribals constituted only 30 per cent while 70 per cent were non-tribals. The total number of tribal PAFs were 17,725, out of these 9,546 were already resettled. It was further the case of the respondents that in Madhya Pradesh the agricultural land of the tribal villages was affected on an average to the extent of 28 per cent whereas in the upper reaches, i.e. Nimad, where the agriculture was advanced, the extent of submergence, on an average, was only 8.5 per cent. The surveys conducted by HMS Gaur University (Sagar), the Monitoring and Evaluation Agency set up by the Government of Madhya Pradesh, reveal that the major resistance to relocation was from the richer, non-tribal families of Nimad who feared a shortage of agriculture labour if the landless labourers from the areas accepted resettlement. In the bi-annual report, 1996 of HMS Gaur University (Sagar), it was observed as follows:

> The pre-resettlement study of submerging villages has revealed many startling realities. Anti-dam protagonists present a picture that tribals and backward people are the worst sufferers of this kind of development project. This statement is at least not true in the case of the people of these five affected villages. Though these villages comprise a significant population of tribals and people of weaker sections, the majority of them will not be a victim of displacement. Instead, they will gain from shifting. The present policy of compensation is most beneficial for the lot of weaker section. These people are living either as labourers or marginal farmers. The status of oustee will make them the owner of two hectares of land and a house. In fact, it is the land-owning class which is opposing the construction of the dam by playing the card of tribals and weaker sections. The land-owners are presently enjoying the benefit of cheap labour in this part of the region. The availability of cheap labour is a boon for agricultural activities. This provides higher returns with less input.

It is apparent that the tribal population affected by the submergence would have to move but the rehabilitation package was such that the living conditions would be much better than what it was before. Furthermore, though 140 villages in Madhya Pradesh would be affected in the plains of Nimad, only 8.5 per cent of the agricultural land of these villages shall come under submergence due to SSP and as such the said project shall have only a marginal impact on the agricultural productivity of the area.

86. While accepting the legal proposition that international treaties and covenants can be read into the domestic laws of the country, the submission of the respondents was that Article 12 of the ILO Convention No. 107 stipulates that '[t]he populations concerned shall not be removed without their free consent from their habitual territories except in accordance with national laws and regulations for reasons relating to national security, or in the interest of national economic development or of the health of the said populations'.

87. The said Article clearly suggested that when the removal of the tribal population is necessary as an exceptional measure, they shall be provided with land of quality at least equal to that of the land previously occupied by them and they shall be fully compensated for any resulting loss or injury. The rehabilitation package contained in the Award of the Tribunal as improved further by the State of Gujarat and the other States *prime facie* shows that the land required to be allotted to the tribals is likely to be equal, if not better, than what they had owned.

88. The allegation that the said project was not in the national or public interest is not correct seeing the need of water for a burgeoning population which is most critical and important. The population of India, which is now one billion, is expected to reach a figure between 1.5 billion and 1.8 billion in the year 2050, would necessitate the need of 2,788 b.cu.m of water annually in India to be above water stress zone and 1,650 b.cu.m to avoid being water-scarce country. The main source of water in India is rainfall which occurs in about 4 months in a year, and the temporal distribution of rainfall is so uneven that the annual averages have very little significance for all practical purposes. According to the Union of India, one third of the country is always under threat of drought not necessarily due to deficient rainfall but many times due to its uneven occurrence. To feed the increasing population, more food grain is required and an effort has to be made to provide safe drinking water, which, at present, is a distant reality for most of the population, specially in the rural areas. Keeping in view the need to augment the water supply, it is necessary that water storage capacities have to be increased adequately to ward off the difficulties in the event of monsoon failure as well as to meet the demand during the dry season. It is estimated that by the year 2050 the country needs to create storage of at least 600 b.cu.m against the existing storage of 174 b.cu.m.

89. Dams play a vital role in providing irrigation for food security, domestic and industrial water supply, hydroelectric power and keeping flood waters back. On full development, the Narmada has a potential of irrigating over 6 million hectares of land and generating 3,000 MW of power. The present stage of development is

very low with only 3 to 4 MAF of water being used by the party States for irrigation and drinking water against 28 MAF availability of water at 75 per cent dependability as fixed by NWDT and about 100 MW power developed. 85 per cent of the water is estimated as flowing in waste to the sea. The project will provide safe and clean drinking water to 8,215 villages and 135 towns in Gujarat and 131 villages in the desert areas of the Jalore district of Rajasthan, though against these, only 241 villages are getting submerged partially and 4 villages fully due to the project.

90. The cost and benefit of the project were examined by the World Bank in 1990 and the following passage speaks for itself:

> The argument in favour of the Sardar Sarovar Project is that the benefits are so large that they substantially outweigh the costs of the immediate human and environmental disruption. Without the dam, the long term costs for people would be much greater and the lack of an income source for future generations would put increasing pressure on the environment. If the waters of the Narmada river continue to flow to the sea unused there appears to be no alternative to escalating human deprivation, particularly in the dry areas of Gujarat. The project has the potential to feed as many as 20 million people, provide domestic and industrial water to about 30 million, employ about 1 million, and provide valuable peak electric power in an area with high unmet power demand (farm pumps often get only a few hours power per day). In addition, recent research shows substantial economic 'multiplier' effects (investment and employment triggered by development) from irrigation development. Set against the futures of about 70,000 project affected people, even without the multiplier effect, the ratio of beneficiaries to affected persons is well over 100:1.

There is merit in the contention of the respondents that there would be a positive impact on the preservation of ecology as a result from the project. The SSP would be making a positive contribution to the preservation of the environment in several ways. The project, by taking water to drought-prone and arid parts of Gujarat and Rajasthan, would effectively arrest ecological degradation which was returning to make these areas inhabitable due to salinity ingress, advancement of desert, groundwater depletion, fluoride and nitrite affected water and vanishing green cover. The ecology of water scarcity areas is under stress and the transfer of Narmada water to these areas will lead to sustainable agriculture and the spread of green cover. There will also be an improvement of fodder availability which will reduce pressure on biodiversity and vegetation. The SSP, by generating clean eco-friendly hydropower, will save the air pollution which would otherwise take place by thermal generation power of similar capacity.

91. The displacement of the tribals and other persons would not per se result in the violation of their fundamental or other rights. The effect is to see that on their rehabilitation at new locations they are better off than what they were. At the rehabilitation sites they will have more and better amenities than they enjoyed in their tribal hamlets. The gradual assimilation in the mainstream of the society will lead to betterment and progress.

Environmental Issues

92. The four issues raised under this head by Shri Shanti Bhushan are as under:

1) Whether the execution of a large project, having diverse and far reaching environmental impact, without the proper study and understanding of its environmental impact and without proper planning and mitigative measures is a violation of fundamental rights of the affected people guaranteed under Article 21 of the Constitution of India?

2) Whether the diverse environmental impacts of the Sardar Sarovar Project have been properly studied and understood?

3) Whether any independent authority has examined the environmental costs and mitigative measures to be undertaken in order to decide whether the environmental costs are acceptable and mitigative measures practical?

4) Whether the environmental conditions imposed by the Ministry of Environment have been violated and if so, what is the legal effect of the violations?

93. It was submitted by Shri Shanti Bhushan that a large project having diverse and far reaching environmental impacts in the concerned States would require a proper study and understanding of the environmental impacts. He contended that the study and planning with regard to environmental impacts must precede construction. According to Shri Shanti Bhushan, when the environmental clearance was given in 1987, proper study and analysis of the environmental impacts and mitigative measures, which were required to be taken, were not available and, therefore, this clearance was not valid. The decision to construct the dam was stated to be a political one and was not a considered decision after taking into account the environmental impacts of the project. The execution of SSP without a comprehensive assessment and evaluation of the environmental impacts and a decision regarding its acceptability was alleged to be a violation of the rights of the affected people under Article 21 of the Constitution of India. It was further submitted that no independent authority has examined vehemently the environmental costs and mitigative measures to be undertaken in order to decide whether the environmental costs are acceptable and mitigative measures practical. With regard to the environmental clearance given in June 1987, the submission of Shri Shanti Bhushan was that this was the conditional clearance and the conditions imposed by the Ministry of Environment and Forests had been violated. The letter granting clearance, it was submitted, disclosed that even the basic minimum studies and plans required for the environmental impact assessment had not been done. Furthermore it was contended that in the year 1990, as the deadline for completion of the studies was not met, the Ministry of Environmental and Forests had declared that the clearance had lapsed. The Secretary of the said Ministry had requested the Ministry of Water Resources to seek extension of the clearance but ultimately no extension was sought or given and the studies and action plans

continued to lag to the extent that there was no comprehensive environmental impact assessment of the project, proper mitigation plans were absent and the costs of the environmental measures were neither fully assessed nor included in the project costs. In support of this contention, Shri Shanti Bhushan relied upon the report of a Commission called the Independent Review or the Morse Commission. The said Commission had been set up by the World Bank and it submitted its report in June 1992. In the report, the Commission had adversely commented on practically all aspects of the project and in relation to environment, it was stated as under:

> Important assumptions upon which the projects are based are now questionable or are known to be unfounded. Environmental and social trade-offs have been made, and continue to be made, without a full understanding of the consequences. As a result, benefits tend to be over-stated, while social and environmental costs are frequently understated. Assertions have been substituted for analysis.
>
> We think that the Sardar Sarovar Projects as they stand are flawed, that resettlement and rehabilitation of all those displaced by the projects is not possible under the prevailing circumstances, and that the environmental impacts of the projects have not been properly considered or adequately addressed.
>
> The history of environmental aspects of Sardar Sarovar is a history of non-compliance. There is no comprehensive impact statement. The nature and magnitude of environmental problems and solutions remain elusive.*

94. Shri Shanti Bhushan submitted that it had become necessary for some independent judicial authority to review the entire project, examine the current best estimates of all costs (social, environmental, financial), benefits and alternatives in order to determine whether the project is required in its present form in the national interest, or whether it needs to be restructured/modified.

95. Shri Shanti Bhushan further submitted that environmental impacts of the projects were going to be massive and a full assessment of these impacts had not been made. According to him the latest available studies show that studies and action plans had not been completed and even now they were lagging behind *pari passu*. It was also contended that a mere listing of the studies does not imply that everything is taken care of. Some of the studies were of poor quality and based on improper data and no independent body had subjected these to critical evaluation.

Re: Environmental clearance

96. As considerable stress was laid by Shri Shanti Bhushan challenging the validity of the environmental clearance granted in 1987 *inter alia* on the ground that it was not preceded by adequate studies and it was not a considered opinion and there was non-application of mind while clearing the project, we first propose to deal with the contention.

---

* The full text of the letter to the President of the World Bank is reproduced at page 331.

97. The events after the Award and up to the environmental clearance granted by the Government vide its letter dated 24 June 1987 would clearly show that some studies, though incomplete, had been made with regard to different aspects of the environment. Learned counsel for the respondents stated that in fact on the examination of the situation, the claim made with regard to the satisfactory progress was not correct. In order to carry out the directions in the Award about the setting up of an authority, the Inter-State Water Disputes Act, 1956 was amended and Section 6-A was inserted to set out how a statutory body could be constituted under the Act. On 10 September 1980, in exercise of the powers conferred by Section 6-A of the Act, the Central Government framed a scheme, constituted the Narmada Control Authority to give effect to the decision of the Award.

98. In January 1980 the Government of Gujarat submitted to the Central Water Commission a detailed project report in 14 volumes. This was an elaborate report and dealt with various aspects like engineering details, canal systems, geology of area, coverage of command area, etc. On 15 February 1980 the Central Water Commission referred SSP to the then Department of Environment in the Department of Science and Technology. At that point of time, environmental clearance was only an administrative requirement. An environmental checklist was forwarded to the Government of Gujarat on 27 February 1980, which sought to elucidate information including the following ecological aspects:

i. Excessive sedimentation of the reservoir,

ii. Waterlogging,

iii. Increase in salinity of the groundwater,

iv. Groundwater recharge,

v. Health hazard-water borne diseases, industrial pollution etc.,

vi. Submergence of important minerals,

vii. Submergence of monuments,

viii. Fish culture and aquatic life,

ix. Plant life-forests,

x. Life of migratory birds,

xi. National park and sanctuaries,

xii. Seismicity due to filling of reservoir.

The Government of Gujarat accordingly submitted information from September 1980 until March 1983. This information was also submitted on physio-social and economic studies for Narmada Command Area, covering cropping pattern, health aspects, water requirement etc. A note of the influence of Navagam Dam on fish yield including impact on downstream fisheries was also submitted.

99. The techno-economic appraisal of the project was undertaken by the Central Water Commission which examined water availability, command area development, construction, etc. The project was considered in the twenty-second meeting of the Technical Advisory Committee on Irrigation, Flood Control and Multi-purpose projects held on 6 January 1983 and found it acceptable subject to environmental clearance.

100. At this point of time, the matter was handled by the Department of Science and Technology which also had a Department dealing with the environment. The Environmental Appraisal Committee of the Department of Environment, then headed by a Joint Secretary, had in its meeting held on 12 April 1983 approved the project, in principle, and required that further data be collected. This Environmental Appraisal Committee dealt with the project on two other occasions, namely on 29 March 1985, when it deferred meeting to await the report of the Dewan Committee on soil conservation, and thereafter on 6 December 1985, when it deferred the meeting to await comments from the Forest Department. As stated hereafter, subsequently the Secretary of the newly constituted Ministry of Environment and Forests took up further consideration of this project along with other higher officials.

101. After the project was approved, in principle, studies and collections of data were continuing. In May 1983 the Narmada Planning Group, Government of Gujarat, after the completion of preliminary surveys, submitted work plans for various activities such as cropping pattern, health aspects, water requirements, distribution system, lay out and operation, development plan of the command, drainage and groundwater development.

102. In July 1983 a study report on 'Ecology and Environment Impact of Sardar Sarovar Dam and its Environs', prepared by MS University, was also submitted by the Government of Gujarat, covering the issues as mentioned below: climate, geology, soil, land use, forest and wildlife, aquatic vegetation, water regime (salinity, tidal movements etc.), fisheries, health, seismicity.

103. A review meeting was convened by the Secretary, Ministry of Water Resources in January 1984 which was attended by a representative of the Department of Environment. During this meeting it was emphasised that the issues regarding catchment area treatment, impact on wildlife, health, waterlogging etc. should be studied in depth for assessment. The issue of charging the cost of catchment area treatment to the project was also discussed. To sort out this matter, a meeting was subsequently convened by the Member, Planning Commission on 23 May 1984 in which the Ministry of Environment and Forests took a stand that there was a need for an integrated approach to basin development covering the catchment and command area. A project report, therefore, should be prepared to cover these aspects. Since the catchment area of Narmada Sagar and Sardar Sarovar was very vast, it was decided that an Inter-Departmental Committee should be set up by the Ministry of Agriculture under the Chairmanship of Dr M.L.

Dewan. This group could submit its report only in August 1985 covering areas of catchment of Narmada and Sardar Sarovar and recommended that at least 25–30 per cent of the area might require treatment for these projects.

104. The consideration of the project in the Ministry, therefore, got deferred for this report on catchment area treatment. During this time, the Government of Madhya Pradesh entrusted the studies on flora for Narmada Valley Project to the Botanical Survey of India and other related surveys were being carried out. Even though there was a request on 10 June 1985 from the Chief Minister of Gujarat to the Minister of State for Environment and Forests for delinking of catchment area treatment works on clearance of the project, this request was not agreed.

105. By this time the approval of SSP was being considered by the Secretary, Ministry of Environment and Forests, who invited other high officials to a review meeting which was held on 31 December 1985 under his Chairmanship. In this meeting, detailed presentations were made by the State officials of Gujarat, Madhya Pradesh and Maharashtra as well as the experts who were involved in the preparation of plans. The Secretary, Ministry of Environment and Forests assessed and reviewed readiness on various environmental aspects like catchment area treatment, compensatory afforestation, rehabilitation, command area development, labour force and health issues, aquatic species, seismicity etc., and discussed the available reports in detail in the presence of the officers of the Central/State governments, Botanical Survey of India, senior officers of the Forest Department, Planning Commission, Agriculture Department, the Additional Inspector of Forests, Government of India, the Deputy Inspector General, Assistant Inspector General of Forests, Government of India, senior officers of the Ministry of Environment and Forests, and the Secretary, Irrigation.

106. As a follow-up, the Government of Maharashtra submitted environmental data regarding affected areas in Maharashtra. This included:

- Impact assessment on wildlife,
- Impact assessment on genetics, specifically identifying the plant types which are likely to be lost as a result of submergence,
- Social anthropological studies on tribals,
- The suitability of alternative land suggested for compensatory afforestation for growing,
- Data regarding alternate land in large blocks,
- Arrangements made for exploitation of mineral resources going under submergence,
- Alternative fuels to the labourers,
- Micro-climatic changes,
- Arrangements made for treatment of catchment area including soil conservation afforestation,
- Steps taken for preserving archaeological and historical monuments,
- Proper land use,

- Actions taken by Government of Maharashtra in pursuance of Dewan Committee Report,
- Arrangements for monitoring for environmental impact for the project,
- Data related to the rehabilitation of project affected persons.

107. The Government of Gujarat also forwarded to the Government of India work plans on the following:

- Forests and wildlife,
- Fish and fisheries,
- Health aspects.

The work plan on forests and wildlife incorporated actions to be taken on the recommendations of the Inter-Departmental Committee headed by Dr Dewan on soil conservation and afforestation works in the catchment area.

108. In March 1986 a meeting was convened by the Ministry of Water Resources in order to discuss the issues of fisheries, flora/fauna, health, archaeology with the officers of the Botanical Survey of India, Zoological Survey of India, Archaeological Survey of India and the officers of the various departments of the State and Centre to gear up the preparation of the environmental work plans. The next meeting was held on 11 April 1986. The Secretary, Ministry of Environment and Forests, who chaired the meeting of senior officials, representatives of States and other agencies, sought additional information to be made available by 30 April 1986 before assessment and management decision.

109. In October 1986 the Ministry of Water Resources prepared and forwarded to the Ministry of Environment and Forests a note on the environmental aspects of the two projects and noted the urgency of the decision. It also considered the importance of the project, should the project be taken at all, environmental aspects of the project and ultimately rehabilitation, compensatory afforestation, fauna and flora, catchment area treatment, public health aspects, and the prevention of waterlogging. It then considered what remained to be done and enumerated the same with a time schedule as follows:

1. Madhya Pradesh to complete the detailed survey of population likely to be affected in all phases of NSP (three years);
2. Maharashtra to prepare a detailed rehabilitation plan for 33 villages under phase 1 of SSP (three years);
3. Madhya Pradesh to identify degraded forest lands twice the forest area to be submerged for compensatory afforestation (six months);
4. Survey of flora in Narmada Valley assigned to Botanical Survey of India (two years);
5. Survey of wildlife by Zoological Survey of India (two years);

6. Aerial photographs and satellite imagery to be analysed by All India Soil and Land Use Survey Organisation and National Remote Sensing Agency and critically degraded areas in catchment. Field surveys (three years);
7. Pilot studies to determine measures for catchment area treatment in 25,000 ha (three years after aerial survey).

110. In this note two options were considered – one to postpone the clearance and the other to clear it with certain conditions with appropriate monitoring authorities to ensure that the action is taken within the time-bound programme. It was concluded that in the light of the position set out, it was necessary that the project should be cleared from the environmental angle, subject to conditions and stipulations outlined.

111. The Department of Environment and Forests made its own assessment though a note of the Secretary, Ministry of Environment and Forests. It took the view that the following surveys/studies as set out therein might take at least 2–3 years. It noted in this regard that:

1. The estimate of the Ministry of Water Resources on analysis of aerial photographs and satellite imageries as 2–3 years;
2. Catchment area treatment programme can be formulated by three years thereafter;
3. Wildlife census by the Zoological Survey of India would take at least three years;
4. Survey by the Botanical Survey of India would take three years.

It further took the view that it was essential that there should be a strong management authority. It finally concluded that if the government should decide to go ahead with the project it should be done with the provision of environmental management authority and adequate power and teeth to ensure that the environment management plan is implemented *pari passu* with engineering and other works. It concluded that effective implementation of the engineering and environmental measures simultaneously will go a long way and that such a project could be implemented by harmonising environmental conservation needs with the development effort.

112. The Ministry of Environment and Forests had not given environment clearance of Narmada Sagar and Sardar Sarovar Dam despite all discussions which had taken place. The documents filed along with the affidavit of Shri P.K. Roy, Under Secretary, Prime Minister's Office, dated 27 April 2000, indicate that there was a difference of opinion with regard to the grant of environmental clearance between the Ministry of Water Resources and the Ministry of Environment and Forests. This led to the matter being referred to the Prime Minister's Secretariat for clearance at the highest level. A note dated 20 November 1986 prepared by the

Ministry of Water Resources was forwarded to the Prime Minister's Secretariat as well as to the Ministry of Environment and Forests after dealing with the environmental aspects relating to rehabilitation, catchment area treatment, command area development, compensatory afforestation, flora and fauna. This note indicated that there were two options with regard to the clearance of the said project. One was to await for two to three years for the completion of the operational plans and other detailed studies, and the second option was that the project should be given the necessary clearance subject to the stipulation with regard to the action to be taken in connection with various environmental aspects and appropriate monitoring arrangements to ensure that the actions were taken in a time bound manner. The Ministry of Water Resources recommended that it should be possible to give environmental clearance of the project and ensure that the conditions are properly met through a process of clear assignment of responsibility and frequent monitoring. The modus operandi for instituting a monitoring system could be discussed at the meeting.

113. On 26 November a meeting took place which was attended, *inter alia*, by the Secretary, Ministry of Water Resources, the Secretary, Ministry of Environment and Forests, the Additional Secretary, Prime Minister Secretariat and representatives of the Governments of Madhya Pradesh and Gujarat regarding the environmental aspects of the Narmada Sagar and Sardar Sarovar Project. The minutes of the meeting, *inter alia*, disclosed 'it was decided that the Government of Gujarat would identify lands for allocation to the project affected persons of Madhya Pradesh within a specified period of time. The meeting also envisaged the arrangement of a monitoring and enforcement authority to monitor the project and to ensure that the actions on the environmental aspects proceed according to the schedule and *pari passu* with the rest of the project'. The authority was not to be mainly an advisory one but was to be given executive powers of enforcement including power to order stoppage of construction activity in the event of its being of the opinion that there was a lack of progress in action on the environment front.

114. On 19 December 1986 the Secretary, Ministry of Environment and Forests sent to the Secretary to the Prime Minister a combined note on the environmental aspects of both the projects, namely Narmada Sagar and Sardar Sarovar Project. In this note, it was, *inter alia*, stated that there was absence and inadequacy on some important environmental aspects even though the Sardar Sarovar Project was in a fairly advanced stage of preparedness. The note also recommended the establishment of the Narmada Management Authority with adequate powers and teeth to ensure that the Environment Management Plan did not remain only on paper but was implemented; and implemented *pari passu* with engineering and other works. In the end, in the note, it stated as follows:

> If, despite the meagre availability of data and the state of readiness on NSP, the Government should decide to go ahead with the project, it is submitted that it should do so only on the basis of providing a Management Authority as outlined above with the hope that the public opposition, not just by vested interests but by

credible professional environmentalists, can be overcome. Effective implementation of the engineering and environmental measures simultaneously would go a long way to prove that even such a project can be implemented by harmonising environmental conservation needs with the development effort.

The choice is difficult but a choice has to be made.

Along with this note was the statement showing the cost and benefits of the Narmada Sagar and Sardar Sarovar dam. The same reads as follows:

Costs	**Narmada Sagar**	**Sardar Sarovar**
Dam construction	Rs 1,400 crores (1981 price level)	Rs 4,240 crores (1982 price level)
Loss of forest	Rs 320 crores	
Environment cost of loss of forests	Rs 30,923 crores	Rs 8,190 crores
Catchment area development	Rs 300 crores	Not available
Command area development	Rs 243.7 crores	Rs 604.0 crores Rs 300.0 crores (conjunctive use)
Loss of mineral reserves	----	---
Diversion of 42 km railway line	----	---
Population affected	1,29,396 (1981 census)	86,572 (Excluding population with land submerged for short period every year)
Land submerged	91,378 ha	39,134 ha
Benefits		
Area irrigated Net culturable land	1,23,000 ha 1,40,960 ha	17,92,000 ha 21,20,000 ha
Power generation	223.5 MW(firm power) 1,000 MW (installed capacity)	300 MW 118.3 MW in 2023 1,450 MW (installed capacity)

115. After a series of meetings held between the Secretary to Prime Minister's office as well as the Ministry of Water Resources, a detailed note dated 15 January

1987 was prepared by Mrs Otima Bordia, Additional Secretary to the Prime Minister. The notes opened by saying that Narmada Sagar and Sardar Sarovar multipurpose projects have been pending approval for a considerable amount of time. The States of Madhya Pradesh and Gujarat have been particularly concerned and have been pressing for their clearance. The main issues of environmental concern related to the rehabilitation of the affected population, compensatory afforestation, treatment of the catchment area, command area development, pertaining particularly to drainage, waterlogging and salinity. The said note mentioned that the Department of Environment and Forests had sent a note with the approval of the Minister for Environment and Forests and had recommended conditional approval to the Narmada Sagar and Sardar Sarovar Projects subject to three conditions:

i. Review of design parameters to examine the feasibility of modifying the height of the dam;

ii. Preparation in due time, detailed and satisfactory plans for rehabilitation, catchment area treatment, compensatory afforestation and command area development;

iii. Setting up of Narmada Management Authority with adequate power and teeth to ensure that environmental management plans are implemented *pari passu* with engineering and other works.

116. It is further stated in the note that the Ministry of Water and the State governments had no difficulty in accepting conditions (ii) and (iii). With regard to the review of design parameters and dam height, the Ministry of Water Resources had examined the same after taking into consideration the comments of the Central Water Commission and concluded that the reduction of the FRL of the NSP would not be worthwhile. The Secretary to the Prime Minister had discussed the matter with the Secretary, Ministry of Water Resources and Secretary, Ministry of Environment and Forests and it was agreed that the recommendation of the Ministry of Environment and Forests, of giving clearance on the condition that items (ii) and (iii) referred to hereinabove, be accepted. The note also stated that in view of the technical report, reduction in the dam height did not appear to be feasible. This note of Mrs Otima Bordia recommended that the Prime Minister's approval was sought on giving conditional clearance. On this note, Mrs Seria Grewal, Secretary to the Prime Minister noted as follows:

> Proposal at paragraph 17 may kindly be approved. This project has been pending clearance of the last seven years and both the Chief Ministers of Gujarat and Madhya Pradesh are keenly awaiting the clearance of the same. The agency, which is proposed to be set up to monitor the implementation of this project, will fully take care of the environmental degradation about which the Prime Minister was concerned. The Ministry of Environment and Forests have recommended clearance of this project subject to conditions which will take care of the Prime Minister's apprehensions. I shall request the Secretary, Water Resources, who will

> be Chairman of the Monitoring Agency, to see that no violation of any sort takes place and that the Prime Minister's office will be kept informed of the progress of this project every quarter. The matter is urgent as last week the Chief Minister of Gujarat had requested the green signal be given before 20 January.
>
> The Prime Minister may kindly approve.

The Prime Minister Shri Rajiv Gandhi, instead of giving the approval, made the following note: 'Perhaps this is a good time to try for a River Valley Authority. Discuss'.

It appears that the Ministry of Environment and Forests gave its clearance to the setting up of an Inter-Ministerial Committee and on 8 April 1987 the following note was prepared and forwarded to the Prime Minister:

> This case has got unduly delayed. The Prime Minister was anxious that speedy action should be taken. As such, since the Ministry of Environment has given its clearance subject to setting up of an Inter-Ministerial Committee as indicated at 'A' above, we may give the necessary clearance. The three Chief Ministers may be requested to come over early next week to give their clearance in principle for the setting up of a River Valley Authority so that simultaneous action can be initiated for giving practical shape to this concept. The clearance of the project, however, should be communicated within two weeks as I have been informed by Shri Shiv Shankar and Shri Bhajan Lal that interested parties are likely to start an agitation and it is better if clearance is communicated before mischief is done by the interested parties.

117. Along with another affidavit of Shri P.K. Roy, Under Secretary, Prime Minister's office, dated 2 May 2000, some correspondence exchanged between Legislature and the Prime Minister has also been placed on record relating to the granting of the environmental clearance by the Prime Minister. On 31 March 1987 Shri Shanker Sing Vaghela, then Member of Parliament, Rajya Sabha, had written a letter to the Prime Minister in which it was, *inter alia*, stated that the foundation stone for the Narmada Project had been laid 25 years ago by the late Pandit Jawahar Lal Nehru and that after the Tribunal's Award, Mrs Indira Gandhi had cleared the project in 1978, but still the environmental clearance had not so far been given. It was also stated in his letter that the project was now being delayed on account of so-called environmental problems. It was further stated in his letter that the Sardar Sarovar Project, when completed, will solve more of the pressing problems of environment than creating them. To this letter of Shri Vaghela, the then Prime Minister sent a reply dated 8 April 1987 stating as follows:

> I have seen your letter of 31 March regarding the Narmada Project. All aspects have to be carefully considered before decisions are taken on a project of this size. This is being done.
>
> The environment and ecological factors cannot be disregarded. We cannot also dismiss the needs of our tribal people. Safeguards are required to ensure that rehabilitation plans are effective.
>
> All these aspects are being examined and a decision will be taken soon.

On 30 April 1987 a press note was released by the Government of India in which it was stated that in a meeting presided over by the Prime Minister, it was agreed by the Chief Ministers of Madhya Pradesh and Gujarat and representatives of the Maharashtra government that, a high level River Valley Authority would be set up for the control and development of the river basin. This press note also stated that the Narmada Sagar and the Sardar Sarovar Project on the river Narmada had been cleared. Soon thereafter Shri Ahmad Pate, Member of Parliament from Gujarat, wrote a letter dated 14 April 1987 to Shri Rajiv Gandhi expressing his gratitude for according clearance to the Narmada multi-purpose project. This letter was replied to on 22 April 1987 by Shri Rajiv Gandhi who thanked Shri Patel for writing his letter dated 14 April 1987 regarding the Narmada project. On 20 April 1987 Shri Shanker Singh Vaghela wrote another letter to the Prime Minister. While thanking him for clearing the project, it was stated that there was apprehension about the environment and ecological factors and also about the needs of the tribal people. The Prime Minister was requested 'to clarify to the people of Gujarat whether or not these aspects have finally been cleared or not and all the doubts on this front have been finally set at rest or not'. On 4 May 1987 the Prime Minister replied to this letter in which it was stated as follows:

> There should be no grounds for any misunderstanding in this regard. The Narmada project has been cleared while at the same time ensuring that environmental safeguards will be enforced and effective measures taken for the rehabilitation of tribals. You could ask the Ministry of Water Resources or the State government for details.

Lastly, we need to make reference to a letter dated 10 June 1987 written by Srimati Chandraben Sureshbhai Shrimali, a Member of the Legislative Assembly of Gujarat and the reply of the Prime Minister thereto. In the said letter dated 10 June 1987, Srimati Shrimali thanked the Prime Minister for clearing the Narmada project and it was stated that the dry land of Gujarat and Saurashtra would be fertilised through Narmada Yojana. To this, the reply dated 30 June 1987 of the Prime Minister was as follows:

> Thank you for your letter of 10 June. The visit to Surendranagar was useful and educative. We are all looking forward to the early implementation of the Sardar Sarovar Project. The question of environmental protection also needs serious attention. I wish you and the people of Surendranagar a good monsoon.

118. From the documents and the letters referred to hereinabove, it is more than evident that the Government of India was deeply concerned with the environmental aspects of the Narmada Sagar and Sardar Sarovar Project. Inasmuch as there was some difference of opinion between the Ministries of Water Resources and Environment and Forests with regard to the grant of environmental clearance, the matter was referred to the Prime Minister. Thereafter, a series of discussions took place in the Prime Minister's Secretariat and the concern of the Prime Minister with regard to the environment and desire to safeguard the interest of the tribals resulted in some time being taken. The Prime Minister gave environmental

clearance on 13 April 1987 and a formal letter was issued thereafter on 24 June 1987.

119. It is not possible, in view of the aforesaid state of affairs, for this court to accept the contention of the petitioner that the environmental clearance of the project was given without application of mind. It is evident, and in fact this was the grievance made by Shri Vaghela, that the environmental clearance of the project was unduly delayed. The government was aware of the fact that a number of studies and data had to be collected relating to the environment. Keeping this in mind, a conscious decision was taken to grant environmental clearance and in order to ensure that environmental management plans are implemented *pari passu* with engineering and other works, the Narmada Management Authority was directed to be constituted. This is also reflected in the letter dated 24 June 1987 of Shri Mudgal giving formal clearance to the project.

Re: Other issues relating to environment

120. Prior to the grant of the environmental clearance on 24 June 1987, sufficient studies were made with regard to different aspects of the environment on the basis of which conditional clearance was granted on 24 June 1987, one of the conditions of clearance being that the balance studies should be completed within a stipulated time frame. According to the Government of Gujarat, the conditions imposed in the environmental clearance granted on 24 June 1987 were:

a) The NCA would ensure that the environmental safeguard measures are planned and implemented *pari passu* with the progress of work on the project.

b) The detailed survey/studies assured will be carried out as per the schedule proposed and details made available to the department for assessment.

c) The catchment area treatment programme and rehabilitation plans be so drawn so as to be completed ahead of reservoir filling.

d) The department should be informed of progress on various works periodically.

It was further submitted by the Government of Gujarat that none of these conditions were linked to any concrete time frame.

a) The first condition casts a responsibility on the NCA to ensure that the environmental aspects are always kept in view. The best way to attain the first and the fourth conditions was to create an environment sub-group headed by the Secretary in the Ministry of Environment and Forest.

b) The second condition – the conducting of surveys by its very nature – could not be made time bound. The surveys related to various activities to undo any damage or threat to the environment not only by the execution of the project but in the long term. Therefore, any delay in the conduct of surveys was not critical. Besides, perusal of the latest status report on the environment shows that a large number of surveys were carried out right from 1983 and also after 1987.

c) The third condition has already stood fully complied with as observed by the Environment Sub-group.

d) The fourth condition again involved keeping the department informed.

121. It was submitted that the concept of 'lapsing' is alien to such conditions. In other words, formal environmental and forest clearances granted by the Ministry of Environment and Forests, Government of India are not lapsed and are very much alive and subsisting.

122. With regard to the lapsing of the clearance granted in 1987, it was contended by Shri Harish Salve that a letter dated 25 May 1992 was written by the Secretary, Ministry of Environment and Forests, Government of India to the Secretary, Ministry of Water Resources stating, *inter alia*, that the conditions of clearance of the project were not yet met and, therefore, a formal request for extension of environmental clearance, as directed by the Review Committee of the Narmada Control Authority, may be made and failing which, a formal notification may be issued revoking the earlier clearance. It is, however, an admitted position that no formal notification has ever been issued revoking and/or cancelling the aforesaid two clearances at any point of time by the Ministry of Environment and Forests, Government of India. The Secretary, Ministry of Environment and Forests has continued to hold and chair the meetings of the Environment Sub-group, Narmada Control Authority closely monitoring the execution of SSP for ensuring that environmental safeguard measures are implemented *pari passu* with the progress of work. On 11 August 1992 a letter was written by the Narmada Control Authority to the Secretary, Ministry of Environment and Forests sending an action plan and status in respect of environmental safeguard measures taken and also stating amongst other details, the following:

> A number of letters were exchanged between the Ministry of Water Resources and MoEF and a great deal of discussion took place both in the Environment Sub-group and NCA as to whether an application for extension of time as above is at all necessary. After a detailed discussion in the last NCA meeting on 25 July 1992, it has been decided that NCA should clearly indicate the additional time required for the completion of the remaining studies, like flora and fauna and some aspects of fisheries, and a revised action plan thereon be also sent expeditiously.
>
> Keeping in view the fact and circumstances mentioned above, I request you to kindly agree to the schedule of the studies and the follow-up actions as presented here. A brief account of the action plan together with bar charts are enclosed, presenting a pictorial view.

On 15 December 1992 a letter was written to the Secretary, Ministry of Environment and Forests, more particularly stating as under, amongst other things:

> The Narmada Control Authority has already prepared an action plan and status on the environmental measures of Sardar Sarovar Project and submitted to the Ministry of Environment and Forests vide their letter No. NCA/EM/683 dated 11 August 1992 for concurrence. As may be seen from their report on action, so far there are no safeguard measures.
>
> During the field season of every year this will be closely reviewed to attain *pari passu* objectives so that the submergence during monsoon is taken care of.
>
> The above actions are scheduled to be completed by June 1993. No doubt, action in Maharashtra is lagging. The matter was taken up with the Chief Secretary of Maharashtra. A copy of his reply dated 7 November 1992 is enclosed. You will observe that the reasons for the lag are largely due to the un-cooperative and agitational approach adopted by some people.
>
> Taking all these into account, you will appreciate that the action plans are adequate.

The Minister for Water Resources, Government of India wrote a letter on 27 January 1993 to the Minister of State for Environment and Forests stating that there had been no violation of environmental safeguard measures. On 7 July 1993 the Secretary, Ministry of Water Resources, Government of India wrote a letter to the Secretary, Ministry of Environment and Forests, Government of India, more particularly stating as under:

> Progress of all the environmental works is summarised in the sheet enclosed herewith. I share your concern for the initial delay in some of the studies but now it seems that the work is in full swing. However, there is a need to keep a close watch and I am advising the NCA for the same.

By the letter dated 17 September 1993, the Minister of State for Environment and Forests, Government of India wrote to the Minister for Water Resources, Government of India appreciating the efforts made by the concerned State governments in making the environmental plans. The exchange of the aforesaid correspondence and the conduct of various meetings of the Environment Sub-group from time to time under the Chairmanship of the Secretary, Ministry of Environment and Forests, dispels the doubt of the environment clearance having been lapsed. In other words, there could not have been any question of environmental clearance granted to SSP being lapsed, more particularly when the Environment Sub-group had been consistently monitoring the progress of various environmental works and had been observing in its minutes of various meetings held from time to time, about its analysis of the works done by the respective States in the matter of the status of studies, surveys and environmental action plans in relation with:

i. Phased catchment area treatment,

ii. Compensatory afforestation,

iii. Command area development,

iv. Survey of flora, fauna, etc,

v. Archaeological and anthropological survey,

vi. Seismicity and rim stability of reservoir,

vii. Health aspects, and

viii. Fisheries development of SSP and NSP reservoirs.

123. Shri Shanti Bhushan, in the course of his submissions referred to the report of the Morse Committee in support of his contentions that the project was flawed in more ways than one.

124. The Morse Committee was constituted, as already noted, by the World Bank. Its recommendations were forwarded to the World Bank. Apart from the criticism of this report from other quarters, the World Bank itself did not accept this report as is evident from its press release dated 22 June 1992 where it was, *inter alia*, stated as follows:

> The Morse Commission provided a draft of its report to the Bank for management comments several weeks prior to the final release of the document. About two weeks before this release, the commission provided a draft of its findings and recommendations. The final version of the report is the sole responsibility of its authors; the report was not cleared by the World Bank.
>
> On resettlement and rehabilitation (R&R), Bank management agrees with the description of the R&R situation in each of the three States and with the report's conclusions about the shortcomings in the preparation and appraisal of the project's R&R aspects. We also agree that work should have been done earlier on the issue of people affected by the canal in Gujarat. However, we do not share the view that resettlement would be virtually impossible even if Maharashtra and Madhya Pradesh adopted the liberal resettlement package provided for displaced people by the State of Gujarat. Given the experience so far, and the fact that most of the impact of submergence on people will not occur until 1997, there is still time to develop meaningful R&R packages and programs in consultation with the affected peoples. Efforts are being intensified to achieve this.
>
> On environment, bank management agrees with the independent review on the need for a more effective central management in the Narmada Basin on environment impact studies and mitigation programmes. Management also agrees on the need to accelerate work on estuary studies and health matters in Gujarat. However, management does not share the review's conclusions about the environmental severity of the study delays. Command area issues are being addressed, including issues of waterlogging and salinity. On water availability (hydrology), Bank Management disagrees with the findings that there is insufficient impoundment of water upstream of the Sardar Sarovar Dam site to make the irrigation system work as designed.

The Government of India vide its letter dated 7 August 1992 from the Secretary, Ministry of Environment and Forests did not accept the report and commented adversely on it.

125. In view of the above, we do not propose, while considering the petitioners' contentions, to place any reliance on the report of the Morse Committee.

126. It was submitted on behalf of the petitioners that the command area development was an important aspect as the benefits of the project depended on this and if proper studies and plans were not done and not implemented, the very areas that were supposed to benefit will end up being rendered unfit for cultivation and the waterlogging and salinisation could render vast areas of the command unproductive. It was also submitted that still there was no integrated command area environment impact assessment. After referring to the status reports and studies regarding the command area development, it was submitted that there was a need for some independent agency to see whether there was ground to believe that the proposed measures will work or not. It was contended that a master plan for drainage and command area development was still not in place and even the full studies had not been done.

127. While refuting the aforesaid contentions it was argued on behalf of the learned counsel for the respondents that the SSP will provide irrigation water for a cultivable command area of 1.9 million hectares in Gujarat and 75,000 hectares in Rajasthan. The introduction of fresh water to the drought-prone areas of Gujarat will create obvious benefits for the farming communities. In order to safeguard these benefits, control and monitoring was suggested by the Secretary, Ministry of Environment and Forests and Chairman of the Environment Sub-group in the following areas from time to time:

- Drainage, waterlogging and soil salinity,
- Water quality,
- Forest loss,
- Potential impact on flora and fauna,
- Effects on public health,
- Socio-economic impacts.

128. Pursuant thereto fifty in-depth studies had been carried out by the State governments of Gujarat and Rajasthan and some of the studies were still in progress. One of the main objectives of carrying out these studies was to prevent excessive use of groundwater and water-logging.

129. There is no reason whatsoever as to why independent experts should be required to examine the quality, accuracy, recommendations and implementation of the studies carried out. The Narmada Control Authority and the Environment Sub-group in particular have the advantage of having with them the studies which had been carried out and there is no reason to believe that they would not be able to handle any problem if and when it arises or to doubt the correctness of the studies made.

130. It was submitted by Shri Shanti Bhushan that the catchment area treatment programme was not to be done *pari passu* but was required to be completed before the impoundment. This contention was based on the terms of the letter dated 24 June 1987 wherein conditional environmental clearance was granted, *inter alia*, on the condition that 'the catchment area treatment programme and rehabilitation plans be drawn so as to be completed ahead of reservoir filling'. Admittedly, the impounding began in 1994 and the submission of Shri Shanti Bhushan was that the catchment area treatment programme had not been completed by then and, therefore, this very important condition had been grossly violated. Reference was also made to the Minutes of the Environment Sub-group meetings to show that there had been slippage in catchment area treatment work.

131. The clearance of June 1987 required the work to be done *pari passu* with the construction of the dams and the filling of the reservoir. The area wherein the rainfall water is collected and drained into the river or reservoir is called catchment area and the catchment area treatment was essentially aimed at the checking of soil erosion and minimising the silting in the reservoir within the immediate vicinity of the reservoir in the catchment area. The respondents had proceeded on the basis that the requirement in the letter of June 1987, that catchment area treatment programme and rehabilitation plans be drawn up and completed ahead of reservoir filling, would imply that the work was to be done *pari passu*, as far as the catchment area treatment programme is concerned, with the filling of the reservoir. Even though the filling of the reservoir started in 1994, the impoundment Award was much less than the catchment area treatment which had been affected. The status of compliance with respect to *pari passu* conditions indicated that in 1999 the reservoir level was 88 metres, the impoundment area was 6,881 hectares (19 per cent) and the area where catchment treatment had been carried out was 1,28,230 hectares being 71.56 per cent of the total work required to be done. The Minutes of the Environment Sub-group as on 28 September 1999 stated that catchment area treatment works were nearing completion in the states of Gujarat and Maharashtra. Though there was some slippage in Madhya Pradesh, however, overall works by and large were on schedule. This clearly showed that the monitoring of the catchment treatment plan was being done by the Environment Sub-group quite effectively.

132. With regard to compensatory afforestation it was contended by Shri Shanti Bhushan that it was being carried out outside the project impact area. Further, it was submitted that the practice of using waste land or lesser quality land for compensatory afforestation means that the forest will be of lesser quality. Both of these together defeated the spirit of the compensatory afforestation. It was contended that the whole compensatory afforestation programme was needed to be looked at by independent experts.

133–4. While granting approval in 1987 to the submergence of forest land and/or diversion thereof for the SSP, the ministry of Environment and Forests had laid down a condition that for every hectare of forest land submerged or diverted for construction of the project, there should be compensatory afforestation on one hectare of non-forest land plus reforestation on two hectares of degraded forest. According to the State of Gujarat, it had fully complied with the condition by raising afforestation in 4,650 hectares of non-forest area and 9,300 hectares in degraded forest areas before 1995–96 against the impoundment area of 19 per cent. The *pari passu* achievement of afforestation in Gujarat was stated to be 99.62 per cent.

135. If afforestation was taking place on wasteland or lower quality land, it did not necessarily follow, as was contended by the petitioners, that the forests would be of lesser quality or quantity.

136. It was also contended on behalf of the petitioners that downstream impacts of the project would include not only the destruction of downstream fisheries, one of the most important ones in Gujarat on which thousands of people are dependent, but will also result in salt water ingress. The project, it was contended, will have grave impacts on the Narmada estuary and unless the possible impacts were properly studied and made public, and mitigation plans demonstrated with the requisite budget, one could not accept the claim that these matters were being looked into. The need to assess the problem was stated to be urgent, as according to the petitioners, rich fisheries downstream of the dam, including the famed hilsa, would be almost completely destroyed. The salinity ingress threatened the water supply and irrigation use of over 210 villages and towns and Bharuch city. All these could not only have serious economic and other impacts but would also directly destroy the livelihood of at least 10,000 fisher families.

137. Again all these contentions were based on the Morse Committee Report which the World Bank and the Union of India had already rejected. That apart, according to the respondents, in 1992 the SSNNL issued an approach paper on environmental impact assessment for the river reach downstream. This provided technical understanding of the likely hydrological changes and possible impact in relation thereto. It was further submitted by the learned counsel for the respondents that the potential for environmental changes in the lower river and estuary had to be seen in the context of the long term development of the basin. The current stage was clearly beneficial. The three stages could be identified as follows:

Stage 1 covers the period roughly from the completion of Sardar Sarovar Dam to the year 2015. Events occurring during this stage include (a) SSP Canal Command will have reached full development and required diversion of some water, (b) the upstream demand will reach about 8 MAF and (c) the Narmada Sagar Dam will have been built and placed in operation.

Stage 2 covers the period from 2015 to 2030 during which the demands upstream of SSP continue to grow and will reach about 12 MAF, still below the volume of 18 MAF that Madhya Pradesh can take in a 75 per cent year.

Stage 3 covers the period up to and beyond full basin development.

The report given by M/s. H.R. Wallingford in March 1993 in respect of the downstream impacts of Sardar Sarovar Dam observers, *inter alia*, as under:

> The overall conclusion of the team undertaking the assessment described in this report is that there are no downstream impacts whose magnitude and effect are such as to cause doubts to be cast over the wisdom of proceeding with the Sardar Sarovar Projects provided that appropriate monitoring and mitigation measures are applied. Much of this work is already in progress under the auspices of the NPG, SSNNL and NCA. The recommendations in this report are intended to provide a synthesis of their work and suggestions as to whether it might be modified to enhance its usefulness.

The said M/s. H.R. Wallingford in the findings of 1995 stated as under:

> It is thought unlikely that any significant negative environmental impacts will occur over the next 30 years as a result of the project. Some possible adverse effects have been identified, the main one being the effect of flood attenuation on hilsa migration. These need to be monitored and more studies undertaken to better understand the conditions which trigger spawning. Beneficial impacts in this period include reduced flooding and more reliable dry season flows as well as an overall improvement of the health and well being of the people to the reliable domestic water supply, improved nutrition and enhanced economic activity.

The above report clearly demonstrates that the construction of the dam would result in a more regulated and perennial flow into the river with an overall beneficial impact. It is also evident that until all the dams are constructed upstream and the entire flow of river is harnessed, which is not likely in the foreseeable future, there is no question of adverse impact including the fishing activity and the petitioner's assertions in this regard are ill-conceived.

138. The area of submergence was stated to be rich in archaeological remains but it still remained to be studied. It was contended that there was a danger of rich historical legacy being lost and even a small increase in the dam height would threaten to submerge many of the sites listed in the report of the Archaeological Survey of India. There were stated to be five monuments which would be affected at dam height of 90 metres or above and no work was stated to have commenced to protect any of the five monuments.

139. According to the State of Gujarat, the Ancient Monuments and Archaeological Sites and Remains Act, 1958 charged the Central and/or State Department of Archaeology with responsibility for the protection of important cultural sites. Under the Act, sites were classified under three categories as follows:

Type 1: Monuments of national importance which are protected by the Central Government;

Type 2: Monuments of religious or cultural importance which are protected by the State government; and

Type 3: Monuments which are neither centrally nor state protected, but which are considered to be an important part of cultural heritage.

Under the same law, authorities charged with the protection of the monuments are permitted to take suitable measures to ensure the preservation of any protected site under threat from decay, misuse or economic activity.

In the case of Sardar Sarovar, where several sites may be submerged, the NWDT Award stipulated that the entire cost of relocation and protection should be chargeable to Gujarat. Relocation work was to be supervised by the Department of Archaeology under the provision of the Ancient Monuments and Archaeological Sites and Remains Act, 1958.

140. The three State governments carried out a complete survey of cultural and religious sites within the submergence zone. The principle of these surveys was to list all archaeological sites, identify and name any site under state protection, and further identify sites of religious or cultural significance which, although not protected under national law, were of sufficient value to merit relocation. So far as the State of Gujarat is concerned the Department of Archaeology surveyed archaeological sites in 19 villages in the submergence zone in Gujarat under the title of Archaeological Survey of Nineteen Villages in Gujarat Submerged by Sardar Sarovar Reservoir, 1989.

In addition to baseline studies on archaeological aspects, work had been carried out on the anthropological heritage of Narmada Basin, including an examination of evidence of ancient dwellings and cultural artefacts. The principal studies in this behalf are described below:

Anthropological Survey of India: **Narmada Salvage Plan**: The Narmada Salvage Plan contains detailed background data on palaeoanthropological, human ecological and other aspects of the Narmada Valley. By May 1992 surface scanning of 17 sample villages coming under submergence had been carried out and 424 specimens including ancient tools, etc. had been collected.

Anthropological Survey of India, Peoples of India: This project entailed a complete survey of 33 tribes of India including those of the Narmada Basin. The study covered all aspects of tribal culture in India and was published in 61 volumes in 1992.

Summary of current situation and progress, Government of Gujarat

Survey of villages in submergence zone	Complete for all items in the State
Identification of cultural sites	Complete for all items in the State
Collection of data and documentation	Complete

of sites	
Selection of appropriate sites	Complete
Action plan	Complete

It was further submitted on behalf of respondents that no centrally or state protected cultural sites were located in the submergence area of the project. In Gujarat, the Department of Archaeology concluded that the temples of Shoolpaneshwar and Hampheshwar were important monuments and should be moved to a higher level. Sites were selected for constructing new Shoolpaneshwar and Hampheshwar temples in consultation with temple trustees. Shoolpaneshwar had been relocated and reconstructed near Gora, about 15 km downstream from the present location. Hampheshwar was also constructed at higher ground in consultation with the temple trustees and pranpratistha also planned on 22 and 24 April 2000, i.e. before the temple was submerged.

141. In relation to flora and fauna studies, it was contended by the petitioners that the studies had finished only recently and the action plans were awaited in many cases. In the meanwhile, extensive deforestation of the submergence zone had taken place, as also part of the area had been submerged, even as the studies have been on. It was also contended that the impact on some of the wild ass sanctuaries in Kachchh would be very severe.

142. The guidelines of the Ministry of Environment and Forests required that while seeking environmental clearance for hydropower projects, surveys should be conducted so that the status of the flora and fauna present could be assessed. A condition of the environmental clearance of 1987, as far as it related to flora, and fauna was that the Narmada Control Authority would ensure in-depth studies on flora and fauna needed for implementation of environmental safeguard measures. It is the case of the respondents that a number of studies were carried out and reports submitted. It was observed that the submergence area and catchment area on the right bank of the proposed reservoir exhibited a highly degraded ecosystem which was in contrast to the left bank area where there was fairly good forest cover which formed part of Shoolpaneshwar Wildlife Sanctuary. With regard to the study of fauna, the said report indicated that a well-balanced and viable eco-system existed in the Shoolpaneshwar Sanctuary. Moreover, with the construction of dam, water availability and soil moisture will increase and support varieties of plants and animals.

143. It was also contended on behalf of the petitioners that the whole project will have serious impacts on health, both around the submergence area and in the command. The preventive aspects had not been given attention. There was no linkage between the studies and work.

144. On behalf of the State of Gujarat, it was contended that large number of studies had been carried out on the health profile of villagers including studies on water related diseases in the SSP command area including the area downstream of the dam. The study of MS University in 1983 and other studies concluded that the most common diseases in the basin were malaria, scabies, dysentery and diarrhoea. Of these only a threat to malaria needed to be of concern. The study concluded that the incidence of hygiene-related diseases other than malaria could be reduced by better water availability. The Gujarat work plan covered villages within a 10 km radius of the reservoir including resettled population and made provision for the monitoring, surveillance and control of malaria. The principal features of the Gujarat Work Plan included establishment of a hospital at Kevadia near the dam site, strengthening of laboratory facility including establishment of mobile unit residual insecticidal spraying operations etc. This showed that the area of public health was in no way being neglected.

145. The petitioner was also critical of the functioning of the Environment Sub-group as it was contended that the claims of the studies and progress report were accepted at face value and without verification. It was also contended that the Ministry of Environment and Forests had grossly abdicated its responsibility. This submission was based on the premise that clearance, which had been granted, had lapsed and the Ministry of Environment and Forests did not insist on the Ministry of Water Resources for its renewal and furthermore the Ministry of Environment and Forests had not taken any cognizance of the criticism about environmental aspects contained in the Morse Committee Report. Lastly the FMG in its first report was critical in many respects and pointed out studies which had remained incomplete but no cognizance was taken by the Ministry of Environment and Forests. The repeated abdication, it was submitted, of responsibility by the Ministry of Environment and Forests indicated that it was not taking the whole issue with the seriousness it deserved.

146. On behalf of the State of Gujarat, it was contended that various alleged dangers relating to the environment as shown by the petitioners were mostly based on the recommendations of the Morse Committee Report and the FMG. The report of the Morse Committee does not require our attention, the same not having been accepted either by the World Bank or the Government of India. Para 4.5.2 of the report of the FMG which relates to creation of the Environmental Sub-group commends its establishment, its observation about its powers is as follows:

> 4.5.2. It must be noted that the Environment Sub-group is not a body which merely observes and reports, but watchdog body which can recommend even the stoppage of work if it feels dissatisfied with the progress on the environment front. The recommendations of the Environment Sub-group will have to be considered by the NCA, and if there is any difference of opinion at that level, it will have to be referred to the Review Committee, which has the Minister of Water and Environment and Forests as a member *It seems doubtful whether any more effective mechanism could have been devised or made to work within the*

> *framework of our existing political and administrative structures, particularly in the context of a federal system.* The Secretary (Environment and Forests) has, in fact, been given a special position in the NCA inasmuch as he can insist on matters being referred to the Review Committee and at the Review Committee the Minister of Environment and Forests forcefully plead the environmental cause; he can also make the environmental point of view heard at the highest level. If in spite all these arrangements, the environmental point of view fails to be heard adequately, and if project construction tends to take an over-riding precedence, that is a reflection of the relative political importance of these two points of view in our system. This can be remedied only in the long term through persuasion and education, and not immediately through institutional arrangements which run counter to the system (emphasis added).

Apart from the fact that we are not convinced that construction of the dam will result in there being an adverse ecological impact, there is no reason to conclude that the Environment Sub-group is not functioning effectively. The group which is headed by the Secretary, Ministry of Environment and Forests is a high powered body whose work cannot be belittled merely on the basis of conjectures or surmises.

147. Shri Shanti Bhushan, learned Senior Counsel, while relying upon *A.P. Pollution Control Board v. Professor M.V. Mayadu* (1999) 2 SCC 718, submitted that in cases pertaining to environment, the onus of proof is on the person who wants to change the status quo and, therefore, it is for the respondents to satisfy the Court that there will be no environmental degradation.

148. In A.P. Pollution Control Board's case this court was dealing with the case where an application was submitted by a company to the Pollution Control Board for permission to set up an industry for production of 'BSS Castor Oil Derivatives'. Though later on a letter of intent had been received by the said company, the Pollution Control Board did not give its no-objection certificate to the location of the industry at the site proposed by it. The Pollution Control Board, while rejecting the application for consent, *inter alia*, stated that the unit was a polluting industry which fell under the red category of polluting industry and it would not be desirable to locate such an industry in the catchment area of Himayat Sagar, a lake in Andhra Pradesh. The appeal filed by the company against the decision of the Pollution Board was accepted by the appellate authority. A writ petition was filed in the nature of public interest litigation and also by the Gram Panchayat challenging the order of the appellate authority, but the same was dismissed by the High Court. On the other hand, the writ petition filed by the company was allowed and the High Court directed the Pollution Control Board to grant consent subject to such conditions as may be imposed by it.

149. It is this decision which was the subject-matter of challenge in this court. After referring to the different concepts in relation to environmental cases like the 'precautionary principle' and the 'polluter-pays principle', this court relied upon the earlier decision of this court in *Vellore Citizens' Welfare Forum v. Union of*

India (1996) 5 SCC 647 and observed that there was a new concept which places the burden of proof on the developer or industrialist who is proposing to alter the status quo and has become part of our environmental law. It was noticed that inadequacies of science had led to the precautionary principle and the said 'precautionary principle' in its turn had led to the special principle of burden of proof in environmental cases where burden as to the absence of injurious effect of the actions proposed is placed on those who want to change the status quo. At page 735, this court, while relying upon a report of the International Law Commission, observed as follows:

> The precautionary principle suggests that where there is an identifiable risk of serious and irreversible harm, including, for example, extinction of species, widespread toxic pollution is a major threat to essential ecological processes, it may be appropriate to place the burden of proof on the person or entity proposing the activity that is potentially harmful to the environment.

150. It appears to us that the 'precautionary principle' and the corresponding burden of proof on the person who wants to change the status quo will ordinarily apply in a case of polluting or other project or industry where the extent of damage likely to be inflicted is not known. When there is a state of uncertainty due to lack of data or material about the extent of damage or pollution likely to be caused then, in order to maintain the ecology balance, the burden of proof that the said balance will be maintained must necessarily be on the industry or the unit which is likely to cause pollution. On the other hand, where the effect on the ecology or the environment of setting up of an industry is known, what has to be seen is that if the environment is likely to suffer, then what mitigative steps can be taken to offset the same. Merely because there will be a change is no reason to presume that there will be an ecological disaster. It is when the effect of the project is known, then the principle of sustainable development would come into play, which will ensure that mitigative steps are and can be taken to preserve the ecological balance. Sustainable development means what type or extent of development can take place which can be sustained by nature/ecology with or without mitigation.

151. In the present case we are not concerned with the polluting industry which is being established. What is being constructed is a large dam. The dam is neither a nuclear establishment nor a polluting industry. The construction of a dam undoubtedly would result in the change of environment but it will not be correct to presume that the construction of a large dam like the Sardar Sarovar will result in ecological disaster. India has experience of over 40 years in the construction of dams. The experience does not show that construction of a large dam is not cost effective or leads to ecological or environmental degradation. On the contrary there has been ecological upgradation with the construction of large dams. What is the impact on the environment with the construction of a dam is well-known in India and, therefore, the decision in A.P. Pollution Control Board's case (supra) will have no application in the present case.

152. Reference was made by Shri Shanti Bhushan to the decision of the United States District Court in the case of *Sierra Club v. Robert F. Froehlke*, 359 F.Supp. 1289 (1973). In that case work had begun on the Wallisville Project which, *inter alia*, consisted of the construction of a low dam. It was the case of the plaintiff that the construction of the project would destroy hundreds of thousands of trees and enormous grain, fish and other wildlife will lose their habitat and perish. It was contended that the defendants were proceeding in violation of law by not complying with the requirements of the National Environment Policy Act, 1969 (NEPA). The plaintiff, *inter alia*, sought an injunction for restraining the undertaking of the project in violation of the said Act. The District Court held that notwithstanding the substantial amount of work that had already been done in connection with the project, the failure to satisfy the full disclosure requirement of the NEPA injunction would be issued to halt any further construction until the requirements of NEPA had been complied with; that even though there was no act like NEPA in India at the time when environmental clearance was granted in 1987, nevertheless by virtue of Stockholm Convention and Article 21 of the Constitution the principles of Sierra Club decision should be applied.

153. In India notification had been issued under Section 3 of the Environmental Act regarding prior environmental clearance in the case of undertaking of projects and setting up of industries including inter-State river projects. This notification has been made effective from 1994. There was, at the time when the environmental clearance which was granted in 1987, no obligation to obtain any statutory clearance. The environmental clearance which was granted in 1987 was essentially administrative in nature, having regard and concern of the environment in the region. Change in environment does not per se violate any right under Article 21 of the Constitution of India especially when ameliorative steps are taken not only to preserve but to improve ecology and environment and in case of displacement, prior relief and rehabilitation measures take place *pari passu* with the construction of the dam.

154. At the time when the environmental clearance was granted by the Prime Minister, whatever studies were available were taken into consideration. It was known that the construction of the dam would result in submergence and the consequent effect which the reservoir will have on the ecology of the surrounding areas was also known. Various studies relating to environmental impact, some of which have been referred to earlier in this judgment, had been carried out. There are different facets of the environment and if in respect of a few of them adequate data was not available it does not mean that the decision taken to grant environmental clearance was in any way vitiated. The clearance required further studies to be undertaken and we are satisfied that this has been and is being done. Care for the environment is an on going process and the system in place would ensure that ameliorative steps are taken to counter the adverse effect, if any, on the environment with the construction of the dam.

155–156. Our attention was also drawn to the case of *Tennessee Valley Authority v. Hiram G. Hill* [437 US 153, 57 L Ed 2d 117, 98 S Ct 2279], where the Tennessee Valley Authority had begun construction of the Tellico Dam and reservoir project on a stretch of the Little Tennessee River. While a major portion of the dam had been constructed the Endangered Species Act, 1973 was enacted wherein a small fish popularly known as the snail darter was declared an endangered species. Environmental groups brought an action in the Unites States District Court for restraining impounding of the reservoir on the ground that such an action would violate the Endangered Species Act by causing the snail darter extinction. The District Court refused injunction but the same was granted by the United States Court of Appeal. On further appeal the US Supreme Court held that the Endangered Species Act prohibited the authority for further impounding the river. The said decision has no application in the present case because there is no such act like the Endangered Species Act in India or a declaration similar to the one which was issued by the Secretary of the Interior under that Act. What is, however, more important is that it has not been shown that any endangered species exists in the area of impoundment. In the Tennessee Valley Authority case it was an accepted position that the continued existence of the snail darter which was an endangered species would be completely jeopardised.

157. Two other decisions were referred to by Shri Shanti Bhushan – *Arlington Coalition on Transportation v. John. A Volpe* [458 F.2d 1323 (1972)] and *Environmental Defense Fund. Inc v. Corps of Engineers of United States Army* [325 F.Supp.749 (1971)]. In both these decisions it was decided that the NEPA would be applicable even in the case of a project which had commenced prior to the coming into force of the said Act but which had not been completed. In such cases there was a requirement to comply with the provisions of NEPA as already noticed earlier. The notification under Section 3 of the Environment Protection Act cannot be regarded as having any retrospective effect. The said notification dated 27 January 1994, *inter alia*, provides as follows:

> Now, therefore, in exercise of the powers conferred by sub-section (1) and clause (v) of sub-section (2) of Section 3 of the Environment (Protection) Act, 1986 (29 of 1986) read with clause (d) of sub-rule (3) of rule 5 of the Environment (Protection) Rules, 1986, the Central Government hereby directs that on and from the date of publication of this notification in the Official Gazette expansion or modernisation of any activity (if pollution load is to exceed the existing one) or a new project listed in Schedule I to this notification, shall not be undertaken in any part of India unless it has been accorded environmental clearance by the Central Government in accordance with the procedure hereinafter specified in this notification.

This notification is clearly prospective and *inter alia* prohibits the undertaking of a new project listed in Schedule I without prior environmental clearance of the Central Government in accordance with the procedure now specified. In the present case, clearance was given by the Central Government in 1987 and at that time no procedure was prescribed by any statute, rule or regulation. The procedure now provided in 1994 for getting prior clearance cannot apply retrospectively to the project whose construction commenced nearly eight years prior thereto.

Relief and rehabilitation

158. It is contended by the petitioner that as a result of the construction of the dam, over 41,000 families will be affected in three States spread over 245 villages. The number of families have increased from 7,000 families assessed by the Tribunal. It was further contended that the submergence area can be broadly divided into two areas. One, the fully tribal area, which covers the initial reach of about 100 or so villages which are almost 100 per cent tribal and hilly. These include all the 33 villages of Maharashtra, all 19 of Gujarat and many of Madhya Pradesh. The second part of the submergence area is the mixed population area on the Nimad plains with a very well developed economy that is well connected to the mainstream. While the tribal areas are stated to be having a rich and diverse resource base and self sufficient economy, the lack of so-called modern amenities like roads, hospitals and schools are far more a reflection of the neglect and disregard by the government over the last fifty years than anything else. Of the 193 villages stated to be affected by the Sardar Sarovar submergence, 140 lie in the Nimad plains. The population of these villages is a mix of caste and tribal and these villages have all the facilities like schools, post offices, bus services etc.

159. It was contended that whereas the project authorities talk only about the families affected by submergence, none of the other families affected by the project are considered PAFs, nor has any rehabilitation package been designed for them. These non-recognised categories for whom no rehabilitation package is given are stated to be those persons living in the submergence area who are not farmers but are engaged in other occupations like petty traders, village shop-keepers who will be affected by submergence; colony affected people whose lands were taken in 1960 to build the project colony, warehouses, etc.; canal affected people who would be losing 25 per cent of their holdings because of the construction of the canals; drainage affected people whose lands will be acquired for drainage; 10,000 fishing families living downstream whose livelihood will be affected; lands of the tribals whose catchment treatment area had been carried out; persons who are going to be affected by the expansion of Shoolpaneshwar Sanctuary; persons going to be affected by the NSP and Garudeshwar Weir. It was contended that there was an urgent need to assess comprehensively the totality of the impact and prepare category specific rehabilitation policies for all of them.

160. It was also submitted that the total number of affected families in all the three States as per the master plan prepared by the Narmada Control Authority is 40,727. According to the petitioner, however, this figure is an underestimate and the estimate of the land required for these PAFs is also too low. The basis for making this submission is:

1. In each village there are many persons left out of the Government list of declared PAFs. These are joint holders (not recognised as landed oustees or PAFs) and the adult sons.

2. Incorrect surveys have been conducted and the affected persons have serious apprehensions about the validity of the surveys since at many places the level markings are suspect, in many cases the people affected at higher levels have been given notices for lower levels, many others at the same levels have been left out and so on. It is also alleged that there have been shortcomings in the policies and if they are corrected many more oustees will be entitled to PAF status. Furthermore the cut-off date for PAFs in Madhya Pradesh including adult sons is linked to the date of issuance of notification. Since land acquisition process is still incomplete the number of adult sons entitled to land would increase with the issuance of fresh Section 4 notifications.

161. From the aforesaid it was contended that the total impact in terms of the number of oustees as well as land entitlement will be much larger than what is considered in the master plan.

162. It is also submitted that there were major lacunae in the said policy like the three States having dissimilar policies for R&R. This difference in rehabilitation packages of different States, with the package of Gujarat being more favourable, is leading to a situation where the oustees are forced to shift to Gujarat. The other lacunae which are stated to have many serious problems are alleged to be the non-provision for fuelwood and grazing land with fodder. No provision for the rehabilitation of people involved in non-agricultural occupation. According to the petitioner the number of affected people even by submergence has been underestimated. The policy regime governing them has many serious lacunae. The increase in the numbers is due to a lack of proper surveys and planning and the provision of just and due entitlements to the PAFs. Since this process of providing just entitlements is still incomplete, and the policies need a thorough review, the numbers and entitlements are likely to go up further. Even the magnitude of the task of R&R cannot be assessed properly until the above are considered and proper policies introduced.

163. It is also contended that before embarking on the Sardar Sarovar Project it was necessary that the master plan for rehabilitation of the families to be affected is completed. According to the petitioner, the master plan which was submitted in the Court cannot be regarded as an acceptable master plan inasmuch as it had no mention of people affected by the SSP other than those affected by submergence and it has no estimate of the resource base of the oustees in their original village. Further the plan makes no estimation of the forest land, grazing land and resources being used by the oustees. The master plan persists with the discriminatory and differential policies which are less than just to the oustees. There is also no planning for community resettlement even though the Award of the Narmada

Tribunal made detailed provision regarding rehabilitation of the oustees which required that there should be village wise community rehabilitation.

164. In support of this contention reliance is placed on the following stipulation for rehabilitation contained in the Award of the Narmada Tribunal: 'That Gujarat shall establish rehabilitation villages in Gujarat in the irrigation command of the SSP on the norms hereafter mentioned for rehabilitation of the families who are willing to migrate to Gujarat'. The submission is that no specific rehabilitation village, as envisaged by the Tribunal's Award, has been established in Gujarat. The issue of community resettlement is stated to be not merely an issue of community facility but is a more fundamental issue. The issue is really one of preserving the social fabric and community relation of the oustees which, it is alleged, is being destroyed due to the dispersal of the community who are being resettled at different sites.

165. Dealing with the situation of those oustees who have been resettled in Gujarat it is established by the petitioner that there are a large number of grievances of the said oustees in 35 resettlement sites. With the passage of time the number of problems overall would become much more, is the contention. The petitioner finds fault with the quality of land which has been given in Gujarat to the oustees contending that a large number of oustees have been given land outside the command area of irrigation and in some resettlement sites there is a serious water-logging problem. It also contends that though some amenities have been provided, they are not adequate. It is also the case of the petitioner that sufficient land for resettlement of the oustees from Madhya Pradesh is not available in Gujarat despite the claim of the State of Gujarat to the contrary.

166. With regard to Maharashtra it is contended by the petitioner that the official figure of the total number of PAFs affected in Maharashtra is not correct and the number is likely to be more than 3,113 PAFs estimated by the State of Maharashtra. Furthermore, adequate land of desired quality has not been made available for resettlement until 90 m and even thereafter. Reference is made to the affidavit of the State of Maharashtra in which it is stated that it proposes to ask for the release of 1,500 hectares of forest land for resettlement and the submission on behalf of the petitioner is that release of such land shall be in violation of the Forest Conservation Act, 1980 and is not in the public interest for forest cover will be further depleted.

167. With regard to the State of Madhya Pradesh it is submitted that as per the Award the PAFs have a right to choose whether to go to Gujarat or to stay in the home State. The State of Madhya Pradesh is stated to have planned the whole resettlement based on the assumption that an overwhelming proportion of oustees entitled to land will go to Gujarat, yet even for the limited number of oustees who are likely to stay in Madhya Pradesh the submission is that no land is available. The petitioner also disputes the averment of the State of Madhya Pradesh that the oustees have been given a choice as to whether they would like to go to Gujarat or

stay in the home State. According to the petitioner, the majority of the oustees would prefer to stay in the home State, that is Madhya Pradesh, but sufficient land for their resettlement in Madhya Pradesh is not available. According to the petitioner the State of Madhya Pradesh has stated that it does not have land for any PAFs above 830 and even for 830 PAFs the land is not available. It is also submitted that the Madhya Pradesh Government cannot wriggle out of its responsibility to provide land for the oustees by offering them cash compensation. The petitioner finds fault with the effort of the State of Madhya Pradesh to push the oustees to Gujarat whose rehabilitation scheme is more attractive and beneficial than that of Madhya Pradesh.

168. The petitioner further contends that one of the fundamental principles laid down is that all the arrangements and resettlement of the oustees should be made one year in advance of submersion. In *B.D. Sharma v. Union of India*'s case this court has held that resettlement and rehabilitation has to be done at least six months in advance of submersion, complete in all respects.* It is therefore, contended that since offers to the Madhya Pradesh oustees affected at 90 m to be settled in Madhya Pradesh has not been made, there cannot be any question of further construction until one year after the resettlement of these PAFs at 90 m.

169. The petitioner is also critical of the functioning of the R&R Sub-group and it is contended that the said Sub-group has not taken any cognizance of the various issues and problems enumerated by the petitioner. It is submitted that in assuring that the relief and rehabilitation arrangements are being done, the said R&R Sub-group merely accepts the assertions of the government rather than verifying the claims independently. There is also a complaint regarding the manner in which the R&R Committee takes decisions on the spot when it makes frequent (sic) visits. It is contended that the decisions which are taken in an effort to solve the grievances of the oustees is done in the most insensitive way. The R&R Sub-group, it is contended, is an official agency of the government, itself being a Sub-group of the NCA, which is pushing the project ahead, and the question raised by the petitioner is, how can the same body which is building a project and executing the R&R be also monitoring it.

170. It is a case of the petitioners that there is a need for an independent monitoring agency in the three States who should be asked to monitor the R&R of the oustees and see to the compliance with the NWDT Award. No construction should be permitted to be undertaken without clearance from this authority. Lastly it is contended that a large number of grievances are persisting even after twenty years and the pace of resettlement has been slow. The petitioner seems to have contended that the relief and rehabilitation can be manageable only if the height of the dam is significantly lessened, which will reduce submersion and the displacement of people.

---

* This document is reproduced at page 265.

171. In order to consider the challenge to the execution of the project with reference to relief and rehabilitation, it is essential to see what is the extent and the nature of submergence.

172. The Sardar Sarovar Reservoir level at 455 ft would affect 193 villages in Madhya Pradesh, 33 villages in Maharashtra and 19 villages in Gujarat. The submergence villages are situated on the banks of the river Narmada having gentle to steep slopes of the Satpura hills. A village is considered affected even when the water level touches the farm/hut at lowest level. It may be noted that only 4 villages (3 villages in Gujarat and 1 village in Madhya Pradesh) are being submerged fully and the remaining 241 villages are affected partially.

173. The state-wise land coming under submergence (category-wise) is given below:

(In hectares)

Sr. No.	Type of Land	States			Total
		Gujarat	Maharashtra	Madhya Pradesh	
1.	Cultivated land	1,877	1,519	7,883	11,279
2.	Forest land	4,166	6,488	2,731	13,385
3.	Other land including river bed	1,069	1,592	10,208	12,869
	Total land	7,112	9,599	20,822	37,533

174. The aforesaid table shows that as much as 12,869 hectares of the affected land is other than agricultural and forest and includes the river bed area.

175. When compared to other similar major projects, the Sardar Sarovar Project has the least ratio of submergence to the area benefited (1.98 per cent only). The ratio of some of the existing schemes is as much as 25 per cent, as can be seen from the table below:

Sr. No.	Name of project	State	Benefit area (in ha)	Submergence area (in ha)	Irrigation benefit per ha submergence	Per-centage of area submerged to area irrigated

1.	Hirakud	Orissa	2,51,150	73,892	3.40	29.42
2.	Shriramsa-gar	Andhra Pradesh	2,30,679	44,517	5.24	19.14
3.	Gandhisa-gar	Madhya Pradesh	5,03,200	66,186	7.60	13.15
4.	Paithan	Maharashtra	2,78,000	35,000	7.94	15.29
5.	Tungbha-dra	Karnataka	3,72,000	37,814	9.84	10.16
6.	Pench	Maharashtra	34,000	7,750	12.13	8.24
7.	Nagarjun-sagar	Andhra Pradesh	8,95,000	28,500	31.40	3.18
8.	Bhakra	Himachal Pradesh	6,76,000	16,800	40.24	2.48
9.	Sardar Sarovar	Gujarat	19,03,500	37,533	50.71	1.97

176. Countering the assertion that the construction of the dam would result in the large-scale relocation and uprooting of tribals, the factual position seems to be that the tribals constitute the bulk of PAFs in Gujarat and Maharashtra, namely, 97 per cent and 100 per cent respectively. In the case of Madhya Pradesh, the tribal PAFs are only 30 per cent while 70 per cent are non-tribals.

177. The tribals who are affected are in indigent circumstances and who have been deprived of the modern fruits of development such as tap water, education, road, electricity, convenient medical facilities, etc. The majority of the project affected families are involved in rain-fed agricultural activities for their own sustenance. There is partial employment in the forestry sector. Since the area is hilly with difficult terrain, they are wholly dependent on the vagaries of the monsoon and normally only a single crop is raised by them. Out of the PAFs of Madhya Pradesh who have resettled in Gujarat, more than 70 per cent are tribal families. The majority of the total tribal PAFs are stated to have already been resettled in Gujarat after having exercised their option. It is the contention of the State of Gujarat that the tribals in large number have responded positively to the resettlement package offered by that State.

178. In Madhya Pradesh the agricultural lands of the tribal villages are affected on an average to the extent of 28 per cent whereas in the upper reaches, i.e. Nimad where the agriculture is advanced, the extent of submergence, on an average, is only 8.5 per cent. The surveys conducted by HMS Gaur University (Sagar), the

Monitoring and Evaluation Agency set up by the Government of Madhya Pradesh, reveals that the major resistance to relocation is from the richer, non-tribal families of Nimad who fear a shortage of agricultural labour if the landless labourers from the areas accept resettlement.

179. The displacement of the people due to major river valley projects has occurred in both developed and developing countries. In the past, there was no definite policy for rehabilitation of displaced persons associated with the river valley projects in India. There were certain project specific programmes for implementation on a temporary basis. For the land acquired, compensation under the provisions of the Land Acquisition Act, 1894 used to be given to the project affected families. This payment in cash did not result in satisfactory resettlement of the displaced families. Realising the difficulties of displaced persons, the requirement of relief and rehabilitation of PAFs in the case of Sardar Sarovar Project was considered by the Narmada Water Disputes Tribunal and the decision and final order of the Tribunal given in 1979 contains detailed directions in regard to the acquisition of land and properties, provision of land, house plots and civic amenities for the resettlement and rehabilitation of the affected families. The resettlement policy has thus emerged and developed along with the Sardar Sarovar Project.

180. The Award provides that every displaced family, whose more than 25 per cent of agricultural land holding is acquired, shall be entitled to and be allotted irrigable land of its choice to the extent of land acquired subject to the prescribed ceiling of the State concerned with a minimum of two hectares of land. Apart from this land-based rehabilitation policy, the Award further provides that each project affected person will be allotted a house plot free of cost and resettlement and rehabilitation grant. The civic amenities required by the Award to be provided at places of resettlement include one primary school for every 100 families, one Panchayat Ghar, one dispensary, one seed store, one children's park, one village pond and one religious place of worship for every 500 families, one drinking water well with trough and one tree platform for every 50 families, an approach road linking each colony to the main road, electrification, water supply, sanitary arrangement etc. The State governments have liberalised the policies with regard to resettlement and have offered packages more than what was provided for in the Award e.g. the Governments of Madhya Pradesh, Maharashtra and Gujarat have extended the R&R benefits through their liberalised policies even to the encroachers, landless/displaced persons, joint holders, tapu land (island) holders and major sons (18 years old) of all categories of affected persons. The Government of Maharashtra has decided to allot one hectare of agricultural land free of cost even to unmarried daughters of all categories of PAFs.

181. In the environmental clearance granted by the Ministry of Environment and Forests vide its letter dated 24 June 1987, one of the conditions stipulated therein

was for information from the project authorities on various action plans including the Rehabilitation Master Plan of 1989.

182. It is the contention of the petitioners that the failure to prepare a master plan constitutes non-compliance with the requirement of the Tribunal's Award as well as environmental clearance. The Tribunal's Award does not use the expression 'Master Plan' but as per Clause XI Sub-clause IV(2)(ii), what is required, is as under:

> The three States by mutual consultation shall determine within two years of the decision of the Tribunal, the number and general location of rehabilitation villages required to be established by Gujarat in its own territory.

183. It is with regard to this clause in the Award that presumably the aforesaid letter of 24 June 1987 granting environmental clearance required the preparation of the new Master Plan.

184. In 1988 when the project was first cleared by the Planning Commission from an investment angle, it was estimated that 12,180 families would be affected in three States. Based on these numbers, the State governments independently prepared their action plans and announced their R&R policies based on the Tribunal's Award. On the basis of the said action plans the Narmada Control Authority submitted a Rehabilitation Master Plan to the Ministry of Environment and Forests along with its letter dated 3/4 May 1989. Out of the total population, which is affected by the submergence, a large number are tribals and hence attention was paid by the State governments to liberalise their policies for protecting the socio-economic and cultural milieu and to extend the R&R benefits even to other categories of persons who were not covered by the Tribunal's Award. This led to the liberalisation of the R&R packages by the three States which packages have been referred to hereinabove. As a result of the liberalisation of the packages, the number of PAFs as estimated in 1992 by the State governments was 30,144. Based on the material available, the three State governments prepared individual action plans in 1993 but those action plans were integrated by the Narmada Control Authority first in 1993 and again in 1995 as an integrated Master Plan to present a holistic picture of the R&R programme. The Master Plan deals with the socio-economic and cultural milieu of PAFs, the legal framework, the R&R policy and procedures, implementation machinery, organisation of R&R, monitoring and evaluation, empowerment of women and youth, special care for vulnerable groups, financial plans for R&R etc. As per the 1990 Master Plan the total PAFs have increased to 40,227 from 30,144 due to the addition of 100 more genuine PAFs in Maharashtra. This Master Plan includes village-wise, category-wise PAFs and their preference in R&R to settle in their home State or in Gujarat.

The reason for increase in the number of PAFs has been explained in the Master Plan and the reasons given, *inter alia*, are:

(a) After CWC prepared backwater level data, the number of PAFs in Madhya Pradesh (MP) increased by 12,000 PAFs as their houses are affected in a 1 in 100 years flood.

(b) Government of Gujarat (GoG) included major sons of the dyke villages as PAFs.

(c) Cut-off date for major sons was extended by GoG and Government of Maharashtra (GoM).

(d) PAFs affected in MP, have increased due to delay in the publication of Section 4 notification under the Land Acquisition Act.

(e) Persons socially or physically cut-off due to the impounding of water in the reservoir, are also considered as PAFs by all three States.

(f) All three States decided to consider encroachers as PAFs.

(g) Major unmarried daughters in Maharashtra are considered as a separate family by the Government of Maharashtra.

(h) Some genuine PAFs were earlier left out (as many stayed in remote areas or used to undertake seasonal migration to towns and developed areas in search of casual work).

185. As far as the State of Gujarat is concerned, its contention is that the task of R&R is not impossible as recognised by the FMG-I in its 1994 report and according to the State, it is fully ready and prepared to resettle in Gujarat all the PAFs up to FRL 455 ft.

186. On 13 November 1996 a meeting of the Review Committee of the NCA chaired by the Union Minister of Water Resources was held. This meeting was attended by the Chief Ministers of all the States including Rajasthan and representatives of the MoEF, the Ministry of Social Justice and Empowerment, Government of India. In the meeting it was unanimously decided that the reviews of the implementation of the resettlement and rehabilitation measures will be undertaken for every five-metre height of the dam jointly by the concerned R&R Sub-group and Environment Sub-group so that work could progress *pari passu* with the implementation measures. In its meeting held on 6 January 1999 the R&R Sub-group of the NCA observed that arrangements made by the States for R&R of the balance families pertaining to the dam height EL 90 metres were adequate, and a meeting of the party States should be convened shortly to finalise the action plan. Pursuant thereto a special inter-State meeting was convened under the chairmanship of the Secretary to the Government of India, Ministry of Social Justice and Empowerment on 21 January 1999 at New Delhi, and an action plan for resettlement and rehabilitation for balanced (sic) families of dam height EL 90 metres was finalised for implementation by the States. It is the case of the State of Gujarat that it had issued notices and made offers in January 1998 to PAFs affected

at RL 90 metres in connection with the selection of land and their resettlement in Gujarat. According to it, even in respect of PAFs affected at RL 95 metre, notices were issued in January 1999 and to the PAFs included in the subsequent list, notices were issued in September 1999. The process of land selection by PAFs who had opted to resettle in Gujarat at RL 95 metres was already started. According to the Union of India, the Master Plan was under implementation and the progress of R&R at various elevations of dam viz. EL 90 metres, EL 95 metres, EL 110 metres and FRL 138.68 metres had been made.

187. The measures which have been implemented for sustainable development with regard to preserving the socio-cultural environment of the displaced persons in the States of Maharashtra, Gujarat and Madhya Pradesh are stated to be as follows:

- Three choices to the people for the selection of relocation sites.
- Integration of the displaced person with the neighbouring villages by organising medical check-up camps, animal husbandry camps, festivals, eye camps, rural development seminars for village workers, etc.
- Establishment of rehabilitation committees at different levels.
- Respect of traditional beliefs, rituals and rights at the starting of house construction, the day and time of leaving the old house and village and the day and time of occupying the new house, etc.
- The sacred places at the native villages are being recreated along with their settlements at new sites.
- Installation of all the religious deities with the due consultation of religious heads.
- Promotion of the cultural milieu viz social festivals, religious rights, rights of passage, presence of priests, shaman, kinsmen, clansmen, etc.
- Special consideration for the preservation of the holistic nature of the culture.
- Proper use of the built-in-mechanism of cultural heritage of the displaced persons.
- Launching of a culturally appropriate development plan.
- Genuine representation of the traditional leader.

188. The Tribunal had already made provision of various civic amenities which were further liberalised by the State governments during implementation. The existing development programmes were strengthened for ensuring sustainable development at the rehabilitation sites. These were Integrated Rural Development Programme (IRDP) for agriculture, business and village industries; Integrated Child Development Scheme (ICDS) for nutrition, health and education; Jawahar Rojgar Yojana (JRY); aids for improved seeds, fertilisers, irrigation, animal husbandry; Training Rural Youth for Self-Employment (TRYSEM); Employment Guarantee Schemes (EGS), Social Assistance; Industrial Training Institute (ITI); Tribal Development Programme (TDP), financial benefits to the backward classes, economically weaker sections, tribals and other backward classes (OBC), eye

camps, subsidies to farmers (seed, tractorisation, fertilisers, diesel, etc.), agricultural prices support subsidy, etc.

189. Other benefits which were extended for improving the quality of life of the resettled PAFs included fodder farm, mobile sale, shop of fodder, seeds cultivation training, initial help in land preparation for agricultural activities, better seeds and fertilisers, access to finance, special programme for women in the traditional skills, entrepreneurship development, employment skill formation, different plantation programmes, special emphasis for pasture management, environment awareness and education programme, programme for bio-gas/smokeless chulhas, safe drinking water supply, electricity, lift irrigation, fertilisers kit distribution, gypsum treatment of soil, etc.

190. The project authorities in these three States of Madhya Pradesh, Gujarat and Maharashtra represented that comprehensive health care was available in tribal areas where the displaced families had been resettled. It was contended that extensive preventive health measures like mass immunisation, anti-malaria programme, family welfare programmes, child development schemes, etc. had been undertaken. What is important is that primary health centres were established at relocation sites for all necessary health facilities to the PAFs.

191. The submission on behalf of the Union of India was that there was a well established mechanism of the Government of India for coordination and monitoring of R&R programmes in the case of the SSP. The R&R Sub-group and Rehabilitation Committee of the NCA are responsible for applying their independent minds on R&R. The Sub-group convenes its meeting regularly to monitor and review the progress of R&R, while the Rehabilitation Committee visits the submergence areas/relocation sites to see whether the rehabilitation is taking place physically and to hear the individual problems of the PAPs. The R&R group, keeping in view the progress of relief and rehabilitation, has not permitted the height to be raised, until and unless it is satisfied that adequate satisfactory progress has been made with regard to R&R. Whereas at an earlier point of time in 1994, the construction schedule had required the minimum block level to be raised to 85 metres, the R&R Sub-group had permitted the same to be raised to EL 69 metres only during that period to match the R&R activity. It was in the meeting of the R&R Sub-group on 6 January 1999 after the R&R Sub-group had reviewed the progress and had satisfied itself that the land for resettlement in Gujarat, Maharashtra and Madhya Pradesh, which was available, was more than required for the resettlement of the balanced PAFs that it cleared the construction up to the dam height EL 90 metres. The action plan for the same had been approved and is under implementation by the States concerned.

192. The petitioners had contended that no proper surveys were carried out to determine the different categories of affected persons as the total number of affected persons had been shown as much lower and that many had been denied PAF status. From what is being stated hereinabove, it is clear that each state has

drawn a detailed action plan and it is after requisite study had been made that the number of PAFs have been identified. The number has substantially increased from what was estimated in the Tribunal's Award. The reason for the same, as already noticed, is the liberalisation of the R&R packages by the State governments. Except for a bald assertion, there appears to be no material on which this court can come to the conclusion that no proper surveys had been carried out for determining the number of PAFs who would be adversely affected by the construction of the dam.

193. Resettlement and rehabilitation packages in the three States were different due to different geographical, local and economic conditions and availability of land in the States. The liberal packages available to the Sardar Sarovar Project oustees in Gujarat are not even available to the project affected people of other projects in Gujarat. It is incorrect to say that the difference in R&R packages, the package of Gujarat being the most liberal, amounts to restricting the choice of the oustees. Each State has its own package and oustees have an option to select the one which is most attractive to them. A project affected family may, for instance, choose to leave its home State of Madhya Pradesh in order to avail the benefits of the more generous package of the State of Gujarat, while other PAFs similarly situated may opt to remain at home and take advantage of the less liberal package of the State of Madhya Pradesh. There is no requirement that the liberalisation of the packages by three States should be to the same extent and at the same time, the States cannot be faulted if the packages offered, though not identical with each other, are more liberal than the one envisaged in the Tribunal Award.

194. Dealing with the contention of the petitioners that there were a large number of persons who were living in the submergence area and were not farmers and would lose their livelihood due to a loss of the community and/or loss of the river and were not being properly rehabilitated, Mr Harish Salve, learned Senior Counsel contended that this averment was not true. According to him, all the families in the 105 hilly tribal villages were agriculturists, cultivating either their own land or government land and all of whom would be eligible for alternative agricultural land in Gujarat. Only a small number of non-agriculturalists, mainly petty shopkeepers, were found in these villages of tribal areas. In Gujarat there were 20 such non-agriculturalist families out of a total of 4,600 affected families and all of these had been resettled as per their choice so that they could restart their business. In Maharashtra out of 3,213 affected families, not a single family was stated to fall under this category. Amongst the affected families of Madhya Pradesh, the figure of such non-agriculturalist families was also stated to be not more than a couple of hundred. In our opinion, it is neither possible nor necessary to decide regarding the number of people likely to be so affected because all those who are entitled to be rehabilitated as per the Award will be provided with benefits of the package offered and chosen.

195. With regard to the colony affected people whose 1,380 acres of land was acquired in six villages for the construction of a colony, most of the landholders had continued to stay in their original houses and about 381 persons were stated to have been provided permanent employment in the project works. At the time, the land was acquired in 1962–63, compensation was paid and in addition thereto, the Government of Gujarat devised a special package in August 1992 providing ex-gratia payment up to Rs 36,000 to the land losers for the purchase of productive assets or land for those who had not received employment in the project.

196. Dealing with the contention of the petitioners that there will be 23,500 canal affected families and that they should be treated on a par with oustees in the submergence area, the respondents have broadly submitted that there is a basic difference in the impacts of the projects in the upstream submergence area and its impacts in the beneficiary zone of the command area. While people who were oustees from the submergence zone required resettlement and rehabilitation, on the other hand, most of the people falling under the command area were in fact beneficiaries of the projects and their remaining land would now be relocated with the construction of the canal leading to greater agricultural output. We agree with this view and that is why, in the Award of the Tribunal, the State of Gujarat was not required to give to the canal affected people the same relief which was required to be given to the oustees of the submergence area.

197. Dealing with the contention of the petitioners that the oustees were not offered a chance to resettle in Gujarat as a community, and that there was a clear requirement of village-wise communication rehabilitation which had not been complied with, the contention of the respondents was that no provision of the Tribunal's Award had been shown which caused any such obligation on the Government of Gujarat. What the Award of the Tribunal required is resettlement of the PAFs in Gujarat at places where civic amenities like dispensary, schools, as already been referred to hereinabove, are available.

198. Subsequent to the Tribunal's Award, on the recommendation of the World Bank, the Government of Gujarat adopted the principle of resettlement that the oustees shall be relocated as village units, village sections or families in accordance with the oustees preference. The oustees' choice has actively guided the resettlement process. The requirement in the Tribunal's Award was that the Government of Gujarat shall establish rehabilitation villages in Gujarat in the irrigation command of the SSP on the norms mentioned for rehabilitation of the families who were willing to migrate to Gujarat. This provision could not be interpreted to mean that the oustees families should be resettled as a homogeneous group in a village exclusively set up for each such group. The concept of community-wise resettlement, therefore, cannot derive support from the above quoted stipulation. Besides, the norms referred to in the stipulation relate to provisions for civic amenities. They vary as regards each civic amenity *vis-à-vis* the number of oustee families. Thus, one panchayat ghar, one dispensary, one

children's park, one seed store and one village pond is the norm for 500 families, one primary school (3 rooms) for 100 families and a drinking water well with trough and one platform for every 50 families. The number of families to which the civic amenities were to be provided was thus not uniform and it was not possible to derive therefrom a standardised pattern for the establishment of a site which had nexus with the number of oustees' families of a particular community or group to be resettled. These were not indicators envisaging resettlement of the oustees families on the basis of tribes, sub-tribes, groups or sub-groups.

199. While resettlement as a group in accordance with the oustees preference was an important principle/objective, the other objectives were that the oustees should have improved or regained the standard of living that they were enjoying prior to their displacement and they should have been fully integrated in the community in which they were resettled. These objectives were easily achievable if they were resettled in the command area where the land was twice as productive as the affected land and where large chunks of lands were readily available. This was what the Tribunal's Award stipulated and one objective could not be seen in isolation of the other objectives.

200. The 1995 Master Plan of the Narmada Control Authority also pointed out that 'the Bhils, who are individualistic people building their houses away from one another, are getting socialised; they are learning to live together'. Looking to the preferences of the affected people to live as a community, the Government of Gujarat had basically relied on the affected families' decision as to where they would like to relocate, instead of forcing them to relocate as per a fixed plan.

201. The underlined principle in forming the R&R policy was not merely of providing land for PAFs, but there was a conscious effort to improve the living conditions of the PAFs and to bring them into the mainstream. If one compares the living conditions of the PAFs in their submerging villages with the rehabilitation packages first provided by the Tribunal's Award and then liberalised by the States, it is obvious that the PAFs had gained substantially after their resettlement. It is for this reason that in the Action Plan of 1993 of the Government of Madhya Pradesh it was stated before this court that 'therefore, the resettlement and rehabilitation of people whose habitat and environment makes living difficult does not pose any problems and so the rehabilitation and resettlement does not pose a threat to the environment'. In the affidavit of Dr Asha Singh, Additional Director (Socio & CP), NVDA, as produced by the Government of Madhya Pradesh in respect of a visit to R&R sites in Gujarat during 21 to 23 February 2000 for ascertaining the status relating to grievances and problems of Madhya Pradesh PAFs resettled in Gujarat. It was, *inter alia*, mentioned that 'the PAFs had informed that the land allotted to them is of good quality and they take the crops of cotton, jowar and tuwar. They also stated that their status has improved from the time they had come to Gujarat but they want that water should start flowing in the canals as soon as possible and in that case they will be able to take three crops in one year as their land is in the

command area'. Whereas the conditions in the hamlets, where the tribals lived, were not good enough the rehabilitation package ensured more basic facilities and civic amenities to the resettled oustees. Their children would have schools and a children's park, a primary health centre would take care of their health and, of course, they would have electricity which was not a common feature in the tribal villages.

202. Dealing with the contention of the petitioners that there was no provision for grazing land and fuelwood for the PAFs, it is rightly contended by the State of Gujarat that grazing land was not mandated or provided for in the Tribunal's Award but nevertheless, the grazing land of six villages was available for use of PAFs. It may be that the grazing land was inadequate but this problem will be faced by the entire State of Gujarat, and not making such land available for them does not in any way violate any of the provisions of the Award.

203. With regard to providing irrigation facilities, most of the resettlement of the project affected families were provided irrigation facilities in the Sardar Sarovar Project command area or in the command areas of other irrigation projects. In many of the out-of-command sites, irrigated lands were purchased. In cases where the irrigation facilities were not functioning, the Government of Gujarat had undertaken the work of digging tubewells in order to avoid any difficulty with regard to irrigation in respect of those oustees who did not have adequate irrigation facilities. It was contended that because of the delay in the construction of the project, the cut-off date of 1 January 1987 for extending R&R facilities to major sons were not provided. The Tribunal's Award had provided for land for major sons as on 16 August 1978. The Government of Gujarat, however, extended this benefit and offered rehabilitation package by fixing the cut-off date of 1 January 1987 for granting benefits to major sons. According to the Tribunal's Award, the sons who had become major one year prior to the issuance of the notification for land acquisition were entitled to be allotted land. The land acquisition notification had been issued in 1981–82 and as per the Award, it was only those sons who had become major one year prior to that date who would have become eligible for the allotment of land. However, in order to benefit those major sons who had attained majority later, the Government of Gujarat made a relaxation so as to cover all those who became major up to 1 January 1987. The Government of Gujarat was under no obligation to do this and would have been quite within its right to merely comply with the provisions of the Tribunal's Award. This being so, the relaxation of the cut-off date so as to give extra benefit to those sons who attained the age of majority at a later date, cannot be faulted or criticised.

204. Dealing with the contention of the petitioners that there is a need for a review of the project and that an independent agency should monitor the R&R of the oustees and that no construction should be permitted to be undertaken without the clearance of such an authority, the respondents are right in submitting that there is no warrant for such a contention. The Tribunal's Award is final and binding on the

States. The machinery of the NCA has been envisaged and consulted under the Award itself. It is not possible accept that the NCA is not to be regarded as an independent authority. Of course, some of the members are government officials but apart from the Union of India, the other States are also represented in this Authority. The project is being undertaken by the government and it is for the government authorities to execute the same. With the establishment of the R&R Sub-group and constitution of the Grievances Redressal Authorities by the States of Gujarat, Maharashtra and Madhya Pradesh, there is a system in force which will ensure satisfactory resettlement and rehabilitation of the oustees. There is no basis for contending that some outside agency or the National Human Rights Commission should see to the compliance of the Tribunal Award.

Monitoring and rehabilitation programme

205. The Ministry of Water Resources, Government of India is the nodal ministry for the SSP and other union ministries involved are the Ministries of Environment and Forests and Social Justice and Empowerment. As a consequence of the Tribunal's Award, the NCA was created to co-ordinate and oversee the overall work of the project and to monitor the R&R activities including environmental safeguard measures. The Review Committee of the Narmada Control Authority consists of the Union Minister of Water Resources as its chairman, the Union Ministry of Environment and Forests and the Chief Ministers of Gujarat, Madhya Pradesh, Maharashtra and Rajasthan as members. This Review Committee may *suo moto,* or on the application of any party State or the Secretary, MoEF, review any decision of the Narmada Control Authority. In the Narmada Control Authority, the R&R Sub-group has been created for closely monitoring the R&R progress. This Sub-group is headed by the Secretary, Government of India, Ministry of Social Justice and Empowerment and is represented by members/invitees of participating States, academic institutions having expertise in R&R, independent socio-anthropological experts and non-governmental organisations. The functions of this Sub-group are as follows:

1. To monitor the progress of land acquisition in respect of the submergence land of the SSP and the ISP.

2. To monitor the progress of implementation of the action plan of rehabilitation of project affected families in the affected villages of SSP and ISP in concerned states.

3. To review the R&R action plan from time to time in the light of results of the implementation.

4. To review the reports of the agencies entrusted by each of the states in respect of the monitoring and evaluation of the progress in the matter of resettlement and rehabilitation.

5. To monitor and review implementation of the resettlement and rehabilitation programmes *pari passu* with the raising of the dam height, keeping in view the clearance granted to ISP and SSP from the environmental angle by the Government of India and the Ministry of Environment and Forests.

6. To coordinate States/agencies involved in the R&R programmes of SSP and ISP.

7. To undertake any or all activities in the matter of resettlement and rehabilitation pertaining to SSP and ISP.

Rehabilitation Committee

206. This Court vide order dated 9 August 1991 in *B.D. Sharma v. Union of India* 1992 Supp (3) SCC 93 directed the formation of a committee under the chairmanship of the Secretary, Ministry of Social Justice and Empowerment, Government of India to visit the submergence areas/resettlement sites and furnish the report of the development and progress made in the matter of rehabilitation. The Rehabilitation Committee headed by the Secretary, Government of India, Ministry of Social Justice and Empowerment, and having representatives of the three State governments as its members, had been constituted. It is the case of the Union of India that this Committee visited regularly the various R&R sites and submergence villages in the three States and submitted reports to this court from time to time. By order dated 24 October 1994, this court in the aforesaid case of B.D. Sharma (*supra*) observed that all the directions issued by the Court from time to time have been complied with and nothing more be done in the petition and the petition was disposed off. Most of the recommendations/observations as made by this Committee are stated to have been complied with fairly by the States concerned.

207. In addition to the above, the officials of the Narmada Control Authority are also stated to be monitoring the progress of R&R regularly by making field visits. The individual complaints of the PAFs are attended and brought to the notice of the respective governments.

Grievances redressal mechanism

208. The appeal mechanism has been established in the policy statements by all the three State governments for the redressal of grievances of the PAFs. According to this mechanism, if a displaced person is aggrieved by the decision of the rehabilitation officers in respect of any R&R process, he may appeal to the concerned agency/officers.

209. Vide Resolution dated 17 February 1999, the Government of Gujarat set up a high-level authority called Grievance Redressal Authority (GRA) before whom the oustees already resettled and to be resettled in Gujarat could ventilate their grievances for redressal after their resettlement until the process of resettlement and rehabilitation is fully completed. The said Grievances Redressal Authority has

Mr Justice P.D. Desai, retired Chief Justice as its chairman. This machinery had been established to:

a) create an authority before whom oustees who have resettled in the State of Gujarat can ventilate their grievances relating to the R&R measures taken by the State of Gujarat;

b) ensure that the oustees already settled and the oustees settled hereinafter in the R&R sites created for resettlement and rehabilitation of the oustees from the State of Madhya Pradesh and Maharashtra receive all the benefits and amenities in accordance with the Award and the various government resolutions from time to time;

c) ensure that Gujarat oustees resettled in Gujarat have received all the benefits and amenities due to them.

210. The Gujarat Rehabilitation Authority has installed a permanent in-house Grievances Redressal Cell (GRC) within the Sardar Sarovar Punarvasavat Agency. The Grievances Redressal Cell deals with the grievances of the PAFs and the grievances redressal is undertaken by it in the following three ways.

i. Grievances Redressal Cell deals with grievances in the regular course on the basis of applications, i.e. by holding enquiries and implementing decisions taken pursuant thereto.

ii. Grievances redressal on the spot through mechanism of Tatkal Fariyad Nivaran Samiti.

iii. Grievances redressal under the mechanism of the Single Window Clearance System.

211. Grievances Redressal Authority has surveyed sites in which PAFs have been resettled and has submitted reports to this court from time to time which disclose substantial compliance with the terms of the Award and the rehabilitation package.

212. In its fourth report dated 15 November 1999, the Grievances Redressal Authority observed 'pursuant to the grievances redressal measures taken by GRC, whose approach is positive and grievance redressal oriented, a considerable number of grievances have been resolved by extensive land improvement work done on agricultural land at different sites within a period of six months i.e. April-September 1999'.

213. The R&R Sub-group in its 20 field visits of the R&R sites in Gujarat on 12/13 January 2000 has noted as follows:

> The Committee after the visit and from interaction with the PAFs, concluded that there is vast improvement in the conditions of PAFs at these R&R sites as compared to the grievances reported in the same sites during previous visits by the Committee/NCA officers. Assessing the perception of PAFs the Committee

observed that the majority of PAFs are happy and joining the mainstream of the country's development.

214. The Grievances Redressal Cell has dealt with and decided a total of over 6,500 grievances.

215. At the instance of the Grievances Redressal Authority, an agricultural cell is set up in Sardar Sarovar Punarvasavat Agency with effect from 1 July 1999. This was done with an objective of enhancing the productivity of agricultural land allotted to PAFs by adopting suitable farm management practices and in assisting in resolving land related grievances. Similarly, with effect from 1 May 1999, medical cells have been set up in the Sardar Sarovar Punarvasavat Agency for ensuring effective functioning of the medical infrastructure and providing an organised system of supervising and monitoring and also for conducting health survey-cum-medical check up activities. The Grievance Redressal Authority has become an effective monitoring and implementing agency with regard to relief and rehabilitation of the PAFs in Gujarat. Apart from resolving independent grievances of PAFs and enforcing the compliance of the provisions of the Award through its exhaustive machinery and mechanism, it is also trying to guide in respect of various other issues not covered by the provisions of the Award such as:

i. vocational training of the oustees;

ii. review of Narmada oustees employment opportunity rules;

iii. issue relating to Kevadia colony;

iv. issue relating to tapu land;

v. development of Kevadia as a tourist centre, etc.

216. In Maharashtra a local committee was constituted comprising of the Additional Collector (SS), Divisional Forest Officer, Resettlement Officer and two representatives of the oustees nominated by the local Panchayat Samities from among the elected members of the village panchayats in the project affected villages/talukas. This committee is required to examine the claims of the PAFs and give directions within a time frame and an appeal from its decision lies to the Commissioner. In addition thereto, vide notification dated 17 April 2000, the Government of Maharashtra has set up a Grievances Redressal Authority in line with that established by the State of Gujarat and Mr Justice S.P. Kurdukar, retired Judge of this court, has been appointed as its Chairman. This authority is expected to be analogous to the Grievances Redressal Authority of Gujarat.

217. In Madhya Pradesh the grievances of the PAFs have first to be made by a claim which will be verified by the patwari and then scrutinised by the tehsildar. PAFs may file an appeal against the decision of the R&R official before the district collector who is required to dispose off the same within a period of three months. In the case of Madhya Pradesh also, by notification dated 30 March 2000, the

Government of Madhya Pradesh has constituted a Grievances Redressal Authority similar to the one in Gujarat with Mr Justice Sohni, retired Chief Justice of Patna High Court, as its Chairman.

Independent monitoring and evaluation agencies

218. The monitoring and evaluation of the rehabilitation programme is also being carried out by the independent socio-anthropological agencies appointed by the State governments of Maharashtra, Madhya Pradesh and Gujarat as well as the Narmada Control Authority. These agencies, which are professional and academic institutes, conduct surveys and in-depth studies relating to PAFs in the submergence and rehabilitation villages. The main object of the monitoring is oriented towards enabling the management to assess the progress, identify the difficulties, ascertaining problem areas, provide early warning and thus call for corrections needed immediately.

219. The Centre for Social Studies, Surat is the monitoring agency for the Government of Gujarat. This institute has prepared 24 six-monthly progress reports in relation to the resettlement of PAFs of submergence villages of Gujarat. Similarly for the project affected families of Madhya Pradesh/Maharashtra who have resettled in Gujarat, the Government of Gujarat has appointed the Gujarat Institute of Development Research, Ahmedabad as the independent monitoring and evaluation agency for monitoring R&R programmes.

220. In Madhya Pradesh the monitoring and evaluation had been carried out by Dr H.S. Gaur University, Sagar and the same has been disengaged now and a new agency is being appointed. The findings of Dr H.S. Gaur University, Sagar indicated that displaced families in Madhya Pradesh are, by and large, happy with the new resettlement in Gujarat and one of the main reasons behind their happiness was that shifting from hamlets had changed their socio-economic status.

221. In Maharashtra the monitoring and evaluation was earlier being done by the Tata Institute of Social Sciences, Mumbai. This agency had reported that the overall literacy rate among project affected persons above six years of age is about 97 per cent, while illiteracy in submergence villages was rampant. Furthermore the report showed that in the submergence villages, the tribals mostly relied on traditional healers for their ailments. Now the current scenario is that at R&R sites, health centres and sub-centres have been established.

222. It is thus seen that there is in place an elaborate network of authorities which have to see to the execution and implementation of the project in terms of the Award. All aspects of the project are supervised and there is a review committee which can review any decision of the Narmada Control Authority and each of the three rehabilitating States have set up an independent grievances redressal authority to take care that the relief and rehabilitation measures are properly implemented and the grievances, if any, of the oustees are redressed.

223. On 9 May 2000 this court directed the State governments of Gujarat, Madhya Pradesh and Maharashtra to file affidavits disclosing the latest status of resettlement and rehabilitation work for the existing as well as prospective oustees likely to be affected by raising the height of the dam.

224. Pursuant to the said direction, affidavits on behalf of the three States have been filed and, in response thereto, the petitioners have also filed an affidavit.

225. On behalf of the State of Gujarat the affidavit of Shri V.K. Babbar, Commissioner (Rehabilitation) and Chief Executive Officer, SSPA has been filed, according to which at FRL 138.68 m the status with regard to PAFs to be resettled is stated to be as follows:

State	Total number of PAFs resettled/ allotted agricultural land in Gujarat	Balance of PAFs to be resettled in Gujarat
Gujarat	4,575	25
Maharashtra	710	290
Madhya Pradesh	3,280	10,450
Total	8,585	10,765

226. It is the case of Gujarat that 8,565 PAFs have been accommodated in 182 R&R sites fully equipped with the requisite civic amenities as provided by the Tribunal Award. The agricultural land allotted to these PAFs is 16,973 hectares.

227. Dealing specifically with the status of PAFs at RL 90 m, 95 m and 110 m, it is averred in the said affidavit that all the PAFs of Gujarat at RL 90 m have been resettled and the balance of PAFs of Madhya Pradesh and Maharashtra affected at RL 90 m have already been offered the R&R package in Gujarat. The process of resettlement is continuing and reliance is placed on the observation of the GRA which has stated in its fourth report, dated 15 November 1999, that:

> There is substantial compliance of the resettlement and rehabilitation measures as mandated by the final report of NWDT, including provision of civic amenities, and also of all the inter-linked provisions of the Government of Gujarat and that, therefore, PAFs from the States of Madhya Pradesh and Maharashtra affected up to the height of RL 90 m can be accommodated as per their choice at these selected 35 sites in Gujarat.

228. With respect to the PAFs affected at 95 m, the affidavit states that the PAFs of Gujarat have already been settled, while the affected PAFs of Madhya Pradesh and Maharashtra have been offered the R&R package in Gujarat in January 1999 and January 2000. The RL 95 m action plan for these PAFs has also been prepared by the Government of Gujarat in consultation with the Governments of Madhya

Pradesh and Maharashtra and has been sent to the NCA. The case of the State of Gujarat, therefore, is that all the PAFs wanting to be resettled in Gujarat have been offered the package but consent of all the PAFs has not so far been received but the Government of Gujarat has sufficient land readily available which can be allotted to the said PAFs as soon as they come and select the same.

229. With regard to the status of PAFs at RL 110 m, all the PAFs of Gujarat have been resettled and 2,761 PAFs (2,642 of Madhya Pradesh and 119 of Maharashtra) remain to be resettled in Gujarat and the R&R package will be offered to them before November 2000. The land which is required to be allotted to them is stated to be around 6,074 hectares and the State of Gujarat has in its possession 8,146 hectares. The civic amenities in 40 new R&R sites are scheduled to be completed by December 2000 and these sites would serve to accommodate not only PAFs between RL 95 m and RL 110 m but would also serve to accommodate PAFs from submergence villages which would be getting affected at levels above RL 110 m. The action plan giving the village-wise details is said to have been sent to NCA in June 2000 for its approval.

230. According to the said affidavit the balance number of PAFs remaining to be resettled in Gujarat at FRL 138.68 m is 10,765. Taking into account that an additional area of 10 per cent towards house plot and civic amenities would be required in addition to the allotment of a minimum 2 hectares of agricultural land, the total land requirement per PAF would be approximately 2.2 hectares. For planning purposes in respect of 10,765 PAFs the land requirement would be about 23,700 hectares. As against this requirement the status of land, as per the said affidavit, under different categories with the Government of Gujarat is stated to be as under:

Sr.No.	Particulars	Land (ha)
1.	Land identified (offers received in respect of private land and Government land)	15,716
2.	Land available (private land for which price is approved by Expert Committee and offer/counter offer conveyed and acceptance of land holder is obtained)	480
3.	Land in possession of SSPA/GoG in 12 districts	8,416
	Total	24,612

It is averred that between March and 21 June 2000 the land in possession as well as the land identified has increased considerably.

231. It has also been explained in the said affidavit that the Government of Gujarat has a well established practice of procuring land for R&R at realistic market prices from willing sellers. Officers hold discussions with prospective resettlers, verify the suitability of land and after the price is settled, the same is procured through the legal process of the Land Acquisition Act and consent awards are passed so that the PAPs are assured of undisputed legal title free from encumbrances. This process of negotiated purchase has been streamlined. At the instance of the GRA, a retired judge of the High Court is now appointed as Chairman of the Expert Committee with retired senior government secretaries as its members. This Expert Committee oversees the exercise of purchase of suitable land at the market price. At the instance of the GRA, PAPs are being issued *Sanads* for the land allotted to them which will ensure provision of a proper legal document in their favour.

232. Dealing with the term of the Award to the effect that Gujarat shall acquire and make available a year in advance of the submergence before each successive stage, land and house sites for rehabilitation of the oustee families from Madhya Pradesh and Maharashtra who are willing to migrate to Gujarat, the affidavit states that the Gujarat Government has already identified sufficient land for accommodating the balance PAFs remaining to be resettled in Gujarat at FRL 138.68 m. In respect of PAFs up to RL 110 m, Gujarat has sufficient land available to meet the R&R requirements but for the PAFs above 110 m, suitable land has already been identified and the same would be acquired and made available one year in advance of submergence before each successive stage. The affidavit gives reason as to why it is not advisable for the State at this stage to acquire the total requirement of land for FRL in one go. What is stated in the affidavit is as follows:

i. Since at present GoG has sufficient land to meet R&R requirement to accommodated PAFs up to RL 110 m, it would not be necessary to acquire further land immediately, especially when the additional land would be required only after the R&R Sub-group and Environment Sub-group give approval for RL 95 m to RL 110 m after examining the preparedness at different stages. This would ensure that public money is not unnecessarily blocked for a long period.

ii. By acquiring land much before it would be required, problems of illegal trespass are likely to arise.

iii. The excess land would, by and large, remain fallow and no agricultural production would take place.

iv. If the land remains fallow for long the overall productivity of the land would be adversely affected.

v. At the time of allotment, the State government would again have to spend a sizeable amount to remove weeds, bushes, small trees etc.

vi. The State government would have to incur a sizeable amount to prevent tampering with the boundary marks, prevent neighbouring farmers removing the top soil or from diverting natural drains passing through their fields toward the land purchased for R&R etc.

233. The affidavit also gives facts and figures showing that all requisite civic amenities have been developed and made available at the R&R sites. Some of the salient features which are highlighted in this behalf are as under:

- A three-room primary school is provided in all MP/MH sites irrespective of the number of families settled.
- A dispensary with examination room, medical equipment, medicines is provided in all MP/MH sites irrespective of the number of resettled families.
- 3439 PAFs (86 per cent) out of the total MP/MH PAFs resettled in Gujarat have availed of the Rs 45,000 financial assistance and built pucca core houses.
- Overhead tanks for drinking water are provided in large R&R sites.
- At the instance of GRA, toilets are being provided in the houses of PAFs with the help of NGOs.

234. The total cost incurred so far by the Government of Gujarat in providing the land and civic amenities up to May 2000 is stated to be 194 crores. The Grievances Redressal Cell is stated to have redressed large number of grievances of the PAFs whether they were related to land, grant of civic amenities or others. The salient features of working of the Grievance Redressal Cell is stated to be as follows:

- At present 2 senior IAS officers with supporting staff are working exclusively for redressal of grievances.
- A reasoned reply is given to the applicants. The applicant is also informed that if he is aggrieved with the decision he may prefer an appeal to GRA within thirty days.
- The single window clearance system's main objective is to proactively resolve grievances and to avoid delays in inter-departmental co-ordination.
- Tatkal Fariyad Nivaran Samitis are held in the R&R sites to resolve grievance of the PAFs in an open forum.
- The PAFs are being involved at every stage of grievance redressal. The works have been carried out in most cases by the PAFs.
- The agriculture officers of the agricultural cell are actively helping, guiding the PAFs in their agricultural operations and upgrading their skills.

235. With a view to effectively rehabilitate and assimilate the PAPs Vasahat Samitis have been constituted in 165 R&R sites, consisting of 5 PAPs, one of

whom is a female. This ensures the participation of the PAFs in the process of development and these Samitis are vested with the responsibility to sort out minor problems. With a view to ensure more effective participation in Panchayat affairs and better integration of PAPs an order under Section 98 of the Gujarat Panchayats Act, 1993 has been issued by the Government of Gujarat providing that there shall be up to two invitees from amongst the PAPs depending upon the number of PAPs at the sites in the village Panchayat within whose jurisdiction the R&R are situated. Pursuant to this 196 PAPs have been inducted as invitees to then village panchayats. The salient features of the rehabilitation programme of the PAPs are as follows:

- PAFs are given productive assets in kind (7,000/PAF) to purchase bullocks, bullock carts, oil engines, etc.
- PAFs are given subsistence allowance (Rs 4,500/PAF) in cash to meet contingency needs in the initial period.
- Vocational training is provided to PAFs for improving their income levels, priority being given to those dependents who are not entitled to be declared as PAFs on their own rights. Tool kits are supplied either free or with 50 per cent subsidy.
- NGOs are actively involved in all the rehabilitation activities such as conducting training classes.
- PAFs are being covered by the ongoing developmental schemes of the Government (DRDA, Tribal Sub Plan, etc.).
- Extension (Agriculture) officers have been appointed for approximately every 150 families to guide them in agriculture operation and assist them in day to day problems (getting ration cards, khedut khatavahis, etc.).
- In recent years focus is on empowering the PAFs, and making them self dependent.

236. Medical cells have been set up for providing services and treatment to PAPs free of cost. The cell is headed by Deputy Director (Medical) and is having a nucleus of medical experts consisting of a physician, a paediatrician, a gynaecologist, 21 MBBS doctors, pharmacists, etc. The salient features of the medical help programmes for the benefit of PAPs is stated to be as follows:

- The medical officers and paramedic staff are making house-to-house visits to motivate the PAPs to come forward to avail of the medical services.
- In all dispensaries, a full time multipurpose health worker (female) is available.

- Multi-specialisation diagnostic/treatment camps are organised fortnightly, where advanced investigations and diagnostic facilities like ECG, X-ray and ultrasound are available.
- Patients requiring further services are brought to government hospitals or any other specialty hospital and necessary treatment given free of cost.
- GoG has placed an order for a mobile medical hospital equipped with diagnostic and treatment equipment.
- A comprehensive health survey and medical check up covering 29,423 PAPs has been completed. A special record system of family health folder and health profile of each PAP is prepared.
- Nutrition supplements are given to children (up to 6 years), expectant and lactating mothers through the Integrated Child Development Scheme (ICDS).
- Special food supplements in the form of 'Hyderabad Mix' is given to malnourished children and vulnerable target groups.
- School-going children are covered under the mid-day meal scheme.
- Under TB Control, all chest symptomatic persons are screened by special examinations like sputum microscopy, X-ray, blood tests and persons found positive for TB are given domiciliary treatment under direct observation of doctors or paramedics. In 77 cases, treatment is completed and patients are cured.
- Under preventive health care, health education material is distributed and health and cleanliness shibirs are organised.
- A special survey covering physically handicapped and mentally retarded persons has been organised and social welfare benefits given.
- Other national health programmes (maternal child health, immunisation, school health check up, family welfare, etc.) are regularly conducted.

237. An agricultural cell has been set up in the SSPA which assists the grievances redressal machinery in resolving the problem relating to the agricultural land. The salient features of this cell are as follows:

- The agriculture cell is involved in purchasing land, supervision of land improvement works and processing land related grievances of the PAFs.
- Agriculture training classes are organised for PAFs in the training institutes of the State government.
- Assistance is given for availing crop-loan credit for banks and extension education is imparted in matters of marketing, cropping pattern, use of improved seeds, insecticides and latest equipment.

- Afforestation was carried out in 33 R&R sites during 1999–2000 by planting 3,500 saplings which are protected by bamboo tree-guards. Plantation is done along the roadside, common plots, school premises, etc. In the remaining sites plantation work is undertaken by NGOs.

238. At the instance of the GRA an educational cell has been set up in the SSPA, the main function of which is to improve the quality of education imparted and to improve the school enrolment. The salient features of this cell are as under.

- School enrolment which was 4,110 in 1998–99, increased to 4,670 in 1999–2000. Out of the 4,670 students enrolled, 2,126 were girls (46.3 per cent).
- The number of schools is 170 and the number of teachers is 384. In the last academic year, 66 schools were upgraded by increasing the number of classes.
- SSPA is regularly sending the teachers for in-service training. So far 120 teachers have been imparted training.
- Every year during the period of June to August, a special drive is taken to increase the school enrolment.
- In the current year 150 adult education classes have been started in the R&R sites with the help of NGOs.
- An advisory committee has been created to make recommendations on how to improve the education being imparted. Members include the faculty of M.S. University, officers of Education Department, Principal of Teacher Training Centre.

239. It is further averred in this affidavit that at the instance of the GRA, a large number of measures have been taken to improve the organisational structure of SSPA so as to effectively meet the challenge of R&R and make the R&R staff accountable. The salient features of this are stated to be as follows:

- A strategic policy decision has been taken to create three separate divisions in SSPA for rehabilitation, resettlement and planning. Each division is in charge of a senior level officer of the rank of Additional/Joint Commissioner.
- Staff strength in SSPA has been considerably augmented especially at the field level.
- To review the structural and functional aspects of SSPA, services of a management consultancy agency (M/s TCS) has been engaged and a draft report has been received and is being examined.
- A demographic survey is to be conducted to comprehensively document information regarding the PAPs with special reference to their family composition, marriage, births, deaths, life expectancy, literacy, customs, culture, social integration, etc.

- Staff is being trained to sensitise them especially with regard to rehabilitation and second generation issues. Senior level officers have been sent for R&R training at the Administrative Staff College of India, Hyderabad.

240. From the aforesaid affidavit it is more than clear that the GRA, of which Mr Justice P.D. Desai is the Chairman, has seen to the establishment of different cells and have taken innovative steps with a view to making R&R effective and meaningful. The steps which are being taken and the assistance given is much more than what is required under the Tribunal's Award. There now seems to be a commitment on the part of the Government of Gujarat to see that there is no laxity in the R&R of the PAPs. It appears that the State of Gujarat has realised that without effective R&R facilities no further construction of the dam would be permitted by the NCA and under the guidance and directions of the GRA meaningful steps are being undertaken in this behalf. In this connection we may take note of the fact that along with the said affidavit Shri V.K. Babbar, again under the directions of the GRA, has given an undertaking to this court, which reads as follows:

1) As per this undertaking, *inter alia*, in respect of scattered pieces or parcels of lands in possession of the SSPA or R&R which do not add up to a contiguous block of 7 hectares by themselves or in conjunction with other lands, steps will be taken to purchase or acquire contiguous lands so that the said small pieces of land become a part of a continuous block of 6 hectares or more. This exercise will be undertaken and completed on or before 31 December 2000. In case it is not possible to have a contiguous block of a minimum of 6 hectares, further directions will be sought from GRA or such piece or parcel of land will be put to use for other public purposes relating to R&R but which may not have been provided for in the NWDT Award.

2) Henceforth, the land which is acquired or purchased for R&R purposes shall be contiguous to each other so as to constitute a compact block of 6 hectares.

3) Henceforth, land to be purchased for R&R will be within a radius of 3 km from an existing or proposed new site and if there is a departure from this policy prior approval of the GRA will be obtained.

4) Demarcation of a boundary of 5,211 hectares of land whose survey has been undertaken by the GRA, and the carving out of individual plots of 2 hectares for allotment to PAFs, will be undertaken and completed on or before 31 December 2000.

5) The other undertakings relate to soil testing and/or ensuring that suitable land is made available to the PAFs after the quality of land is cleared by the agriculture experts of the Gujarat Agriculture University. With regard to the lands in possession of the SSPA which are low-lying and vulnerable to waterlogging during monsoon, an undertaking has been given that the land has been deleted from the inventory of lands available for R&R unless such lands

are examined by the agriculture cell of SSPA and it is certified that access to these lands is clear and unimpeded and that they are suitable for R&R. A compliance report in this regard is to be submitted to the GRA on or before 31 December 2000.

241. In addition to the aforesaid undertaking of Shri V.K. Babbar, undertakings of the Collectors of Khedr, Vadodara, Ahmedabad, Narmada, Panchmahal and Bharuch districts have also been filed. Apart from reiterating what is contained in the undertaking of Shri V.K. Babbar, in these undertakings of the collectors, it is stated that necessary mutation entries regarding entering the name of SSPA/SSNNL in the village records of right in respect of the land in possession of R&R or PAFs likely to be resettled in Gujarat have been made, but the certification of these entries will be completed and matter reported to the GRA before 31 August 2000. If this is not done the land is to be deleted from the inventory of land available for R&R. Necessary mutation entries in the village records or rights regarding removal of encumbrances of original landholders shall also be completed by that date.

242. From what is noticed hereinabove, this court is satisfied that more than adequate steps are being taken by the State of Gujarat not only to implement the Award of the Tribunal to the extent it grants relief to the oustees, but the effort is to substantially improve thereon and, therefore, continued monitoring by this court may not be necessary.

243. On behalf of the State of Madhya Pradesh, response to this court's order dated 9 May 2000, an affidavit of Shri H.N. Tiwari, Director (TW), NVDA has been filed. It is stated therein that with a view to arrange resettlement of the PAFs to be affected at different levels detailed instructions to the field officers of the submergence area were issued by Shri Tiwari vide letter dated 20 May 2000 in respect of all the aspects of resettlement of the PAFs. This is related to the identification of land, processing of land acquisition cases and passing of the Award, taking of PAFs to Gujarat for selection of land, allotment of land to the PAFs who decide to remain in Madhya Pradesh and development of sites. There are 92 sites for resettlement of the PAFs which are required to be established and out of these 18 are stated to be fully developed, development in 23 sites is in process, 18 sites are such where location has been determined and land identified but development work has not started, and 33 sites are such where the location of the land for development is to be decided by the task force constituted for this purpose.

244. Dealing specifically with the status of PAFs to be affected at different levels this affidavit, *inter alia*, states that with regard to PAFs to be affected at EL 85 m, for those who have opted to go to Gujarat, land has been offered to them by the Government of Gujarat, while for those PAFs who have changed their mind and now want to remain in Madhya Pradesh, land is being shown to them in Madhya Pradesh.

245. It has not been categorically stated whether the PAFs who are so affected have been properly resettled or not. On the contrary, it is stated that no Awards in land acquisition cases have been passed in respect of six villages and it is only after the Awards are passed that house plots will be allotted and compensation paid. The provision of financial assistance for the purchase of productive assets will be released when PAFs shift and start construction of the houses. The reason for not making the payment in advance rightly is that if the grants are paid to the oustees before they shift they may possibly squander the grant and the State government may be required to pay again to establish them on some self-employment venture. For the resettlement of PAFs in Madhya Pradesh, out of ten relocation sites mentioned in the affidavit only five have been fully developed. It is also stated that 163 PAFs are resisting resettlement to Gujarat under the influence of anti-dam activists, though they have been given notices containing an offer of the land and house plots by the Government of Gujarat. In addition thereto 323 PAFs who were earlier resisting have now been persuaded and arrangements for the selection of land for them in Gujarat has been initiated.

246. With regard to the R&R status of PAFs to be affected at EL 95 m, it is, *inter alia*, stated that those losing 25 per cent of their holdings are entitled to be allotted cultivable land, and notices were given to them to identify the land which can be allotted. In the said notice it was stated that the development process will be undertaken with regard to the said land only after it is selected by the PAFs. There is also a mention in the affidavit filed in the name of Narmada Bachao Andolan, the petitioner herein, not allowing the State government to conduct survey for demarcation of the submergence area and identification of the PAFs to be affected at EL 132.86 m (436 ft). Six out of 25 relocation sites required to be developed have been fully developed.

247. The affidavit on behalf of the State of Madhya Pradesh draws a picture of rehabilitation which is quite different from that of Gujarat. There seems to be no hurry in taking steps to effectively rehabilitate Madhya Pradesh PAFs in their home State. It is indeed surprising that even Awards in respect of six villages out of 33 villages likely to be affected at 90 m dam height have not been passed. The impression which one gets after reading the affidavit on behalf of the State of Madhya Pradesh clearly is that the main effort of the said State is to try and convince the PAFs that they should go to Gujarat, whose rehabilitation package and effort is far superior to that of the State of Madhya Pradesh. It is, therefore, not surprising that the vast majority of the PAFs of Madhya Pradesh have opted to be resettled in Gujarat, but that does not by itself absolve the State of Madhya Pradesh of its responsibility to take prompt steps so as to comply at least with the provisions of the Tribunal's Award relating to relief and rehabilitation. The State of Madhya Pradesh has been contending that the height of the dam should be lowered to 436 ft so that a lesser number of people are dislocated, but we find that even with regard to the rehabilitation of the oustees at 436 ft the R&R programme of the State is nowhere implemented. The State is under an obligation to effectively

resettle those oustees whose choice is not to go to Gujarat. Appropriate directions may, therefore, have to be given to ensure that the speed in implementing the R&R picks up. Even the interim report of Mr Justice Soni, the GRA for the State of Madhya Pradesh, indicates a lack of commitment on the State's part in looking after the welfare of its own people who are going to be under the threat of ouster and who have to be rehabilitated. Perhaps the lack of urgency could be because of a lack of resources, but then the rehabilitation even in Madhya Pradesh is to be at the expense of Gujarat. A more likely reason could be that, apart from electricity, the main benefit of the construction of the dam is to Gujarat and to a lesser extent to Maharashtra and Rajasthan. In a federal set up like India, whenever any such inter-State project is approved and work undertaken, the States involved have a responsibility to cooperate with each other. There is a method of settling the differences which may arise amongst the like, for example, in the case of an inter-State water dispute the reference of the same to a tribunal. The award of the tribunal being binding, the States concerned are duty bound to comply with the terms thereof.

248. On behalf of the State of Maharashtra affidavit in response to this court's order dated 9 May 2000, the position regarding the availability of land for distribution to the PAFs was stated to be as follows:

i) Total land available by the Forest Department	4,191.86 ha
ii) Land which could not be allotted at present to PAFs	
a) Gaothan land (used for residential purposes)	209.60 ha
b) Land occupied by river/nallah/hills	795.62 ha
c) Land under encroachment by third parties	434.13 ha
Therefore, the net land available at present for allotment was 4,191.86 (-) 1,439.35	2,752.51 ha
Total area of land allotted to 1,600 PAFs	2,434.01 ha
Remaining cultivable land available with the State 2,752 (-) 2,434.01	318.50 ha

It is further stated in this affidavit that out of 795.62 ha of forest land which was reported to be uncultivable, the State has undertaken a survey for ascertaining whether any of these lands can be made available for cultivation and distribution by resorting to measures like bunding, terracing and levelling. It is estimated that 30 to 40 hectares of land would become available. In addition thereto the affidavit states that the Government of Maharashtra has decided to purchase private land in nearby villages for the resettlement of PAFs and further that the GRA has been established and Justice S.P. Kudukar, a retired judge of this court has been appointed as its Chairman. It is categorically stated in this affidavit that the State government would be in a position to make these lands available to all the concerned project affected families.

Conclusion

249. Water is one element without which life cannot be sustained. Therefore, it is to be regarded as one of the primary duties of the government to ensure availability of water to the people.

250. There are only three sources of water. They are rainfall, groundwater and from river. A river itself gets water either by the melting of the snow or from rainfall, while the groundwater is again dependent on rainfall or from the river. In most parts of India, rainfall takes place during a period of about 3 to 4 months known as the monsoon season. Even at the time when the monsoon is regarded as normal, the amount of rainfall varies from region to region. For example, North-Eastern States of India receive much more rainfall than some of other states like Punjab, Haryana or Rajasthan. Dams are constructed not only to provide water whenever required but they also help in flood control by storing extra water. Excess of rainfall causes floods while deficiency thereof results in drought. Studies show that 75 per cent of the monsoon water drains into the sea after flooding a large land area due to the absence of storage capacity. According to a study conducted by the Central Water Commission in 1998, surface water resources were estimated at 1,869 cu. km and rechargeable groundwater resources at 432 cu. km. It is believed that only 690 cu. km of surface water resources (out of 1,869 cu. km) can be utilised by storage. At present the storage capacity of all dams in India is 174 cu. km which is incidentally less than the capacity of Kariba Dam in Zambia/Zimbabwe (180.6 cu. km) and only 12 cu. km more than Aswan high dam of Egypt.

251. While the reservoir of a dam stores water and is usually situated at a place where it can receive a lot of rainfall, the canals take water from this reservoir to distant places where water is a scare commodity. It was, of course, contended on behalf of the petitioner that if the practice of water harvesting is restored and some check dams are constructed, there would really be no need for a high dam like Sardar Sarovar. The answer to this given by the respondent is that water harvesting serves a useful purpose but it cannot ensure an adequate supply to meet all the requirements of the people. Water harvesting means to collect, preserve and use the rain water. The problem of the area in question is that there is deficient rainfall and small water harvesting projects may not be adequate. During the non-rainy days, one of the essential ingredients of water harvesting is the storing of water. It will not be wrong to say that the biggest dams to the smallest percolating tanks meant to tap the rainwater are nothing but water harvesting structures to function by receiving water from the common rainfall.

252. A dam serves a number of purposes. It stores water, generates electricity and releases water throughout the year and at times of scarcity. Its storage capacity is

meant to control floods and the canal system which emanates therefrom is meant to convey and provide water for drinking, agriculture and industry. In addition thereto, it can also be a source of generating hydro-power. A dam has, therefore, necessarily to be regarded as an infrastructural project.

253. There are three stages with regard to the undertaking of an infrastructural project. One is the conception or planning, second is the decision to under take the project and the third is the execution of the project. The conception and the decision to undertake a project is to be regarded as a policy decision. While there is always a need for such projects not being unduly delayed, it is at the same time expected that as thorough a study as is possible will be undertaken before a decision is taken to start a project. Once such a considered decision is taken, the proper execution of the same should be taken expeditiously. It is for the government to decide how to do its job. When it has put a system in place for the execution of a project and such a system cannot be said to be arbitrary, then the only role which a court may have to play is to see that the system works in the manner envisaged.

254. A project may be executed departmentally or by an outside agency. The choice has to be of the government. When it undertakes the execution itself, with or without the help of another organisation, it will be expected to undertake the exercise according to some procedure and principles. The NCA was constituted to give effect to the Award, various sub-groups have been established under the NCA and to look after the grievances of the resettled oustees, and each State has set up a grievance redressal machinery. Over and above the NCA is the Review Committee. There is no reason now to assume that these authorities will not function properly. In our opinion the courts should have no role to play.

255. It is now well settled that the courts, in the exercise of their jurisdiction, will not transgress into the field of policy decision. Whether to have an infrastructural project or not, what type of project to be undertaken, and how it has to be executed are part of policy making process and the courts are ill-equipped to adjudicate on a policy decision so undertaken. The court, no doubt, has a duty to see that in the undertaking of a decision, no law is violated and people's fundamental rights are not transgressed upon except to the extent permissible under the Constitution. Even then any challenge to such a policy decision must be before the execution of the project is undertaken. Any delay in the execution of the project means over-run in costs and the decision to undertake a project, if challenged after its execution has commenced, should be thrown out at the very threshold on the ground of laches if the petitioner had the knowledge of such a decision and could have approached the Court at that time. Just because a petition is termed as a PIL does not mean that ordinary principles applicable to litigation will not apply. Laches is one of them.

256. PIL was an innovation essentially to safeguard and protect the human rights of those people who were unable to protect themselves. With the passage of time the PIL jurisdiction has been ballooning so as to encompass within its ambit subjects

such as probity in public life, granting of largess in the form of licences, protecting the environment and the like. But the balloon should not be inflated so much that it bursts. PIL should not be allowed to degenerate to becoming Publicity Interest Litigation or Private Inquisitiveness Litigation.

257. While exercising jurisdiction in PIL cases, this court has not forsaken its duty and role as a court of law dispensing justice in accordance with law. It is only where there has been a failure on the part of any authority in acting according to law or in non-action in violation of the law that the Court has stepped in. No directions are issued which are in conflict with any legal provisions. Directions have, in appropriate cases, been given where the law is silent and inaction would result in violation of the fundamental rights or other legal provisions.

258. While protecting the rights of the people from being violated in any manner utmost care has to be taken that the Court does not transgress its jurisdiction. There is in our constitutional framework a fairly clear demarcation of powers. The Court has come down heavily whenever the executive has sought to impinge upon the Court's jurisdiction.

259. At the same time, in exercise of its enormous power, the Court should not be called upon to undertake governmental duties or functions. The courts cannot run the government nor the administration indulge in abuse or non-use of power and get away with it. The essence of judicial review is a constitutional fundamental. The role of the higher judiciary under the Constitution casts on it a great obligation as the sentinel to defend the values of the Constitution and rights of Indians. The courts must, therefore, act within their judicially permissible limitations to uphold the rule of law and harness their power in public interest. It is precisely for this reason that it has been consistently held by this court that in matters of policy the Court will not interfere. When there is a valid law requiring the government to act in a particular manner the Court ought not to, without striking down the law, give any direction which is not in accordance with law. In other words the Court itself is not above the law.

260. In respect of public projects and policies which are initiated by the government, the courts should not become an approval authority. Normally such decisions are taken by the government after due care and consideration. In a democracy, welfare of the people at large, and not merely of a small section of the society, has to be the concern of a responsible government. If a considered policy decision has been taken, which is not in conflict with any law or is not mala fide, it will not be in the public interest to require the Court to go into and investigate those areas which are the function of the executive. For any project which is approved after due deliberation the Court should refrain from being asked to review the decision just because a petitioner, in filing a PIL, alleges that such a decision should not have been taken because an opposite view against the undertaking of the project, which view may have been considered by the government, is possible. When two or more options or views are possible and after

considering them the government takes a policy decision, it is then not the function of the Court to go into matter afresh and, in a way, sit in appeal over such a policy decision.

261. What the petitioner wants the Court to do in this case is precisely that. The facts enumerated hereinabove clearly indicate that the Central Government had taken a decision to construct the dam as that was the only solution available to it for providing water to water-scarce areas. It was known at that time that people will be displaced and will have to be rehabilitated. There is no material to enable this court to come to the conclusion that the decision was mala fide. A hard decision need not necessarily be a bad decision.

262. Furthermore environment concern has not only to be of the area which is going to be submerged and its surrounding area. The impact on environment should be seen in relation to the project as a whole. While an area of land will submerge but the construction of the dam will result in multifold improvement in the environment of the areas where the canal waters will reach. Apart from bringing drinking water within easy reach, the supply of water to Rajasthan will also help in checking the advancement of the Thar Desert. Human habitation will increase there which, in turn, will help in protecting the so far porous border with Pakistan.

263. While considering Gujarat's demand for water, the government had reports that with the construction of a high dam on the river Narmada, water could not only be taken to the scarcity areas of northern Gujarat, Saurashtra and parts of Kachchh but some water could also be supplied to Rajasthan.

264. Conflicting rights had to be considered. If for one set of people, namely those of Gujarat, there was only one solution, namely construction of a dam, the same would have an adverse effect on another set of people whose houses and agricultural land would be submerged in water. It is because of this conflicting interest that considerable time was taken before the project was finally cleared in 1987. Perhaps the need for giving the green signal was that while for the people of Gujarat, there was no other solution but to provide them with water from Narmada, the hardships of oustees from Madhya Pradesh could be mitigated by providing them with alternative lands, sites and compensation. In the governance of the State, such decisions have to be taken where there are conflicting interests. When a decision is taken by the government after due consideration and full application of mind, the Court is not to sit in appeal over such decision.

265. Since long the people of India have been deriving the benefits of river valley projects. At the time of Independence, foodgrain was being imported into India but with the passage of time and the construction of more dams, the position has been reversed. The large-scale river valley projects per se all over the country have made India more than self-sufficient in food. Famines which used to occur have now become a thing of the past. Considering the benefits which have been reaped

by the people all over India with the construction of the dams, the government cannot be faulted with deciding to construct the high dam on the river Narmada with a view to provide water not only to the scarcity areas of Gujarat but also to the small areas of the State of Rajasthan where there has been shortage of water since time immemorial.

266. In the case of projects of national importance where the Union of India and/or more than one State(s) are involved, and the project would benefit a large section of the society, and there is evidence to show that the said project had been contemplated and considered over a period of time at the highest level of the states and the Union of India, and more so when the project is evaluated and approval granted by the Planning Commission, then there should be no occasion for any court carrying out any review of the same or directing its review by any outside or 'independent' agency or body. In a democratic set up, it is for the elected government to decide what project should be undertaken for the benefit of the people. Once such a decision has been taken, unless and until it can be proved or shown that there is a blatant illegality in the undertaking of the project or in its execution, the Court ought not to interfere with the execution of the project.

267. Displacement of people living on the proposed project site and the areas to be submerged is an important issue. Most of the hydrology projects are located in remote and inaccessible areas, where local population is, like in the present case, either illiterate or having marginal means of employment and the per capita income of the families is low. It is a fact that people are displaced by projects from their ancestral homes. Displacement of these people would undoubtedly disconnect them from their past, culture, custom and traditions, but then it becomes necessary to harvest a river for the larger good. A natural river is not only meant for the people close by but it should be for the benefit of those who can make use of it, being away from it or near by. Realising the fact that the displacement of these people would disconnect them from their past, culture, custom and traditions, the moment any village is earmarked for takeover for a dam or any other developmental activity, the project implementing authorities have to implement R&R programmes. The R&R plans are required to be specially drafted and implemented to mitigate problems whatsoever relating to all, whether rich or poor, landowner or encroacher, farmer or tenant, employee or employer, tribal or non-tribal. A properly drafted R&R plan would improve the living standards of displaced persons after displacement. For example residents of villages around Bhakra Nangal Dam, Nagarjun Sagar Dam, Tehri, Bhilai Steel Plant, Bokaro and Bala Iron and Steel Plant and numerous other developmental sites are better off than people living in villages in whose vicinity there was no development project. It is not fair that tribals and the people in undeveloped villages should continue in the same condition without ever enjoying the fruits of science and technology for better health and have a higher quality of lifestyle. Should they not be encouraged to seek greener pastures elsewhere, if they can have access to it, either through their own efforts due to information exchange or due to outside compulsions? It is

with this object in view that the R&R plans which are developed are meant to ensure that those who move must be better off in the new locations at government cost. In the present case, the R&R packages of the States, specially of Gujarat, are such that the living conditions of the oustees will be much better than what they had in their tribal hamlets.

268. Loss of forest because of any activity is undoubtedly harmful. Without going into the question as to whether the loss of forest due to a river valley project because of submergence is negligible, compared to deforestation due to other reasons like the cutting of trees for fuel, it is true that large dams cause submergence leading to a loss of forest areas. But it cannot be ignored and it is important to note that these large dams also cause the conversion of waste land into agricultural land and making the area greener. Large dams can also become instruments in improving the environment, as has been the case in western Rajasthan, which is transformed into a green area because of Indira Gandhi Canal, which draws water from Bhakra Nangal Dam. This project not only allows the farmers to grow crops in deserts but also checks the spread of the Thar desert in the adjoining areas of Punjab and Haryana.

269. Environmental and ecological consideration must, of course, be given due consideration but with proper channellisation of developmental activities ecology and environment can be enhanced. For example, Periyar Dam Reservoir has become an elephant sanctuary with thick green forests all round, while at the same time wiped out famines that used to haunt the district of Madurai in Tamil Nadu before its construction. Similarly Krishnarajasagar Dam which has turned Mandya district which was once covered with shrub forests with wild beasts into a prosperous one with green paddy and sugarcane fields all round.

270. So far a number of such river valley projects have been undertaken in all parts of India. The petitioner has not been able to point out a single instance where the construction of a dam has, on the whole, had an adverse environmental impact. On the contrary the environment has improved. That being so, there is no reason to suspect, with all the experience gained so far, that the position here will be any different and there will not be overall improvement and prosperity. It should not be forgotten that poverty is regarded as one of the causes of degradation of the environment. With improved irrigation systems the people will prosper. The construction of Bhakra Dam is a shining example for all to see how the backward area of erstwhile undivided Punjab has now become the granary of India with an improved environment and what was there before the completion of the Bhakra Nangal project.

271. The Award of the Tribunal is binding on the States concerned. The said Award also envisages the relief and rehabilitation measures which are to be undertaken. If for any reason, any of the State governments involved lag behind in providing adequate relief and rehabilitation then the proper course, for a court to take, would be to direct the Award's implementation and not to stop the execution

of the project. This Court, as a Federal Court of the country specially in a case of inter-State river dispute where an Award had been made, has to ensure that the binding Award is implemented. In this regard, the Court would have the jurisdiction to issue necessary directions to the State which, though bound, chooses not to carry out its obligations under the Award. Just as an ordinary litigant is bound by the decree, similarly a State is bound by the Award. Just as the execution of a decree can be ordered, similarly, the implementation of the Award can be directed. If there is a shortfall in carrying out the R&R measures, a time-bound direction can and should be given in order to ensure the implementation of the Award. Putting the project on hold is no solution. It only encourages a recalcitrant State to flout and not implement the Award with impunity. This certainly cannot be permitted. Nor is it desirable in the national interest that where the fundamental right to life of the people who continue to suffer due to a shortage of water, to such an extent that even the drinking water becomes scarce, non-cooperation of a State results in the stagnation of the project.

272. The clamour for the early completion of the project and for the water to flow in the canal is not by Gujarat but is also raised by Rajasthan.

273. As per Clause 3 of the final decision of the Tribunal published in the Gazette notification of India dated 12 December 1979, the State of Rajasthan has been allocated 0.5 MAF of Narmada water in the national interest from Sardar Sarovar Dam. This was allocated to the State of Rajasthan to utilise the same for irrigation and drinking purposes in the arid and drought-prone areas of Jalore and Barmer districts of Rajasthan situated on the international border with Pakistan, which have no other available source of water.

274. Water is the basic need for the survival of human beings and is part of the right of life and human rights as enshrined in Article 21 of the Constitution of India, and can be served only by providing a source of water where there is none. The Resolution of the UN in 1977 to which India is signatory, during the United Nations Water Conference resolved unanimously *inter alia* as under:

> All people, whatever their stage of development and their social and economic conditions, have the right to access to drinking water in quantum and of a quality equal to their basic needs.

275. Water is being made available by the State of Rajasthan through tankers to the civilians of these areas once in four days during the summer season in quantity, which is just sufficient for their survival. The districts of Barmer and Jalore are part of the Thar Desert and on account of the scarcity of water the desert area is increasing every year. It is a matter of great concern that even after half a century of freedom, water is not available to all citizens even for their basic drinking necessity violating the human right resolution of the UN and Article 21 of the Constitution of India. Water in the rivers of India has great potentiality to change the miserable condition of the arid, drought-prone and border areas of India.

276. The availability of drinking water will benefit about 1.91 lakh people residing in 124 villages in arid and drought-prone border areas of Jalore and Barmer districts of Rajasthan who have no other source of water and are suffering grave hardship.

277. As already seen, the State of Madhya Pradesh is keen for the reduction of the dam's height to 436 ft. Apart from Gujarat and Rajasthan the State of Maharashtra also is not agreeable to this. The only benefit from the project which Rajasthan is to get is its share of hydel power from the project. The lowering of the height from 455 ft to 436 ft will take away this benefit even though 9,399 hectares of its land will be submerged. With the reduction of height to 436 ft not only will there be a loss of power generation but it would also render the generation of power seasonal and not throughout the year.

278. One of the indicators of the living standard of people is the per capita consumption of electricity. There is, however, a perennial shortage of power in India, and, therefore, it is necessary that the generation increases. The world over, countries having rich water and river systems have effectively exploited these for hydel power generation. In India, the share of hydropower in the total power generated was as high as 50 per cent in the year 1962–63 but the share of hydropower started declining rapidly after 1980. There is more reliance now on thermal power projects. But these thermal power projects use fossil fuels, which are not only depleting fast but also contribute towards environmental pollution. Global warming due to the greenhouse effect has become a major cause of concern. One of the various factors responsible for this is the burning of fossil fuel in thermal power plants. There is, therefore, international concern for the reduction of greenhouse gases which is shared by the World Bank resulting in the restriction of sanction of funds for thermal power projects. On the other hand, the hydel power's contribution in the greenhouse effect is negligible and it can be termed ecologically friendly. Not only this but the cost of generation of electricity in hydel projects is significantly less. The Award of the Tribunal has taken all these factors into consideration while determining the height of the dam at 455 ft. Giving the option of generating eco-friendly electricity and substituting it by thermal power may not, therefore, be the best option. Perhaps the setting up of a thermal plant may not displace as many families as a hydel project, but at the same time the pollution caused by the thermal plant and the adverse effect on the neighbourhood could be far greater than the inconvenience caused in shifting and rehabilitating the oustees of a reservoir.

279. There is, and has been in the recent past, protests and agitations not only against hydel projects but also against the setting up of nuclear or thermal power plants. In each case reasons are put forth against the execution of the proposed project either as being dangerous (in the case of nuclear), or causing pollution and ecological degradation (in the case of thermal), or rendering people homeless and causing adverse environment impacts as has been argued in the present case. But

then electricity has to be generated and one or more of these options exercised. What option to exercise, in our constitutional framework, is for the government to decide, keeping various factors in mind. In the present case, a considered decision has been taken and the Award made whereby a high dam having an FRL of 455 ft with the capability of developing hydel power to be constructed. In the facts and circumstances enumerated hereinabove, even if this court could go into the question, the decision so taken cannot be faulted.

Directions

280. While issuing directions and disposing of this case, two conditions have to be kept in mind, (i) the completion of the project at the earliest, and (ii) ensuring compliance with conditions on which clearance of the project was given including completion of relief and rehabilitation work and taking of ameliorative and compensatory measures for environmental protection in compliance with the scheme framed by the government thereby protecting the rights under Article 21 of the Constitution. Keeping these principles, in view, we issue the following directions.

1. Construction of the dam will continue as per the Award of the Tribunal.

2. As the Relief and Rehabilitation Sub-group has cleared the construction up to 90 metres, the same can be undertaken immediately. Further raising of the height will be only *pari passu* with the implementation of the relief and rehabilitation and on the clearance by the Relief and Rehabilitation Sub-group. The Relief and Rehabilitation Sub-group will give clearance of further construction after consulting the three Grievances Redressal Authorities.

3. The Environment Sub-group under the Secretary, MoEF, Government of India will consider and give, at each stage of the construction of the dam, environment clearance before further construction beyond 90 metres can be undertaken.

4. The permission to raise the dam height beyond 90 metres will be given by the NCA, from time to time, after it obtains the above-mentioned clearances from the Relief and Rehabilitation Sub-group and Environment Sub-group.

5. The reports of the GRAs, and of Madhya Pradesh in particular, shows that there is a considerable slackness in the work of the identification of land, acquisition of suitable land and the consequent steps necessary to be taken to rehabilitate the project oustees. We direct the States of Madhya Pradesh, Maharashtra and Gujarat to implement the Award and give relief and rehabilitation to the oustees in terms of the packages offered by them, and these States shall comply with any direction in this regard which is given either by the NCA or the Review Committee or the Grievances Redressal Authorities.

6. Even though there has been substantial compliance with the conditions imposed under the environment clearance, the NCA and the Environment Sub-group will continue to monitor and ensure that all steps are taken not only to protect but to restore and improve the environment.

7. The NCA will within four weeks from today draw up an action plan in relation to further construction and the relief and rehabilitation work to be undertaken. Such an action plan will fix a time frame so as to ensure relief and rehabilitation *pari passu* with the increase in the height of the dam. Each State shall abide by the terms of the action plan so prepared by the NCA and in the event of any dispute or difficulty arising, representation may be made to the Review Committee. However each State shall be bound to comply with the directions of the NCA with regard to the acquisition of land for the purpose of relief and rehabilitation to the extent and within the period specified by the NCA.

8. The Review Committee shall meet whenever required to do so in the event of there being any unresolved dispute on an issue which is before the NCA. In any event the Review Committee shall meet at least once in three months so as to oversee the progress of construction of the dam and implementation of the R&R programmes. If for any reason serious differences in implementation of the Award arise, and the same cannot be resolved in the Review Committee, the Committee may refer the same to the Prime Minister whose decision, in respect thereof, shall be final and binding on all concerned.

9. The Grievances Redressal Authorities will be at liberty, in case the need arises, to issue appropriate directions to the respective States for due implementation of the R&R programmes, and in case of non-implementation of its directions, the GRAs will be at liberty to approach the Review Committee for appropriate orders.

10. Every endeavour shall be made to see that the project is completed as expeditiously as possible.

281. This and connected petitions are disposed off in the aforesaid terms.

Narmada Bachao Andolan v. Union of India, Minority Judgment of Justice S.P. Bharucha, October 2000

Source: *Narmada Bachao Andolan v. Union of India*, Writ Petition (Civil) No. 319 of 1994, Supreme Court of India, Judgment of 18 October 2000.

1. Bharucha, J. I have read the judgment proposed to be delivered by my learned brother, the Hon'ble Mr Justice B.N. Kirpal. Respectfully, I regret my inability to agree therewith.

2. I do not set out the facts here: they are detailed in brother Kirpal's judgment.

3. I take the view that the Sardar Sarovar Project does not require to be re-examined, having regard to its cost effectiveness or otherwise, and that the seismicity aspect of the project has been sufficiently examined and no further consideration thereof is called for. I do not accept the submission on behalf of the petitioner that those ousted by reason of the canals emanating from the reservoir in the project must have the same relief and rehabilitation benefits as those ousted on account of the reservoir itself; this is for the reason that the two fall in different classes.

4. Having said this, I turn to the aspect of the environmental clearance of the project. The Planning Commission accorded provisional sanction to the project subject to the environmental clearance thereof being obtained. At the relevant time, the responsibility for giving environmental clearance lay with the Department of Environment in the Ministry of Environment and Forests of the Union Government. The department had in January 1985 issued Guidelines for Environmental Impact Assessment of River Valley Projects. The preface thereof stated that environmental appraisal was an important responsibility assigned to the department. It involved the evaluation of the environmental implications of, and the incorporation of necessary safeguards in, activities having a bearing on environmental quality. While river valley projects were a basic necessity to a country whose economy was largely based on agriculture, over the years the realisation had dawned that river valley projects had their due quota of positive and adverse impacts which had to be carefully assessed and balanced for achieving sustained benefits. Therefore, it had been decided in the late 1970s that all river valley projects should be subjected to a rigorous assessment of their environmental impact so that necessary mitigative measures could be duly incorporated therein at the inception stage. The Guidelines set out the procedure to be adopted for carrying out environmental impact assessments. In the chapter headed Relevance of Environmental Aspects for River Valley Development Projects, the Guidelines stated,

> Concern for environmental pollution is rather a recent phenomenon which has been triggered mainly by the backlash effect of accelerated industrial growth in the developed countries. The two major criteria – the project should maximise economic returns and it should be technically feasible – are no longer considered adequate to decide the desirability or even the viability of the project. It is now widely recognised that the development effort may frequently produce not only sought for benefits, but other – often unanticipated – undesirable consequences as well which may nullify the socio-economic benefits for which the project is designed.

After reference to the strong feelings that were often expressed in favour of measures that would provide the provision of adequate food and shelter to the millions, the Guidelines stated:

> Such strong feelings are easy to understand in the context of the prevailing economic stagnation. It does not, however, follow that the arguments advanced are valid. The basic flaw in these arguments is that they presume incompatibility between environmental conservation and the development effort.

Apart from some selected cases where the uniqueness of the natural resources, like wildlife, flora and genetic pool, which demanded exclusive earmarking of a given region for their specific use, the majority of cases did not call for a choice between development projects and preservation of the natural environment; but in all cases there was great need to consider the environmental aspects along with other feasibility considerations. It was imperative to analyse whether the adoption of environmental measures was going to result in any short or long-term social or economic benefits. A careful study of the direct costs involved, which would be caused by the absence of environmental mitigative measures on river valley projects, was an eye opener. These included effects on health, plant genetic resources, aquatic resources, water-logging and salinity of irrigated soils, deforestation and soil conservation. During the planning and feasibility assessment stages, several factors had to be taken into account, including short and long-term impact on population and human settlements in the inundated and watershed areas, impact on flora and fauna (wildlife) in the vicinity, impact on wildlife, including birds, impact on national parks and sanctuaries, on sites and monuments of historical, cultural and religious significance and on forests, agriculture, fisheries and recreation and tourism. Requisite data for impact assessment was not readily available, this being relatively a new discipline, and it had to be generated through such field surveys as:

- pre-impoundment census of flora and fauna, particularly the rare and endangered species, in submergence areas;
- census of animal population and available grazing areas;
- land-use pattern in the area with details of extent and type of forest;
- pre-impoundment survey of fish habitat and nutrients levels;
- groundwater level, its quality, and existing water use pattern;
- mineral resources, including injurious minerals, in the impoundment;
- living conditions of affected tribals/aboriginals, etc.

The cost of proposed remedial and mitigative measures to protect the environment had to be included in the project cost. Mitigative measures included, among other things, compensatory afforestation. Only when the incorporation of environmental aspects in the project planning was made a part and parcel of all river valley projects would there be hope to protect and preserve

> our natural environment and fulfil the objective of rapid economic development on the sustained basis while safeguarding the natural resources including the air, water, land, flora and fauna for the benefit of present and future generations.

The necessary data that was required to be collected for impact assessment was set out in the Guidelines. A chart of the impact assessment procedure was also contained in the Guidelines.

5. It appears that, though it ought rightly to have been taken by the Ministry of Environment and Forests, the decision whether or not to accord environmental clearance to the project was left to the Prime Minister.

6. A Note was prepared by the Ministry of Water Resources in or about October 1986 on the environmental aspects of the Sardar Sarovar and the Narmada Sagar Multi Purpose Projects. It stated that a decision on the clearance of these projects from the environmental angle and under the Forest Conservation Act, 1980 had become a matter of urgency. Delays had occurred which had necessitated a recasting of the schedule. The MoEF had been doing its best to expedite the process of examination and clearance 'but have been finding the material submitted inadequate and unsatisfactory'. While the State governments had done their best to meet the requirements, 'some of the information and action will necessarily take time and will have to proceed *pari passu* with the implementation of the project, which in any case will take a decade or more to complete'. The Note stated that the Ministry of Water Resources shared the concerns and anxieties of the MoEF, as also the sense of urgency of the Governments of Gujarat and Madhya Pradesh, who felt that it was urgently necessary to take a decision in regard to the clearance. Under the sub-heading, 'Should the projects be taken up at all?', the Note stated that the abandonment of the projects would mean the abandonment of the generation of 2,450 MW of power and of the possibilities of economic development which that quantum of power would bring, as also increased agricultural production resulting from the creation of an irrigation potential of 2.041 million hectares. No effective alternatives to the two projects were available. Reference to the adverse environmental impact of the projects carried the implicit assumption that if the projects were not sanctioned the status quo would remain and there would be no deterioration of the environment. Such an assumption was not warranted. Despite the submergence of land and displacement of people and livestock, there was no case for the abandonment of the projects. What needed to be done was to take appropriate and adequate counter measures to off-set the environmental impact of the projects. In respect of the flora and fauna, it said 'Quantified data not yet available'. In respect of the possibility of soil erosion from the catchment leading to excessive siltation of the reservoir it said, 'Extent of critically degraded area needing treatment to be identified'. Specifically in respect of the Sardar Sarovar Project, the Note said that for the area to be submerged in Maharashtra, the Maharashtra Government had proposed compensatory afforestation over an area of 6,490 hectares and that Madhya Pradesh was preparing an action plan to reforest about 5,500 hectares of the denuded forest in the impact area. In respect of fauna, the Note said that the NSP authorities had commissioned a wildlife census of the areas by the Zoological Survey of India and were negotiating terms with the Indian Institute of Wildlife Management,

Dehradun, for carrying out detailed wildlife studies for relocation purposes. They proposed to undertake all necessary steps to minimise the adverse impact of the project on wildlife. Gujarat and Maharashtra were also taking similar action with the help of specialised agencies. In respect of the projects' flora, the Note said that the first preliminary survey in the area by the Botanical Survey of India was started in December 1985 and it was estimated that the survey would take two to three years to be completed. In respect of catchment area treatment, the Note said that field surveys were likely to be started shortly. The project authorities had identified three representative pilot project areas. The biological and engineering measures to be adopted in the treatment of the balance of the catchment area would be designed on the basis of the experience to be gained from these pilot projects. Under the sub-heading, 'What still remains to be done', the Note stated,

> *While some plans have been made, studies undertaken and action initiated, it will be clear from the preceding paragraphs that much still remains to be done. Indeed, it is the view of the Ministry of Environment, Forests and Wildlife that what has been done so far whether by way of action or by way of studies does not amount to much, and that many matters are as yet in the early and preliminary stages* (emphasis supplied).

What was then set out was an enumeration of what remained to be done. *The survey of flora*, to assess if there were any rare or threatened plant species, had been assigned to the Botanical Survey of India, which was *expected to be completed in a period of two years. The Wildlife survey* undertaken by the Zoological Survey of India *was also likely to take two years*. The Indian Institute of Wildlife Management, Dehradun was to consider and assess the impact on wildlife of the destruction of their habitat, and to prepare a project report for their re-location. *After all these reports became available, a master plan had to be prepared*. Field surveys for the *identification of the critically eroding areas* was necessary and *would take three years. The results from pilot studies would be available only after three years*. Then, under the sub-heading, 'Options in regard to the clearance of the projects', the Note stated:

> *There are two options*:
>
> 1. As a number of studies, census, field surveys, mapping of areas, etc., are likely to take between 2 and 3 years, one possibility is that all these should be completed; detailed operational plans for catchment treatment, compensatory afforestation, rehabilitation and resettlement of affected population, and remedial or relocation measures for planned species, wildlife, etc., formulated; the responsibility for their implementation clearly identified; and then the projects should be given a clearance from the environmental and forest angles. This will mean a postponement of the clearance of projects by about 3 years.
>
> 2. The other option is that the projects should be given the necessary clearance now, with clear conditions and stipulations in regard to the actions to be taken on the various environmental aspects and appropriate monitoring arrangements to ensure that the actions are taken in time-bound manner.

> 13.2 The arguments against a postponement of clearance by three years are very strong (emphasis supplied)

The postponement of the decision at this stage seemed, to the writers of the Note, 'scarcely conceivable'. A postponement would lead to substantial increases in project costs and the benefits expected from the projects would be delayed. Also the work that had already been done would be rendered infructuous. The deferment of clearance by three years would put the organisational set-up that had been built up into a state of uncertainty, retard the momentum that had been gathered, and sap the organisational morale and motivation. The Note added,

> Finally, the numerous studies, surveys, data collection exercises, plans for remedial measures, etc., which have been enumerated earlier would involve time, money and organisational commitment. With the project decision postponed for three years, and with no assurance that at the end of that period, the decision will be positive, it is difficult to believe that all these studies, surveys and plans relating to the environmental aspects will be pursued with energy and enthusiasm, and the necessary resources devoted to them. In other words, the postponement of the decision in the interest of collecting the information relating to the environment aspects and completing the formulation of the necessary operational plans, may in fact prove to be self defeating exercise. On the other hand, if the project decisions are taken now, subject to firm conditions and stipulations regarding the environmental aspects, there is greater likelihood of these conditions being met (…). A possible argument against the immediate clearance of the projects could be that once the projects are cleared, the management would concentrate on the engineering and construction aspects and would not pay adequate attention to the environmental and human aspects. There seems to be no need for such apprehensions. It should be entirely possible to give a conditional clearance and ensure that the conditions are properly met through a process of clear assignment of responsibility and frequent monitoring (...). Moreover, even assuming that the postponement of a decision by three years will improve the availability of detailed information and the state of preparedness on environmental matters, there can be no greater assurance at that stage than there is now regarding the whole-hearted and effective implementation of the remedial and ameliorative measures. We would still have to depend on proper monitoring.

In conclusion, the Note urged that clearance from the environmental angle and under the Forest Conservation Act, 1980 be given immediately, subject to conditions and stipulations relating to the various environmental and related aspects outlined in the Note.

7. Another Note was prepared by the Ministry of Water Resources and forwarded to the Additional Secretary to the Prime Minister on 20 November 1986. Insofar as catchment area treatment was concerned, it concluded that it was certain that the catchment area treatment programme could not be realistically formulated and assessed for at least another three years. Therefore, it was premature to comment on the efficacy or otherwise of the catchment area treatment programme which was still to be formulated. The action programme for command area development was yet to be made available. The lining of the canal network and the digging of

tubewells in the command could not be considered to be adequate. A lot of fieldwork and planning was needed to be done to arrive at a workable and effective command area development programme. As to compensatory afforestation, the land for the same was yet to be identified and procured before it could be evaluated for the purpose. In regard to the loss of flora and fauna, the following studies were considered absolutely essential to determine the adequacy or otherwise of the left over habitat to sustain wildlife:

> 'A wildlife census of the area' (ZSI will take at least 2–3 years to complete the survey);
>
> 1. Preparation of a master plan showing all protected areas, national parks, wildlife reserves, reserve and protected forests, etc. on which should be superimposed the area to be taken up for various reservoirs, roads, canals, settlement colonies, etc.
>
> 2. Study of the carrying capacity of the surrounding areas where the wildlife from the submergence area will disperse.

In the circumstances, it was not considered possible to assess the impact of the loss of habitat on the wildlife and the overall loss of biological diversity. The absence and inadequacy of data on the following environmental aspects persisted:

1. rehabilitation,
2. catchment area treatment,
3. command area development,
4. compensatory afforestation, and
5. flora and fauna.

Considering the magnitude of rehabilitation, involving a large percentage of tribals, loss of extensive forest area rich in biological diversity, enormous environmental cost of the project and *considering the fact that the basic data on vital aspects was still not available 'there could be but one conclusion, that the project(s) are not ready for approval*'. 'There were two options in regard to the clearance. As a number of studies, censuses, field surveys, mapping of areas, etc. was likely to take between two and three years, one possibility was that all these should be completed; detailed operational plans for catchment treatment, compensatory afforestation, rehabilitation and resettlement of affected population and remedial or relocation measures for plant species, wildlife, etc. formulated; the responsibility for their implementation clearly identified; and then the projects should be given clearance from the environmental and forest angles. This would mean postponement of the clearance of projects by about three years'. The other option was that the project should be given the necessary clearance with conditions and stipulations in regard to the actions to be taken on the various environmental aspects with appropriate monetary arrangements. The Note recommended the latter option (emphasis supplied).

8. On 19 December 1986, the Ministry of Environment and Forests sent to the Secretary to the Prime Minister a Note on the environmental aspects of the Narmada Sagar and the Sardar Sarovar Projects.* The Note stated that it covered the major environmental issues which included the rehabilitation of the affected population, catchment area treatment, command area development, compensatory afforestation, and the loss of flora and fauna. It explained the then status of each of these aspects in terms of availability of data and plans and the readiness to execute them. It said that other components of the environmental aspects like the higher incidence of water borne diseases and loss of mineral reserves were important but were not dealt with in detail in the note. It stated that in respect of catchment area treatment, the requirement was of demarcation of critically degraded areas on the basis of aerial photographs, satellite imagery and ground checks; creation of a chain of nurseries of suitable species for biological treatment of the catchment area; and preparation of phased action programme for biological and engineering treatment of the degraded catchment area. Considering that catchment area treatment on an intensive scale was imperative, both to reduce silt load and to maintain ecological balance, and keeping in view the fact that the interpretation of the aerial photographs and satellite imagery would take at least one year for completion, to be followed by ground truth checks; the detailed land and soil surveys would take three years to be completed; the geo-morphological studies to suggest the engineering and biological treatment for the eroded areas were still to be taken up and the chain of nurseries needed to provide the necessary saplings in adequate quantity along with manpower and other infrastructure requirements were still to be mobilised, it was 'reasonable to conclude that the catchment area treatment programme can be realistically formulated only after three years when these data become available'. Command area development was to achieve the prevention of waterlogging and salinity, the optimisation of water utilisation and the maintenance of water quality. A detailed survey of the command area was required on priority to prepare a package of the nature and quantity of development and drainage and on farm works to fully utilise the irrigation potential. An action programme was yet to be detailed. The Ministry of Water Resources was preparing an evaluation report covering the extent of likely water-logging and salinity problems and the effectiveness of measures proposed or likely to be proposed to combat these problems 'as per the action programme to be formulated'. In so far as compensatory afforestation was concerned, the project authorities had not been able to identify non-forest land for compensatory afforestation and had proposed to undertake afforestation on double the extent for degraded forest land, which proposal was fairly detailed and seemed satisfactory. In the matter of loss of flora and fauna the Note stated that

> [t]he forest area specially affected by the NSP represents areas harbouring rich heritage of genetic resources as well as wildlife. The preliminary study carried out by the Environmental Planning and Coordination Organisation, Bhopal, as well as

* The full text of this document can be found at http://www.ielrc.org/content/c8601.pdf.

the observations made by the World Bank, clearly underline the need for preparing a master plan showing not just the present status but also the likely scenario after the project is implemented. The prime concern is to ascertain the loss of biological diversity and whether the wildlife will be able to sustain itself after the destruction of its habitat, specially on the Southern side which is surrounded by agriculture fields. The following studies were considered absolutely essential to determine the loss of flora and the adequacy or otherwise of the left-over habitat to sustain the wildlife:

- A wildlife census of the area (ZSI will take at least 2–3 years to complete the survey).
- Preparation of a master plan showing all protected areas i.e. national parks, wildlife reserves, reserve and protected forests, etc. on which should be superimposed the areas that cannot be taken up for various reservoirs, roads, canals, settlement colonies, etc.
- Study of the carrying capacity of the surrounding areas where the wildlife from the submergence area will disperse.

These studies are considered specially important in the case of NSP. The work initiated by the Botanical Survey of India and ZSI at the request of the project authorities will be completed only by 1989. The other studies have not been initiated. Under the circumstances, it is not possible to assess the impact of the loss of habitat on the wildlife and the overall loss of biological diversity and genetic reserves.

Even if one were to assume that the forests to be destroyed do not contain genetic resources, which in any case cannot be valued, the simple loss of these forests would have an environmental cost estimated at several thousand crores of rupees as per norms developed by the FRL. The environmental cost is thus colossal.

The Note concluded:

1. Taking note of the fact that the project formulation has been in progress for more than three decades and the active interaction of the project authorities with the Department of Environment has been going on for almost three years, the absence and inadequacy of data on important environmental aspects still persists.
2. In an objective sense, the NSP is not ready for clearance from the environmental angle. Even though SSP is in a fairly advanced stage of preparedness, it is neither desirable nor recommended that the SSP should be given approval in isolation on technical and other grounds.
3. The state of readiness in the case of NSP is such that it gives just an outline of the intention plan. The fact that this intention plan will be converted to an action plan and thereby effectively implemented has to be taken on trust. In case of Sardar Sarovar Project (SSP), readiness to execute is reasonably good except on the issue of rehabilitation of oustees, specially from MP and Maharashtra.
4. Holding up of the projects even for the next few months is not likely to improve the level of preparedness on most of the environmental aspects, specially in the case of NSP. In the meanwhile, further studies will not

perhaps pick up speed and thus at no time will the requisite information be fully available.

5. A large amount of money has already been invested in SSP which is critically linked – on technical and operational aspects – to NSP. However, *it may not be too late even now to modify some of the parameters of NSP and SSP to minimise environmental damage while at the same time ensuring optimal utilisation of water resources.*

(…)

The choice is difficult but a choice has to be made (emphasis supplied).

9. A Note was prepared on 15 January 1987 in the Prime Minister's Office. It noted that the main issues on environmental concerns were the rehabilitation of the affected population, compensatory afforestation, treatment of the catchment area and command area development, pertaining, particularly, to drainage, waterlogging and salinity. The Department of Environment and Forests had raised the point that the rehabilitation plan was not ready, land had not been surveyed, areas of land use capability and water availability had not been identified and the land being suggested for rehabilitation, prima facie, appeared to be infertile. Detailed meetings with the State governments revealed that they were seriously undertaking surveys, land identification and preparation of a rehabilitation plan, of which the first phase was more or less ready. The catchment area treatment preparation would take time. A compensatory afforestation programme could be chalked out without difficulty. The issue was whether detailed plans should be made fully ready before giving environmental clearance or whether there could be a conditional clearance so that the project could start. The Secretary to the Prime Minister had discussed the matter with the Secretary, Water Resources and the Secretary, Environment and Forests and it had been agreed that clearance might be given on the following conditions:

> Preparation in due time of detailed and satisfactory plans for rehabilitation, catchment area treatment, compensatory afforestation and command area development.
>
> Setting up of the Narmada Management Authority with adequate powers and teeth to ensure that environmental management plans are implemented *pari passu* with engineering and other works.

Below the aforesaid Note, the Secretary to the Prime Minister sought his approval to conditional clearance of the project from the environmental angle. The project, she said, had been pending clearance for seven years and the Chief Ministers of Gujarat and Madhya Pradesh were keenly awaiting it. The Chief Minister of Gujarat had requested a 'green signal' before 20 January 1987.

10. On 19 January 1987 the Prime Minister made a handwritten endorsement on the aforesaid Note, 'Perhaps this is a good time to try for a River Valley Authority. Discuss'. But it appears that a River Valley Authority was not found feasible and

the sanction to the project from the environmental angle was issued by the Ministry of Environment and Forests on 24 June 1987.

11. The environmental sanction to the project reads thus:*

12. Even in 1987, when the environmental clearance to the project was given, it had been found necessary by the Union of India to rigorously assess the environmental impact of river valley projects. This was to determine whether the uniqueness of the natural resources, like wildlife, flora and fauna and the genetic pool in the region, demanded its exclusive earmarking for that purpose, in which event the river valley project would not be accorded clearance. Even otherwise, it was imperative to consider the project's environmental aspects, such as its effect on health, plant genetic resources, aquatic resources, water-logging and salinity of irrigated soils, deforestation and soil conservation. Its short and long-term impact on population, on flora and fauna, on wildlife, on national parks and sanctuaries, on historical, cultural and religious monuments, on forests, agriculture, fisheries and recreation and tourism had to be taken into account. Field surveys were necessary for generating the requisite data for the impact assessment. The cost of the proposed remedial and mitigated measures had to be included in the project cost. The necessary data that was required to be collected for the purposes of the assessment of a project's environmental impact was set out in the Guidelines for the purpose issued by the Ministry of Environment and Forests of the Union Government (which have been referred to above).

13. The contemporaneous notes prepared by the Ministry of Water Resources and the Ministry of Environment and Forests, also referred to above, leave no manner of doubt that the requisite data for assessment of the environmental impact of the project was not available when the environmental clearance thereof was granted. In the words of one of the Notes, 'While some plans have been made, studies undertaken and action initiated, it will be clear from the preceding paragraphs that much still remains to be done. Indeed it is the view of the Ministry of Environment, Forests and Wildlife that what has been done so far, whether by way of action or by way of studies, does not amount to much and that many matters are yet in the early and preliminary stages'. The Notes make it clear that the studies, censuses, mapping of areas and field surveys for the collection of data for assessment of the environmental impact of the project were likely to take a further 2 to 3 years. An environmental clearance based on next to no data in regard to the environmental impact of the project was contrary to the terms of the then policy of the Union of India in regard to environmental clearances and, therefore, no clearance at all.

14. The environmental clearance of 24 June 1987 stated that details had been sought from the project authorities in respect of the rehabilitation master plan, phased catchment area treatment scheme, compensatory afforestation plan,

---

* Text omitted, see page 77 for the text of the Office Memorandum.

command area development, survey of flora and fauna, carrying capacity of surrounding area, seismicity and health aspects; field surveys had yet to be completed and complete details had been assured by 1989. Clearly, therefore, the necessary particulars in regard to the environmental impact of the project, as required by the Guidelines, were not available when the environmental clearance was given, and it, therefore, could not have been given.

15. The conditions upon which the environmental clearance was given were that detailed surveys and studies would be carried out and the Narmada Control Authority, whose terms of reference had been amplified, would ensure that 'environmental safeguard measures' were planned and implemented *pari passu* with the progress of work on the project. No further assessment of the environmental impact of the project was contemplated by the environmental clearance, nor, indeed, was it ever carried out.

16. What the environmental safeguard measures the Narmada Control Authority was to ensure were, and what their cost would be, was not known when the environmental clearance was given. There was, therefore, no way in which this cost could be included in the cost of the project, which was a requirement of the Guidelines.

17. While environmental safeguard measures were to be planned and implemented *pari passu* with the progress of the work on the project, the catchment area treatment programme and the rehabilitation plans were required to be 'so drawn as to be completed ahead of reservoir filling'. This condition clearly required that before any water was impounded in the reservoir the catchment area treatment programme was not only to be drawn up but also to be completed; so also the rehabilitation plans. If, as the project authorities interpreted this clause, only the drawing of the catchment area treatment programme and the rehabilitation plans were to be completed ahead of reservoir filling, the clause would have read: 'The catchment area treatment programme and the rehabilitation plans shall be drawn ahead of reservoir filling'. What the clause as drawn required was that the catchment area treatment programme and the rehabilitation plans should be drawn in such a manner that the catchment area treatment and rehabilitation works would be completed ahead of impoundment in the reservoir. This, plainly, was intended to offset, so far as was possible in the circumstances, the adverse effect of the impoundment of water in the reservoir upon the catchment and those who were required to be settled elsewhere. In fact, the impoundment began much before.

18. Learned counsel for the Union of India submitted that most of the necessary surveys and studies had been carried out in regard to the environmental impact of the project before the environmental clearance was given, and he invited our attention to what had been done. The short answer to the submission on behalf of the Union of India is that the two concerned ministries of the Union of India thought otherwise at the relevant time. To quote the Note of one ministry again: 'While some plans have been made, studies undertaken and action initiated, it

would be clear from the preceding paragraph that much still remains to be done. Indeed it is the view of the Ministry of Environment, Forests and Wildlife that what has been done so far, whether by way of action or by way of studies, does not amount to much and that many matters are yet in the early and preliminary stages'.

19. The fact that the environmental clearance was given by the Prime Minister and not by the Ministry of Environment and Forests, as it would ordinarily have been done, makes no difference at all. Under its own policy, as indicated by the Guidelines, the Union of India was bound to give environmental clearance only after a) all the necessary data in respect of the environmental impact of the project had been collected and assessed; b) the assessment showed that the project could proceed; and c) the environmental safeguard measures, and their cost, had been worked out.

20. An adverse impact on the environment can have disastrous consequences for this generation and generations to come. This Court has in its judgments on Article 21 of the Constitution recognised this. This Court cannot place its seal of approval on so vast an undertaking as the project without first ensuring that those best fitted to do so have had the opportunity of gathering all necessary data on the environmental impact of the project and of assessing it. They must then decide if environmental clearance to the project can be given, and, if it can, what environmental safeguard measures have to be adopted, and their cost. While surveys and studies on the environmental aspects of the project have been carried out subsequent to the environmental clearance, they are not, due to what are euphemistically called 'slippages', complete. Those who now examine whether environmental clearance to the project should be given must be free to commission or carry out such surveys and studies and the like as they deem necessary. They must also, of course, consider such surveys and studies as have already been carried out. Given that the construction of the dam and other work on the project has already commenced, this factor must play a part in their deciding whether or not environmental clearance should be accorded. Until environmental clearance to the project is accorded by them, further construction work on the dam shall cease.

21. The Union of India has issued a notification on 27 January 1994 called the Environmental Impact Assessment Notification 1994 (and amended on 4 May 1994). Its terms are not applicable to the present proceedings, but its provisions are helpful in so far as they prescribe who is to assess the environmental impact assessment reports and environmental management plans that are submitted by applicants for new projects, including hydro-electric projects. The notification says:

> The reports submitted with the application shall be evaluated and assessed by the Impact Assessment Agency, and if deemed necessary, it may consult a Committee of Experts, having composition as specified in Schedule III of this notification. The Impact Assessment Agency (IAA) would be the Union Ministry of Environment and Forests. The Committee of Experts mentioned above shall be

constituted by the IAA or such other body under the Central Government authorised by the IAA in this regard.

Schedule III of the notification reads thus:

1. Composition of the Expert Committees for Environmental Impact Assessment

    i. The Committees will consist of experts in the following disciplines:

    ii. Eco-system management,

    iii. Air/water pollution control,

    iv. Water resource management,

    v. Flora/fauna conservation and management,

    vi. Land use planning,

    vii. Social sciences/rehabilitation,

    viii. Project appraisal,

    ix. Ecology,

    x. Environmental health,

    xi. Subject area specialists,

    xii. Representatives of NGOs/persons concerned with environmental issues.

2. The Chairman will be an outstanding and experienced ecologist or environmentalist or technical professional with wide managerial experience.

3. The representative of IAA will act as Member-Secretary.

4. Chairman and members will serve in their individual capacities, except those specifically nominated as representatives.

5. The membership of a Committee shall not exceed 15.

The Environmental Impact Agency of the Union Ministry of Environment and Forests shall now appoint a committee of experts composed of experts in the fields mentioned in Schedule III of the notification and that committee of experts shall assess the environmental impact of the project as stated above.

When the writ petition was heard at the admission stage, this court was most concerned about the distressing state of the relief to and rehabilitation of those ousted on account of the project. The proper implementation of relief and rehabilitation measures was the aim of the Court at that time, but it was not contemplated that the other issues in the writ petition would not to be considered at the stage of its final hearing.

23. The many interim orders that this court made in the years in which this writ petition was pending show how very little had been done in regard to the relief and rehabilitation of those ousted. It is by reason of the interim orders, and, in fairness, the co-operation and assistance of the learned counsel who appeared for the States,

that much that was wrong has now been redressed. The States have also been persuaded to set up GRAs and it will be the responsibility of these authorities to ensure that those ousted by reason of the project are given relief and rehabilitation in due course.

24. The States are lagging behind in the matter of the identification and acquisition of land upon which the oustees are to be resettled. Having regard to the experience of the past, only the GRAs can be trusted by this court to ensure that the States are in possession of vacant lands suitable for the rehabilitation of the oustees. During the time that it takes to assess the environmental impact of the project, the States must take steps to obtain, by acquisition or otherwise, vacant possession of suitable lands upon which the oustees can be rehabilitated. When the project obtains environmental clearance, assuming that it does, each of the GRAs of the States of Gujarat, Madhya Pradesh and Maharashtra must certify, after inspection, before work on the further construction of the dam can begin, that all those ousted by reason of the increase in the height of the dam by 5 m from its present level have already been satisfactorily rehabilitated and also that suitable vacant land for rehabilitating all those who will be ousted by the increase in the height of dam by another 5 m is already in the possession of the respective States; and this process must be repeated for every successive proposed 5 m increase in the dam height.

25. Only by ensuring that relief and rehabilitation is so supervised by the GRAs can this court be assured that the oustees will get their due.

26. It is necessary to provide for the contingency that, for one or other reason, the work on the project, now or at any time in future, does not proceed and the project is not completed. Should that happen, all oustees who have been rehabilitated must have the option to continue to reside where they have been rehabilitated or to return to where they were ousted from, provided such place remains habitable, and they must not be made at all liable in monetary or other terms on this account.

27. When the writ petition was filed the process of relief and rehabilitation, such as it was, was going on. The writ petitioners were not guilty of laches in that regard. In the writ petition they raised other issues, one among them being related to the environmental clearance of the project. Given what has been held in respect of environmental clearance, when the public interest is so demonstrably involved, it would be against the public interest to decline relief only on the ground that the Court was approached belatedly.

28. I should not be deemed to have agreed to anything stated in Brother Kirpal's judgment for the reason that I have not traversed it in the course of what I have stated.

29. In the premises,

1. The Environment Impact Agency of the Ministry of Environment and Forests of the Union of India shall forthwith appoint a Committee of Experts in the

fields mentioned in Schedule III of the notification dated 27 January 1994, called the Environmental Impact Assessment Notification, 1994.

2. The Committee of Experts shall gather all necessary data on the environmental impact of the project. They shall be free to commission or carry out such surveys and studies and the like as they deem necessary. They shall also consider such surveys and studies as have already been carried out.

3. Upon such data, the Committee of Experts shall assess the environmental impact of the project and decide if environmental clearance to the project can be given and, if it can, what environmental safeguard measures must be adopted, and their cost.

4. In so doing, the Committee of Experts shall take into consideration the fact that the construction of the dam and other work on the project has already commenced.

5. Until environmental clearance to the project is accorded by the Committee of Experts as aforesaid, further construction work on the dam shall cease.

6. The Grievance Redressal Authorities of the States of Gujarat, Madhya Pradesh and Maharashtra shall ensure that those ousted by reason of the project are given relief and rehabilitation in due measure.

7. When the project obtains environmental clearance, assuming that it does, each of the Grievance Redressal Authorities of the States of Gujarat, Madhya Pradesh and Maharashtra shall, after inspection, certify, before work on the further construction of the dam can begin, that all those ousted by reason of the increase in the height of the dam by 5 metres from its present level have already been satisfactorily rehabilitated and also that suitable vacant land for rehabilitating all those who will be ousted by the increase in the height of the dam by another 5 metres is already in the possession of the respective States.

8. This process shall be repeated for every successive proposed 5-metre increase in the dam height.

9. If for any reason the work on the project, now or at any time in the future, cannot proceed and the project is not completed, all oustees who have been rehabilitated shall have the option to continue to reside where they have been rehabilitated or to return to where they were ousted from, provided such place remains habitable, and they shall not be made at all liable in monetary or other terms on this account.

30. The writ petition is allowed in the aforementioned terms. The connected matters are disposed of in the same terms.

31. No order as to costs.

Narmada Bachao Andolan v. Union of India, Review Petition, November 2000

Source: Review Petition under Order 40 of the Supreme Court Rules Read with Order 47 of CPC, *Narmada Bachao Andolan v. Union of India and Others*, Writ Petition (Civil) No. 319 of 1994, Supreme Court of India, 17 November 2000.

This Review Petition is being filed against the judgment and order of this Hon'ble Court dated 18 October 2000 by which this Hon'ble Court disposed of the above Writ Petition of the petitioners with certain directions, findings and observations (Annexure A). The petitioners are seeking this review on *inter alia* the following grounds which are set out below. It is the petitioner's submission that a review of those parts of the judgment which are assailed herein below are required since those parts of the judgment not only contain errors which are apparent on the face of the record but would also lead to violations of the fundamental rights of the oustees from the Sardar Sarovar Project guaranteed by Article 14 and Article 21 of the Constitution.

'Order' and 'Judgment' refers to Majority Order and Judgment respectively, unless specifically qualified otherwise.

1. Order Will Lead to Violation of the Tribunal Award

Violation of Clause XI, Sub-clause IV(2)(iv) Read with Sub-clause IV(6)(i)

The order/permission for immediate construction of the project up to 90 m needs to be reviewed as construction of 90 m at this stage can take place only in clear violation of the Tribunal order. The order of the Tribunal states, *inter alia*:

> Clause XI, Sub-clause IV(2)(iv): Gujarat shall acquire and make available a year in advance of the submergence before each successive stage, irrigable lands and house sites for rehabilitation of the oustee families from Madhya Pradesh and Maharashtra who are willing to migrate to Gujarat. Gujarat shall in the first instance offer to rehabilitate the oustees in its own territory.
>
> Clause XI, Sub-clause IV(6)(i): In the event of Gujarat being unable to resettle the oustees or the oustees being unwilling to occupy the area offered by Gujarat, Madhya Pradesh and Maharashtra shall make such provisions for rehabilitation, civic amenities, etc. on the lines mentioned in Clauses IV(1) to (4) above.

Together the above two clauses clearly require that lands in all the states for those who will be submerged by construction up to 90 m in July 2001 had to be available in July 2000 itself. It is the admitted position of GoMP that no land is available for PAFs up to 90 m who want to settle in MP.

The detailed position regarding land in Madhya Pradesh is as follows.

The Government of MP claims it has identified lands to be allotted to PAFs. GoMP has identified land through the following ways (Affidavit of GoMP of April 2000, page1):

i. By reducing grazing land areas of villages, and land found suitable for development 'subject to development process',

ii. Forest land,

iii. Land bought from prospective sellers.

None of the lands to be purchased have actually been purchased or even the process for the same initiated. Neither is there any assurance of quality of these lands. Further, of their own admission, most of the lands are mostly pasture lands which are absolutely uncultivable. GoMP has stated in their affidavit dated 11 April 2000, which has been placed before the Supreme Court, that:

> All the aforesaid parcels of land whether pertaining to grazing land, or under encroachment, or under forest will require to be bought under development process, like tractorisation, land shaping, levelling, ripping, extraction of root stumps, creation of drainage system, examination of availability of underground water for irrigation, etc. before allotment to the PAFs.

Obviously, they mean that ALL the identified lands are uncultivable as of now. There is no guarantee about these lands 'becoming cultivable' in the future. Further in the affidavit of GoMP dated 6 July 2000, they have stated that these lands are being shown to those PAFs affected at EL 85 m and 90 m. The notice also states (GoMP Affidavit of July 2000, page 99):

> The land will be developed for making it culturable and irrigation will also be provided. Land will be allotted only after all the developments and the works for provision of irrigation are completed.

The notice is self-explanatory in that the land being offered is presently uncultivable and after selection the land will be made cultivable after it is subjected to the development works as stated in the same affidavit. (Page 2–3, paragraph (iv), GoMP Affidavit of July 2000).

> The development process like tractorisation, bunding, ripping, development of drainage system, etc. may be carried out only after the PAFs select the land.

Thus this 'offer' is clearly invalid. Obviously the PAFs will refuse and reject these lands, as all those shown land have done. The GoMP in their affidavit has admitted to this (Affidavit of GoMP, July 2000, page 103):

> The oustees of tehsil Badwani (...) getting affected at 90 m, were shown government lands (...) but none of the oustees selected these lands, stating that these lands are non-culturable, stony, sandy and are hilly; such land, even if developed by the government, is not liked by them.

Obviously they have rejected these lands since they are uncultivable and the PAFs have no belief in the hollow vague promises of the government of making them cultivable. Shockingly, many of the PAFs affected at 90 m have been offered forest land – and even this was uncultivable! (Affidavit of GoMP July 2000, pages 115–116). This is ridiculous considering that even the proposal for the release of these forest lands is yet to be made.

Further, the lands 'identified' in MP are not even in large chaks. Thus the PAFs do not have the choice to settle as a community, a choice that is at the heart of the principles of the R&R policy. To quote (Affidavit of GoMP, 11 April 2000, page 3):

> The State of Madhya Pradesh submits that since the land suitable for agriculture is available in small parcels in a large number of villages and not in large 'chaks', the PAFs will have to be offered land and resettled as per the choice wherever the land is available.

In sum, as of today, all identified land is mostly uncultivable and ill fit for agriculture. This is evident from the above admissions of GoMP as also the joint surveys conducted by NBA and government officials. Also, if the government had any cultivable land available, then this would have been the land they would have shown to the PAFs.

The GoMP, in its affidavit of July 2000 submitted to this court, states that (page 4) at 90 m level, there are 1,034 PAFs of MP yet to be resettled, out of which 435 are to be settled in MP and out of which 228 (page 5) are eligible for land. Thus, GoMP acknowledges that there are 228 PAFs at 90 m who need to be given land in MP but it does not have a single hectare of land to give to them, nor has it made the mandatory offers to them.

This is clearly a violation of the NWDTA that irrigable agricultural lands will be made available for rehabilitation one year in advance [NWDTA Clause XI, Sub-clause IV(2)(iv)].

Violation of Clause XI, Sub-clause IV(6) (ii)

This clause reads:

> In no event shall any areas in Madhya Pradesh and Maharashtra be submerged under the Sardar Sarovar unless all payment of compensation, expenses and costs as aforesaid is made for acquisition of land and properties and arrangements are made for the rehabilitation of the oustees therefrom in accordance with these directions and intimated to the oustees [NWDTA Clause XI, Sub-clause IV(6)(ii)].

This clause requires that before submergence:

A. All payment of compensation, etc. for acquisition of land has been made;

B. Arrangements are made for rehabilitation of oustees;

C. Oustees are duly intimated of the same.

We have already seen above that one of the key components of the arrangements for rehabilitation, namely land, is not yet ready. Regarding the other aspects, the Court itself has noted in its judgment:

> The affidavit on behalf of the State of Madhya Pradesh draws a picture of rehabilitation which is quite different from that of Gujarat. There seems to be no hurry in taking steps to effectively rehabilitate the Madhya Pradesh PAFs in their home State. It is indeed surprising that even Awards in respect of six villages out of 33 villages likely to be affected at 90 m dam height have not been passed (paragraph 247).

Further, referring to the status of R&R at 90 m in Madhya Pradesh, the judgment states:

> It has not been categorically stated whether the PAFs who are so affected [at EL 85 m and 90 m] have been properly resettled or not. On the contrary, it is stated that no Awards in land acquisition cases have been passed in respect of six villages and it is only after the Awards are passed that house plots will be allotted and compensation paid. (...) For the resettlement of PAFs in Madhya Pradesh, out of ten relocation sites mentioned in the affidavit only five have been developed (paragraph 245).

Thus, in terms of the arrangements, it is an admitted fact that:

1. Land acquisition in 6 villages have not been done,

2. Five out of ten resettlement sites are not ready.

While petitioners submit that many other problems remain with the arrangement, the above are the admitted facts recorded by the Court itself. To allow the construction to proceed to 90 m, in the hope that the State will somehow make sure all this is done by the time submergence comes is totally against the spirit and letter of the Tribunal. It turns the condition precedent into a condition subsequent. Further, the performance of the governments, especially the MP Government does not warrant any faith that the arrangements will be complete.

The report of the GRA for MP, quoted in the Court judgment itself, too shows this:

> Even the interim report of Mr Justice Soni, the GRA for the State of Madhya Pradesh, indicates a lack of commitment on the State's part in looking to the welfare of its own people who are going to be under the threat of ouster and who have to be rehabilitated (paragraph 247).

Given this, any construction allowed before all the conditions required by the Clause XI, Sub-clause IV(6)(ii) are completed is a clear violation of the clause.

Violation of the Orders of this Court

In the case of *Shri B.D. Sharma v. Union of India* (1201 of 1990),* this court has passed an order stating that all resettlement and rehabilitation processes should be

---

* This case is reproduced at page 265.

completed 6 months before the lands are submerged. This means that the rehabilitation of those affected by the 90 m level dam should be complete by 31 December 2000. However, the admitted position, as noted above in detail, is that there is no land for those affected up to 90 m, and even the resettlement sites are not ready, nor even the land acquisition complete. There is no way that the rehabilitation can be completed six months before submergence. Thus, the construction up to 90 m is leading to a clear violation of the order of this court.

Thus it is necessary for the Court to review the permission to go up to 90 m and instead order the State governments to first complete all the arrangements for rehabilitation, including land, one year in advance, make sure that the rehabilitation is complete six months in advance, order the GRAs to verify the same, and only then allow any further construction.

2. Clearance of the R&R Sub-group up to 90 m

The majority order states: (paragraph 280, Direction No. 2)

> As the Relief and Rehabilitation Sub-group has cleared the construction up to 90 metres, the same can be undertaken immediately.

It may be noted, however, that the R&R Sub-group has nowhere given an explicit and formal clearance for the construction to proceed to 90 m. It has only indicated that the arrangements are complete to resettle all PAFs affected to 90 m.

As the majority judgment notes (paragraph 186):

> In its meeting held on 6 January 1999, the R&R Sub-group of the NCA observed that arrangements made by the states for R&R of the balance families pertaining to the dam height EL 90 metre were adequate. (...) Pursuant thereto (...) on 21 January 1999 (...) an action plan for resettlement and rehabilitation for balanced families of dam height EL 90 metres was finalised.

However, even this assertion of the R&R Sub-group was ill-conceived and not based on any cross-checking of the ground realities. Possibly, it was based only on statements of the State governments. The arrangements were, in fact, not ready – not in January 1999, nor, one and a half years later, in July 2000.

It has already been shown above how the arrangements with regards the land acquisition, as also the resettlement sites, are not ready even today (hence not ready in January 1999 too). It has also been shown above that no land is available in MP and hence even this arrangement is not ready. Clearly, the R&R Sub-group has shown great irresponsibility in assessing the 'adequacy' of arrangements. In fact, this persisted even after the fact that lands are not available had been pointed out to them by the GoMP.

Situation of Land for Resettlement

Soon after the R&R Sub-group meeting of January 1999, this court permitted construction on the dam up to 88 m from 80.3 m (85 m exclusive of humps by

Order of this court dated 18 February 1999 and later humps of 3 m). In this connection, the NCA wrote to Madhya Pradesh Government regarding the very 'arrangements' that the R&R Sub-group had found 'adequate'. In a letter from Shri Afroze Ahmed, Director (Rehabilitation), Narmada Control Authority to Government of Madhya Pradesh dated 12 March 1999 (placed before the Supreme Court in GoMP Affidavit, 14 July 1999, page 18), he writes:

> I would like to draw your kind attention regarding the Government of MP's submission during the 43rd R&R Sub-group meeting of NCA held on 6 January 1999 and also during the inter-State meeting on 21 January 1999 for the finalisation of the R&R action (sic) for the dam height EL 90 m, that 1,973 ha has been identified in Dhar and Jhabua districts for the allotment of MP-PAFs in MP including PAFs of dam height EL 85 m, and EL 90 m who are entitled for the land in MP. (...) I shall be grateful if you could kindly arrange to send the details of the land available, indicating the name of the place/village.

Shri Mazumdar, Member (Rehabilitation), Narmada Valley Development Authority in his reply dated 26 March 1999 also placed before the Supreme Court in GoMP's affidavit dated 14 July 1999 (page 19) stated:

> At the outset, it may be stated that the GoMP, neither in their affidavits submitted to the Honourable Supreme Court nor in the meetings of RCNCA, NCA, SSCAC, R&R Sub-group including the 43rd meeting held on 6 January 1999, ever said that an area of 1,973 ha of land identified by the State of MP can be made available to or can be allotted to those PAFs who are being affected at EL 85 m and/or 90 m.

Thus it was clear that even the arrangements regarding land availability were not in place in January 1999.

The position as of today has already been given above. The GRAs had been asked by the courts to report on the ground realities of the land availability in all the three States. Petitioners do not have access to the GRA reports submitted to this court. However, the findings of the GRA as noted in the order itself are quite evident:

> (5) The reports of the Grievances Redressal Authorities, and of Madhya Pradesh in particular, shows that there is a slackness in the work of identification of land, acquisition of suitable land and the consequent steps necessary to be taken to rehabilitate the project oustees (paragraph 280, Direction No. 5).
>
> Even the interim report of Mr Justice Soni, the GRA for the State of Madhya Pradesh, indicates a lack of commitment on the State's part in looking to the welfare of its own people who are going to be under the threat of ouster and who have to be rehabilitated (paragraph 247).

Given this, it is indisputable that the arrangements for resettlement of PAFs up to 90 m were not complete in January 1999 (when the R&R said they were) and are not complete to date, R&R Sub-group thus showed complete irresponsibility when it recorded 'adequacy' where there was none. It is shocking that it did not bother to change its recordings even after MP pointed it out to the NCA that they do not have land. Hence the noting in the judgment with regards the findings of the R&R Sub-group may be suitably modified and construction of the dam be ordered to be immediately stopped. Further, the R&R Sub-group should not be trusted with any major responsibility.

3. Clearance for further construction

Clearance from R&R angle

The court has ordered that further construction (over 90 m) will be subject to clearance and monitoring by the R&R Sub-group (paragraph 280, Direction No. 2). The NCA will give the final clearance based on the clearance of the R&R and Environment Sub-group. In view of the facts given above it is seen that the R&R Sub-group has found arrangements adequate even when the arrangements were not ready. The petitioner has already demonstrated similar irresponsibility at earlier points. For instance, the R&R Sub-group has never monitored the progress of R&R with respect to specific clauses of the Tribunal Award. Even when the Sub-group itself decided that lands for resettlement should be acquired in minimum chunks of 200 ha, they allowed the states to acquire lands in much smaller pieces not suitable for community resettlement. Hence, the R&R Sub-group cannot be relied upon to impartially monitor the readiness and implementation of rehabilitation in terms of the provisions of the Award, orders of this Court and State policies. Hence this very critical function needs to be placed in the hands of a more independent and responsible machinery.

The court order asks the R&R Sub-group only to 'consult' the GRAs. In view of the fact that the GRAs are headed by retired judges enjoying the confidence of the Court, they would be far more appropriate agencies to carry out this function of clearing further construction. However, the GRAs, as the name suggests, are essentially created for grievance redressal. More over, the GRAs themselves have the limitation that their entire machinery is essentially on deputation from the State governments. Further, since there are 3 different GRAs in three States, it would be difficult for the GRAs to take a holistic point of view regarding the project.

In view of the above, the most appropriate arrangement would be for the three GRAs to accord clearance for further construction based on the situation in their respective States, and a central independent authority (independent from the project), like the NHRC (National Human Rights Commission), to give the final clearance for construction based on the State-wise clearance of the three GRAs and after consulting the R&R Sub-group.

Clearance for construction from environmental angle

Similarly, an independent expert group/agency should be created to certify that environmental mitigation measures are progressing as per the norms. Based on this the agency would give permission for further construction at each stage. This permission would be given by the agency after consulting the Environment Sub-group. Even this should happen only after the project has received fresh clearance after a comprehensive environmental impact assessment.

Such an independent agency is required because the Environment Sub-group consists of the implementing agencies themselves, and to leave the monitoring to the Sub-group can lead to a conflict of roles and interests.

The judgments of the court may be suitably reviewed in view of the above.

4. Linkage of construction to R&R

The Tribunal Award clearly lays down a linkage between the extent of resettlement and the height of the dam. This is done through a number of clauses, mainly Clause XI, Sub-clause IV(2)(iv) and Clause XI, Sub-clause IV(6)(ii). The first of these specifies that land must be made available one year in advance of submergence. The second states that in no event can the areas be submerged unless full compensation is paid and arrangements for rehabilitation are in place and intimated to the oustees. The use of the words 'in no event' indicates the spirit of the Tribunal order in unequivocal terms. The order of the Supreme Court in the B.D. Sharma case goes further and makes it mandatory for rehabilitation to be complete in all respects six months prior to submergence.

Apart from these specific linkages, the Tribunal Award also provided for an overall linkage between the construction and submergence. Reading the relevant clauses of the Tribunal together shows that the Tribunal required the land as well as the location of the rehabilitation villages in all the three States to be identified within two years of the Tribunal Award. This clearly shows the Tribunal's intention that all arrangements for rehabilitation including land had to be in place before construction progressed to a point where it could submerge any village.

The monitoring of the rehabilitation by the R&R Sub-group has never explicitly considered these requirements of the Tribunal and the orders of this court. Hence, it is not sufficient for the Court to direct that the dam be built as per the Tribunal Award and that the resettlement be monitored by the R&R Sub-group. There is a need for the Sub-group (or any other agency that is mandated to monitor the R&R) to be directed to monitor and record explicit findings on the various provisions of the Tribunal Award including the abovementioned clauses and the order of this court in the B.D. Sharma case. The absence of such directions has led, and will continue to lead, to submergence without rehabilitation. The directions in the majority order may therefore be suitably reviewed.

5. Situation of Land and Resettlement

Situation in MP

The status of the land availability and other arrangements like resettlement sites, land acquisition in the State of MP have already been outlined above.

Situation in Maharashtra

The government claims that at EL 90 m 220 PAFs remain to be resettled. This is a gross underestimate. The ground reality is that the government has chosen to ignore thousands of those to be affected. This gross underestimation of PAFs has been pointed out by petitioners. This underestimation is due to:

A. A large number of PAFs have been omitted due to sketchy and faulty baseline and level surveys.

B. Although some of the 'tapu' villages and those that will be rendered socially unviable have been declared affected, the adivasis residing in those villages have not been counted as declared PAFs! For instance, Maal, Bamana, Teenismaal, Khardi, Savaria, Bilgaon, Atti and other villages (a handful of the PAFs here have been declared but most omitted).

C. There are a number of adivasis who are eligible for PAF status, being adult sons or other reasons, and whose claims are pending. There are at least 757 such claims in 24 affected villages and 477 adivasis with similar claims in the R&R sites. These claims are yet to be verified.

D. Another crucial issue is the issue of cut-off date for adult sons. If a realistic and just cut-off date were adopted instead of the arbitrary 1 January 1987 then a further significant increase in the number of oustees would ensue.

Given the above, a large number of PAFs are not even counted as affected at 90 m though they would be. This situation has been presented by the petitioner to GRA Maharashtra. However, since the report is not available to the petitioners, it is not clear what GRA has to say on this issue.

Related to this is the issue of land availability in Maharashtra. While Maharashtra claims that some land is available and that this is sufficient to resettle the PAFs up to 110 m, the ground reality is totally different. There is virtually no land in Maharashtra. This is evident most clearly from the fact that the government still does not have land to allot to the 201 families who were resettled here 4–6 years ago but are yet without land.

The petitioners do not have access to the report of the GRA which was supposed to verify the land availability in Maharashtra. However, the one reference to the GRA report in the Court Order does not augur well. As the Court notes (paragraph 280, direction No. 5):

> (5) The reports of the Grievances Redressal Authorities, and of Madhya Pradesh in particular, shows that there is a considerable slackness in the work of identification of land, acquisition of suitable land and the consequent steps necessary to be taken to rehabilitate the project oustees.

The reference to the reports of all authorities and the finding that there is slackness in the work of identification of land is worrying. The ground reality is that there is no land, and even when the PAFs affected at 90 m have asked the government to show land to them, the government has not been able to do so. The GRA report seems to support this. It is imperative that the report of the GRA on land availability be made public so that the real situation is clear.

At this point, the situation is that a large number of PAFs in Maharashtra are to be affected by the 90 m level of dam and there is no land to resettle them – a clear violation of the Award. Thus, this is an additional ground to review the judgment's directions permitting construction to 90 m.

6. Rationalisation of Polices In Various States

The petitioners have pointed out in their submissions that the rehabilitation policies in the three States are dissimilar. Not only that, but the policy regime in MP and Maharashtra violate the spirit and letter of the NWDT Award requirements. Also, some of the policies have shortcomings which violate the fundamental rights of the people. For instance, a large number of tribals do not have title to their land simply because the land is still held on the name of the grandfather or great-grandfather and division has not been recorded in the land records. This is their custom. However, these are then classified as landless and in MP they are not entitled to land. Petitioners have given a detailed analysis of this in written submissions (written submissions of petitioners February 2000, Displacement, Resettlement and Rehabilitation, Vol. 147, pages 16–24, especially pages 21–22). These shortcomings in polices, if not rationalised, will violate the fundamental rights of the affected people. Note that the government call this 'liberalisation' of policies, but the more appropriate term would be 'rationalisation'. The Court has recorded in the judgment (paragraph 193):

> There is no requirement that the liberalisation of the packages by the three States should be to the same extent and at the same time, the States cannot be faulted if the package offered, though not identical with each other, are more liberal than the one envisaged in the Tribunal Award.

However, petitioners have pointed out that the package in Maharashtra and MP is not fully as per the Tribunal Award, and also that some rationalisation will be needed to ensure that the letter and spirit of the Tribunal Award as well as fundamental rights are ensured. Therefore, this particular observation needs to be reviewed and the States directed to rationalise the packages as petitioners have submitted. This is also important, as the land requirement would go up at each stage once the rationalisation takes place and hence the amount of land required for clearance for further construction would change.

7. Other categories of oustees (non-submergence affected people)

The majority judgment does not consider it necessary to order any relief to the non-submergence oustees, including the canal affected people, the downstream affected people, the people with non-agricultural occupations, etc. Only in terms of the colony affected people, it notes that the GRA, Gujarat 'is trying to guide in respect of (...) issue relating to Kevadia Colony' (paragraph 215).

This judgment needs to be reviewed. In general, as the FMG has recommended (paragraph 5.8.1):

> complete census of all categories (...) affected in any manner whatsoever, including canal affected persons, communities downstream of the dam, groups and individuals providing supplies and services to others. (...) A number of category-specific rehabilitation packages should be worked out.*

Two points are very important in this recommendation. One is that the FMG has called for a census/survey of all those affected in any manner whatsoever. It is implicit in this that the nature and extent of impact on these categories will also be included in the survey. The other significant point in this is it calls for category-specific rehabilitation packages, and not for the same package for each. Again, implicit in this is that the package developed would depend on the nature and extent of impact. In declining to grant the relief to these affected people, the Court possibly has assumed that the impact is marginal (e.g. on the people who would lose their livelihoods due to dispersion of the community), or that it is too far in the future (e.g. downstream affected), or that indeed those affected would actually benefit (e.g. canal affected people). However, these are based on assertions by the State governments and not necessarily on any ground surveys.

Hence, it is imperative that the Court orders a survey, as recommended by the FMG, to carry out a census of the nature and extent of impact on various categories, and then develop suitable packages for those affected depending on this nature and extent. Land-based rehabilitation would be necessary where the lost livelihood depended on land; in other cases, appropriate packages would need to be evolved. It may ask some expert body to carry this out, under the supervision of the GRA, with input from the affected people themselves.

With respect to specific categories, the following can be stated.

Colony affected people

The Court has not mentioned anything in this regard except that the GRA, Gujarat is guiding in the issue of Kevadia Colony affected people. But the Court has declined to give any relief to these people. In this context, it is important to note that the majority judgment accepts the proposition that ILO Convention has to be read into the domestic legal regime. To quote:

> 86. While accepting the legal proposition that international treaties and covenants can be read into the domestic laws of the country the submission of the respondents was that Article 12 of the ILO Convention No. 107 stipulates that '[t]he populations concerned shall not be removed without their free consent from their habitual territories except in accordance with national laws'.
>
> 87. The said Article clearly suggested that when removal of the tribal population is necessary as an exceptional measure, they shall be provided with lands of quality at least equal to that of the land previously occupied by them (...).

The judgment then goes on to state that the rehabilitation package contained in the Award of the Tribunal fulfils this requirement.

---

* Document partly reproduced below at page 357.

However, the petitioners had pointed out that the ILO Convention 107 would apply to the colony-affected people also since they were tribals and were displaced in 1961 (Note in rejoinder of petitioners on R&R, May 2000, page 3). In their case however, lands of equal quality have NOT been given and they do not come under the ambit of the Award. Thus, their basic right to life under Article 21 with the ILO Convention read into it has been violated. The current 'package' offered to them – merely a small sum of money – is nowhere near any kind of rehabilitation, a fact that even the FMG recognised. This needs to be reviewed and the Court should issue directions to the Gujarat Government to immediately provide them with land-based rehabilitation.

Canal affected people

The Court has said that: (paragraph 196):

> [M]ost of the people falling under the command area were in fact beneficiaries of the projects and their remaining land would now get relocated with the construction of the canal leading to greater agricultural output.

However, there are a large number of people who would be rendered landless, or marginal by the canal. Further, there are likely to be a number of people whose remaining land would not be in the command area. None of these people would be then the 'beneficiaries of the project'. The official surveys themselves point out some of the above. Surveys regarding whether remaining land will be in the command or not have to be carried out. Thus, clearly, there will be many who do not benefit on the whole but would lose significantly their sources of livelihood. Hence, it is necessary to carry out some surveys, and then develop appropriate packages for those affected.

Other occupations

There are a large number of persons living in the submergence area who would lose their livelihood due to a loss of community and/or loss of river. For example, fisher people, shopkeepers, carpenters, etc. would lose their livelihoods but were not being rehabilitated. NBA had pointed out that while no surveys were done, its estimates of these were several thousand families, mostly in MP. NBA called for an immediate survey of the same and the development of rehabilitation packages.

The Court Judgment quotes Gujarat in saying that the number of such families in MP was 'not more than a couple of hundred' (paragraph 194) (no survey is mentioned by Gujarat for this) and then (paragraph 194):

> In our opinion, it is neither possible nor necessary to decide regarding the number of people likely to be so affected because all of those who are entitled to be rehabilitated as per the Award will be provided with benefits (…).

But the very argument was that the Tribunal Award leaves out these people. Also, why is it not possible to decide the number of people so affected? A simple survey would suffice. Such a survey would put to rest the question of whether there are a few hundreds or a few thousands of such families. Even if there are only a few thousands, Article 21 of the Constitution certainly guarantees their right to be rehabilitated and livelihoods restored.

In sum, carrying out surveys and developing appropriate packages will take care of all the grounds that the Court has raised. The petitioners too had asked only for the development of appropriate packages and not the same package for all categories.

Hence, the Court may review the judgment appropriately and order the carrying out of the survey and the development of the appropriate packages for all those affected in any manner whatsoever by the project, to safeguard the rights of the affected people under Article 21 and Article 14 of the Constitution. This issue is extremely critical because it pertains to the fundamental rights of the people.

8. Environment impact assessment and environmental clearance

The central premise of the majority judgment vis-à-vis the environment impact assessment and the 1987 environmental clearance is that there was a proper application of mind by the authorities (paragraph 119). The arguments of the Gujarat and the Union Governments is that the government was aware that the critical studies remained to be done and this was considered while according the clearance. The majority judgment has accepted this contention.

> Even though complete data with regard to the environment was not available, the government did in 1987 finally give environmental clearance (paragraph 77).

and,

> [It] is not possible (...) for this court to accept the contention of the petitioner that the environmental clearance of the project was given without application of mind. It is evident, (…) that the environmental clearance of the project was unduly delayed. The government was aware of the fact that a number of studies and data had to be collected related to the environment. Keeping this in mind, a conscious decision was taken to grant environmental clearance (paragraph 119).

This contention cannot stand the test of accepted and established legal principles. In short, application of mind cannot take place if there is nothing to apply one's mind to. Hence, the 1987 clearance must be held as null and void as the minority judgment has held.

Even after the conditional environmental clearance of 1987, there has been no comprehensive environmental impact assessment of this project which is necessary to ensure that there are no serious adverse environmental impacts of the project which either cannot be mitigated or the cost of whose mitigation would be too high. This assessment is necessary for safeguarding the environment which has been held to be part of the fundamental right of the citizens under Article 21 of the Constitution. A majority judgment inasmuch as it finds such a comprehensive assessment unnecessary, is erroneous and contrary to the law on the subject laid down by this Hon'ble Court. The majority judgment, insofar as it relies on the

Environment Sub-group of the Narmada Control Authority as an adequate safeguard against potential damage of the environment caused by this project, is erroneous since the Environment Sub-group is a Sub-group of the dam construction authority which is the Narmada Control Authority, and, therefore, cannot be relied upon to act as an adequate regulatory agency. The very fact that the project has been allowed to go ahead for so long without a comprehensive environmental impact assessment is an adequate ground for not trusting the Environment Sub-group alone as a safeguard against potential environmental damage to be caused by the project.

The majority judgment insofar as it holds that the conditional environmental clearance is not illegal, even if it violates the MoEF's own guidelines for such clearance, since it is merely an administrative clearance, is also an error on the face of the records. Whether the clearance is administrative or statutory, the fact that the clearance violates the established guidelines of the MoEF itself is an indication of the fact that the government's own standards for environmental safeguards have not been followed and, therefore, the project could be environmentally hazardous, and allowing such a project to go ahead in violation of such administrative guidelines would be a violation of Article 21 of the Constitution. This part of the majority judgment also needs to be reviewed and the project authorities asked to carry out a proper environmental impact assessment and seek a clearance based on the same. The minority judgment and order in this aspect – in particular minority order paragraph 280, points 1–5 of Directions – may be adopted in the final judgment and order of this Court on this matter.

9. Laches (delay)

The majority judgment finds the petitioners guilty of laches as the petitioners were active in 1986 and did not bring up the matter to the Court until 1994. It may be pointed out that the petitioners had taken a step-by-step approach and had gone to the Executive (Government) with all the issues. It had hoped that the Government would consider the issues raised by it. This period was marked by two reviews of the project, namely the Morse Committee and the FMG. It was only natural that the NBA would wait for the reports of these committees and give the government enough time to implement their recommendations. It was only when the government's response was bureaucratic and indifferent to these findings and recommendations that the NBA finally decided to move the Supreme Court.

In particular, with reference to the clearance of the Ministry of Environment and Forests, it was difficult for the NBA to obtain the official documents, as is the case always. Even when the NBA eventually obtained the clearance letters it decided to wait until the period mentioned therein for studies to be completed. Even after this, as the Ministry of Water Resources and the MoEF debated the issue back and forth, it was difficult for the NBA to know and obtain conclusive documents about the status of the clearance. Therefore, the NBA decided to wait for the report of the FMG which it hoped would shed light on the situation. However when the government tried to undermine the review by the FMG, the NBA decided to move the Court.

It may further be pointed out that the NBA had filed cases in the district and high courts on specific issues related to the project. It is only when these courts did not address the concerns raised that the NBA had to move the Supreme Court. Also, the implementation of the R&R programme was then in full swing in Gujarat and moving in a somewhat sluggish manner in other States. More and more cases of serious violations of basic human rights of the affected people were coming to light at this point, which the existing machinery (NCA and its sub-groups) was neither willing to nor capable of addressing. This too prompted the petitioners to come to this court. It may also be pointed out that an NGO had approached the Gujarat High Court and subsequently this very Court in the 1980s itself to safeguard the rights of the affected people.

Given all this, the petitioners cannot be said to be guilty of laches.

The majority judgment, insofar as it holds that a public interest petition should also be dismissed if it is delayed on the ground of laches, is also erroneous and liable to be reviewed. Such a principle is incorrect in law and will have serious deleterious consequences on public interest. The principle of laches is akin to a principle of limitation which is based on the principle that a person who does not challenge the violation of his rights for a sufficient length of time is deemed to have acquiesced to that violation which then gives a vested right to the party which has caused such violation. Such a principle obviously cannot apply to a public interest case where a party has come to Court not in its own interest but in the interest of the general public. The fact that such a party has moved the Court late cannot mean that the public interest should be compromised merely because of any delay on the part of the person who has moved the Court not in his own interest but in the general public interest. In public interest cases, the yardstick for determining whether a matter be entertained or not must always be in the public interest and cannot be a technical plea of laches on the part of the person who has moved the case in the public interest. In a case like the present one, in deciding whether the petition should be entertained or not on the ground of environmental impact, costs/benefits, etc., the Court can certainly take into consideration the fact that the construction of the project had begun and had proceeded significantly. Yet if the petitioner has come forward with the case as it has in the present case, that even the remaining costs and adverse consequences of the project would greatly outweigh its benefits and, therefore, the project needs to be reviewed even at this stage, the Court must not reject such a plea merely on account of laches. For instance, if the facts which emerged were that the dam was being built in a seismically active and high risk zone, and that there were very high chances of a dam collapse within the next few years resulting in a loss of lakhs of lives, could the Court refuse to entertain such a plea on the ground of laches? This principle laid down in the majority judgment will have serious deleterious consequences to public interest if it were to be followed, as it would have to be by all other courts in such cases.

10. Amendment/Review of the Tribunal Award

The majority judgment states that:

> Once the Award is binding on the States, it will not be open to a third party like the petitioners to challenge the correctness thereof. ... We therefore, do not propose to deal with any contention which, in fact, seems to challenge the correctness of an issue decided by the Tribunal (paragraph 82).

The above principle laid down by the majority judgment is erroneous for at least three reasons: Firstly, It is incorrect to hold that the Award of a Tribunal is binding itself on the State and cannot be challenged even in the Supreme Court. It is submitted that every order of any Court or Tribunal must be subject to judicial review by a higher Court or reviewable by the Tribunal itself. Even under order 47 of the CPC, a review is allowed against any decree or order from which no appeal is allowed on *inter alia* the ground of an error apparent on the face of the record, or due to a discovery of new matter or evidence which was not within the knowledge of the petitioner when the decree was passed or for any other sufficient reason. Even if the order happened to be an Award of a Tribunal, this principle must clearly be held to apply to it, otherwise it would lead to a situation in which even the subsequent facts clearly show the Award of the Tribunal to be erroneous and grossly against public interest, such an order would still have to be enforced. For example, if it was discovered after the Award that the construction of the dam permitted by an Award would lead to its immediate collapse and would thereby kill lakhs of people, could it be said that there was no possibility of a review of such an order. In the present case, the petitioners had shown that the assumptions on which the Tribunal's Award was based, had been found to be grossly incorrect by discovery of facts which went to the very root of the matter and, therefore, the Award needed to be reviewed. These subsequently discovered facts included facts regarding the number of people displaced, the facts regarding environmental impact which were not considered by the Tribunal, the hydrology of the river, etc.

That even otherwise, the petitioners or the affected people were not parties to the proceedings of the Narmada Tribunal. Thus, if the Award violated the fundamental rights of the petitioners, it was surely open to them to challenge the Award on that ground. The Constitution clearly empowers and makes it the responsibility of the Hon'ble Court to protect the fundamental rights of the citizens of this country even if those fundamental rights are violated by the Award of a Tribunal. This Court would still have the power and responsibility to make suitable orders to prevent that violation. It is, thus, totally erroneous as a matter of law for this Hon'ble Court to have held that third parties like the petitioners cannot challenge the correctness of an Award even if it violates their fundamental rights. In fact, passing of an Award which affects the rights of the petitioners without giving them a hearing is a violation of natural justice and it has been held by this Hon'ble Court that this court can entertain a challenge on that ground even to a judgment of this Hon'ble Court passed in a case in which that party was represented. However, in a case where an Award of any Tribunal is made in violation of natural justice and which affects fundamental rights of some citizens, it would surely be open to challenge in this Hon'ble Court on the ground of violation of natural justice as well as violation of fundamental rights. This part of the majority judgment, therefore, also needs to be reviewed.

11. Large dams – Benefits and costs

The majority judgment deals with the issue of large dams and their overall benefits to the country. This issue was not an issue before the Court – only the Sardar Sarovar Project was under discussion. The issue of the impacts of large dams does form a background to the issue of SSP, but there was no material presented or discussed related to the same, nor was the issue itself under adjudication. Moreover, the issue of large dams has been extensively discussed, debated and a large number of studies have been carried out on this all over the world. In 1998 the World Bank and International Union for Conservation of Nature initiated the setting up of the World Commission on Dams (WCD) to assess the impacts and experience of dams all over the world. This Commission consists of eminent experts from all over the world representing both the dam-building sector and those who oppose large dams as well as some independent academics. The Report of this Commission is to be released on 16 November by Nelson Mandela. The 'India Country Study' prepared for the WCD is already public. This study is a devastating indictment of large dams in India, not only from the angle of social and environmental impacts but also from the viewpoint of economic and financial performance.

Given the complexity of the issue, it is not proper for the Court to conclude on this without giving due consideration to the large knowledge-base including the India Country Study. It is understandable that the Court did not go into all this material since this (performance of large dams in general) was not an issue before the Court. However, it then follows that the Court cannot give any findings on the same. Particularly, the judgment contains observations about the experiences specific to certain projects, namely Bhakra Nangal, Nagarjun Sagar Dam, Tehri, Bhillai Steel Plant, Bokaro and Bala Iron and Steel plants (paragraph 267) Regardless of whether the ground reality in these areas is reflected in the said observations or not (experience from these areas actually runs contrary to these observations), the fact is that there was no material before the Court on these areas and hence these observations should not from a part of the judgment.

Hence, the judgment may be reviewed in light of the above and the general discussion on the benefits of dams presented as a part of conclusions of the Court may be deleted.

12. PILs and intervention in infrastructure projects

The Court judgment observes that judicial interventions in policy decisions and in particular with regards to the infrastructure projects should be made only at the time such decisions are being taken and once the construction/implementation of the project begins the judiciary should not interfere. This raises certain serious issues and it may be necessary for the judgment to be reviewed in light of these. These issues are :

a) The working of the government is largely non-transparent and non-participatory. The common people and even people directly affected by policy decisions related to infrastructure projects are not informed that the government is considering any such thing. The related studies, documents, factors under consideration and so on are not available to the people as a matter of right. In most of the cases people get to hear of the project only after decisions have been taken, a considerable amount of money spent and work initiated. Thus it is very difficult for the people to intervene in any meaningful manner at this stage.

b) In certain cases when people have approached the courts at the policy-making stage itself, the courts have rejected their claims on the grounds that the cause of action has not yet taken place and it is premature.

c) In many cases the implementation of the projects results in violations of basic rights of the people. At this point, the people certainly have a right to challenge this and ask that projects be implemented in a manner that the basic rights are not violated. It may happen in such cases that it is found there is reasonable ground to believe that it would be impossible to implement the project without violation of the fundamental rights of the people. In such cases, certainly, the right to challenge the project itself would remain even though the project may be in an advanced stage.

In any case, the overall public interest in general and the fundamental rights of the people in particular cannot be superseded on the grounds of delay, given especially the absence of right to information.

The part of the judgment related to these issues may be suitably modified in the light of the above and in particular directions or at least observations with regard to the need for right to information be included in the same.

Public Interest Litigation (PIL)

PIL is a unique contribution of the Indian judiciary and needs to be strengthened further. It is not anybody's contention that frivolous and even mala fide cases of PIL do not exist. However the percentage of such frivolous cases are very small and the judiciary has enough powers and discretion to handle these. Indeed the percentage of frivolous and mal-intentioned cases would be far more in the case of regular litigation as compared to PIL.

Given this situation the remarks with regard to PIL made in the judgment are likely to be interpreted, especially by the lower courts, as an indictment of PIL in general. Even a hint of displeasure of the Supreme Court towards PILs can cause an enormous setback to this special innovation of the judiciary which has greatly empowered the common people. It would therefore be appropriate for the said remarks to be deleted from the judgment. In fact, unless the Court wanted to say that this petition by the NBA was in any manner a 'Publicity Interest Litigation' or 'Private Inquisitiveness Litigation', these remarks (paragraph 256) would be out of place in the judgment. It is clear that the Court has not directly cast any such aspersions on the current petition because if this were the case, the Court would have said so directly. Hence, it is important that these remarks be removed to prevent any negative impacts on PILs.

Prayers

It is therefore respectfully prayed that this court may be pleased to:

A) Review and recall the majority judgment and order in the light of the above;

B) Considering the importance of the matter, grant oral hearing to the petitioners in this review petition;

C) Until the review petition is disposed off, stay and suspend the operation of the judgment dated 18 October 1999;

D) Pass any other order that this court may deem fit and proper.

Narmada Bachao Andolan v. Union of India, Review Orders, March 2001

Source: Supreme Court of India, Civil Original Jurisdiction Review Petition (C) No. 1259 of 2000 in Writ Petition (C) No. 319 of 1994, *Narmada Bachao Andolan v. Union of India & Ors*, 29 March 2001.

Order (CJI & B.N. Kirpal, Justice)

This petition has been filed for the 'review and recall' of the majority judgment of this court in Writ Petition (C) No.319 of 1994.

Prayer for oral hearing is declined.

We have carefully gone through the review petition and the connected record but we do not find any error apparent on the face of the record which may call for a review of the majority judgment. The contentions raised in the review petition, in effect, seek to challenge the correctness of the said judgment and that is impermissible being outside the parameters of review. The only ambiguity, if any, has been clarified with the recording of the statement of the counsel for the Union of India and the State of Gujarat on 23 November 2000 in IA No. 17 (filed by the petitioner) that even if humps are required to be constructed, the effective height of the dam shall not be raised beyond 90 metres for purposes of submersion until further construction is permitted as per the conditions laid down in the judgment.

The review petition is accordingly dismissed.

Order (S.P. Bharucha, Justice)

The review petition seeks the review of the majority judgment. The learned Judges who wrote the majority judgment have found no case for a review. The review petition must, therefore, be dismissed.

I stand by the view expressed by me in the minority judgment.

The review petition is dismissed.

Chapter 6

Resettlement

B.D. Sharma v. Union of India, 1991

Source: *B.D. Sharma v. Union of India*, Writ Petition (Civil) No. 1201 of 1990, Order of 9 August 1991.

Order

1. We have heard Mr Sharma in person, learned Additional Solicitor General for Union of India and counsel for the States of Gujarat, Madhya Pradesh and Maharashtra, Mr Bagla, the present Commissioner who is present in Court has also been heard.

2. A letter received from Mr Sharma, while he was in office as Commissioner of Scheduled Castes and Scheduled Tribes, has been treated as a writ petition under Article 32 of the Constitution. His letter essentially raised questions with reference to the relationship between the Commissioner and the Union of India as also the State governments, the effective nature of the reports made from time to time by the Commissioner, implementation and non-implementation of the recommendations and the consequences arising out of the same and the constitutional methods which should have been generated for the purpose of treating the scheduled castes and scheduled tribes for the purpose of monitoring their welfare. That letter apparently appeared to raise constitutional issues of importance and, therefore, we considered it appropriate that it should be examined. It is true that in the letter reference was made to certain institutions where, according to Mr Sharma, there was infraction of the obligation and, therefore, certain action should be taken.

3. The interlocutory application now filed is in relation to a question which perhaps requires more of regulation than of constitutional questions to be looked into. We are told that writ petitions are already pending before the High Courts of Bombay and Gujarat filed by people who are affected by the developments that are taking place in those States with reference to the SSP. Apart from the fact that writ petitions are pending and the high courts had made certain interim orders, the nature and particulars are yet not known. It is very difficult for this court to

regulate the details of the working of a scheme of this nature, but we are of the view that the rehabilitation of the oustees as a result of the project coming up can be examined.

4. Counsels for the three States have supplied the figures on the basis of which we find that in Maharashtra there will be 2,468 oustees, in Gujarat the number would be 4,500 and in Madhya Pradesh it would be 6,800 or so. The Award indicated certain figures but there has been a change in the position of oustees.

5. From the affidavit of the Gujarat Government we find that out of 4,500 oustee families about 3,100 families have already accepted rehabilitation. The remainder obviously is yet to be attended to. Mr Sharma has also told us that in view of the fact that the State of Gujarat has already been taking rehabilitatory steps, there is not much agitation against the scheme in Gujarat. We have been told that in the other two States there is some amount of agitation.

6. Sardar Sarovar is an inter-State project, the feeder being Narmada an inter-State river. This is financed by the World Bank and assistance is forthcoming from some foreign countries. As it is, completion is behind schedule. It is, therefore, difficult to look for enforcement of what had been contemplated either in the agreement or in the Award. While we agree that the rehabilitation should be done as far as possible in a methodical and meticulous way, to enforce terms and conditions stipulated in the agreement such as eighteen months' notice before effecting evacuation in terms may be difficult and may not be beneficial for the ultimate purpose. We understand that there is a Committee headed by the Secretary, Social Welfare, as a Sub-Committee under Narmada Control Authority. We would require this Committee where the Secretary, Social Welfare is the Chairman, to be activised so as to ensure rehabilitation. We would direct that this Committee should move in the areas where there is rehabilitation to be undertaken and directly ensure that rehabilitative process is undertaken. The rainy season is on but within a month or six weeks the weather would improve. We would, therefore, require rehabilitation to be personally supervised at intervals by the Committee in all the areas likely to be submerged when water is stored.

7. It was submitted to us that the first storage of water in Gujarat area is to be done in 1992, and within two years it would be done in the State of Madhya Pradesh. It is, therefore, necessary that before April 1992, rehabilitation should be effected in regard to the oustees who are said to be the remainder out of 4,500. Rehabilitation should be so done that at least six months before the area is likely to be submerged, rehabilitation should be complete and should be in respect of homestead substitution of agricultural property and such other arrangements which are contemplated under the rehabilitation scheme. This Court would require a report to be furnished of the developments and progress made in the matter of rehabilitation once in every month. We would, therefore, suggest to the Committee to meet at least once after they have visited the areas which they consider necessary, and give

their views with particular details of rehabilitation to be placed before the Court for directions.

8. We make it clear that it is not our intention to hold up the progress of the work. On the other hand, we would like it to be completed expeditiously so that the time lag may not affect the construction of the project.

9. Mr Bagla wanted to make a point about his difficulties in functioning. To meet it, learned Additional Solicitor General made a statement that government had decided to set up the National Commission before the end of September 1991.

Letter from the Secretary, National Commission for Scheduled Castes and Scheduled Tribes, 1993

Source: Letter from S.K. Basu, Secretary, National Commission for Scheduled Castes and Scheduled Tribes to M.S. Reddy, Chairman, Narmada Control Authority – Secretary, Ministry of Water Resources, 1 December 1993.

Dear Shri Reddy,

1. It is understood that the Government of Gujarat have asked for closure of sluice gates of the Sardar Sarovar dam by mid-December 1993. This will cause immediate and permanent submergence of the villages which were temporarily submerged during the 1993 monsoon. It is apprehended that this action will lead to submergence of a much larger area during the next monsoon.

2. From various reliable sources, e.g., the official report of the R&R Monitoring and Evaluation Agency for the Sardar Sarovar affected villages of Maharashtra, i.e., the Tata Institute of Social Sciences, it seems all rehabilitation measures undertaken so far are seriously flawed and arrangements for the future rehabilitation to be done this year are inadequate. The tribals of Madhya Pradesh do not want to go to Gujarat and the Narmada Water Dispute Tribunal Award too offers them a choice to stay in their home State. Yet, no land has been identified in Madhya Pradesh so far for R&R purposes and there are grave doubts about the amount and quality of land available for this purpose in Gujarat.

3. This Commission feels that it will be prudent and advisable that until the various issues of R&R are resolved, the work of eviction of the people should be suspended and the closure of the sluice gates put off. I trust the NCA will duly consider our suggestion urgently and apprise us of its decision.

Pradip Prabhu v. State of Maharashtra, 1995

Source: *Pradip D. Prabhu and Ors v. State of Maharashtra and Ors*, Supreme Court of India (Kuldip Singh, N. Venkatachala and S. Saghir Ahmad, JJ), Writ

Petition (C) No. 4647, 5564–7 and 7339–43 of 1982, 13696–700 of 1983, 15206–8 of 1984 and 1778 of 1986, Decision of 7 March 1995.

Order

1. We have heard learned counsel for the parties. There is almost a consensus between the learned counsel that these matters be referred back to the Government of Maharashtra for a fresh decision, keeping in view the merit of each case. These petitions, being public interest litigation, have been filed on behalf of landless adivasis occupying various lands in various districts of Maharashtra. It is not disputed that the adivasis have encroached upon the land and further that the land under their possession is the forest land. It is alleged in the petitions that the adivasis have been in possession of the land since prior to 1978. The claim of the petitioners is that the Government of Maharashtra has issued instructions from time to time whereunder it has been provided that for persons who are in possession of lands for a particular period of years, their possession may be regularised in terms of the government instructions. Apart from that, it is not denied that the Government of India have also issued instructions, dated 18 September 1990 laying down guidelines in this respect. The petitioners' only claim is that their cases for regularisation be considered in terms of the instructions issued by the State of Maharashtra from time to time and also the above-mentioned instructions of the Government of India. We see considerable force in the contention of the learned counsel for the petitioners.

2. We direct the State of Maharashtra to appoint responsible officers in different districts to examine the claims of adivasis who are in possession of land and decide their claims for regularisation in accordance with law and the above-mentioned instructions.

3. Needless to say that while determining the rights of the adivasis for regularisation, they shall be given an opportunity to be heard by the officers concerned and also to adduce evidence in support of their claims. We further direct that until the cases of adivasis concerned are finally disposed of, they shall not be dispossessed from the lands which are in their possession. The writ petitions are disposed of in the above terms. No costs.

Writ Petitions (C) No. 13696–700 of 1983

4. We have heard learned counsel for the State of Madhya Pradesh. These petitions, being PIL, have been filed on behalf of landless adivasis occupying various lands in various districts of the State of Madhya Pradesh.

5. We direct the State of Madhya Pradesh to appoint responsible officers in different districts to examine the claims of the adivasis who are in possession of the various lands and decide their claims for regularisation in accordance with the instructions of the Government of India.

6. Needless to say that while determining the rights of the adivasis for regularisation, they shall be given an opportunity to be heard by the officers concerned. We further direct that until the cases of adivasis concerned are finally disposed of, they shall not be dispossessed from the lands which are in their possession. The writ petitions are disposed of. No costs.

Narmada Bachao Andolan v. Union of India (Petition 328 of 2002) Supreme Court Order, September 2002

Source: *Narmada Bachao Andolan v. Union of India and Others*, Supreme Court, Writ Petition (Civil) No. 328 of 2002, Order of 9 September 2002.

The Grievance Redressal Authority having been put in place, there is no reason for this court to interfere. As far as the dispute raised in this petition is concerned, that is over and finalised with the earlier decision of this court. In case an oustee or a person affected by the project has any grievance, it is open to him to approach the Grievance Redressal Authority.

It is also contended that land for land has not been given. If there is any person so aggrieved or who has a justifiable grievance, it is open to that person to approach the GRA, failing which this court.

It is made clear that full assistance will be rendered by the NCA as well as the State governments to the GRA in the discharge of their respective functions.

This writ petition is disposed of in the aforesaid terms.

Narmada Bachao Andolan v. Union of India (Petition 328 of 2002) Supreme Court Order, April 2004

Source: *Narmada Bachao Andolan v. Union of India and Others*, Supreme Court, Writ Petition (Civil) No. 328 of 2002, Order of 16 April 2004.

Upon hearing counsel the Court made the following Order:

IA Nos 10 & 11. The prayer in these applications is to direct the respondents not to proceed with further construction of the Narmada Dam by raising its height to 110 metres until all affected people are rehabilitated. The main grievance of the applicants is the subject matter of two IAs (IA Nos 4 & 7).

Having heard learned counsel for the parties at considerable length, we are of the view that for the present, no case has been made out to stop the ongoing construction raising the height of the dam. At the same time, it is necessary to note that the matter relating to the rehabilitation of oustees is required to be examined by all concerned and implemented in terms of the Award made by the Narmada Water Disputes Tribunal, orders and directions issued by this court from time to time, the orders and directions of the NCA and that of the GRAs of the State concerned. Further, it has been brought to our notice by the learned counsel appearing for the applicants that the land in village Matanya, Tehsil Mahaeshwar and some of the other villages being proper cultivable land can be obtained/acquired by the State governments and given to the oustees as a rehabilitation measure. The applicants may give concrete suggestions in this respect to the respondent State of Madhya Pradesh. On receipt of such suggestions, the matter shall be expeditiously examined by the State government. In case the applicants/oustees are still not satisfied with the decision of the State government, it would be open to them to approach the GRA. The said Authority shall also examine the grievance of the oustees as expeditiously as possible as the matter pertaining to rehabilitation of the oustees cannot be taken up or treated lightly. It may also be noticed that a positive approach has been adopted by the oustees that what they are concerned with is the proper rehabilitation and not the stoppage of the construction of the dam, and they have approached this court seeking a stay of construction as they were not satisfied about the proper offer having been made to all affected parties for their rehabilitation and other related aspects of rehabilitation.

For the present, having noticed aforesaid issues, we express no opinion. We however direct that these two applications along with IA Nos 4 & 7 shall be listed for hearing in the third week of July 2004. Mr Muralidhar, learned counsel appearing for the applicants in IAs 7 & 11 shall file a brief synopsis on the reopening of the Court after vacation.

The matter is adjourned in the above terms.

Order of the Grievance Redressal Authority, Madhya Pradesh, 2004

Source: Grievance Redressal Authority, Madhya Pradesh (Sardar Sarovar Project), Case No. 234 of 2004, Order of 11 September 2004.

Order

The applicants in IA Nos 4 & 7 in Writ Petition (C) No. 328 of 2002 filed before the Hon'ble Supreme Court seeking their rehabilitation, as mandated by the provisions of the Narmada Water Disputes Tribunal Award, hereinafter referred to as the Award, have approached this Authority as directed by the Hon'ble Supreme Court by orders passed on 16 April 2004 and 23 July 2004.

2. To appropriate the matter arising for consideration before this Authority, it would be useful to refer to the following relevant facts:

(i) Acting under Section 4 of the Inter-State Water Disputes Act, 1956, hereinafter referred to as the said Act, the Government of India constituted a Tribunal and made a reference to it relating to an inter-State water dispute referred by the State of Gujarat regarding the inter-State river Narmada and the river valley thereof. The States of Gujarat, Madhya Pradesh, Maharashtra and Rajasthan were made parties.

(ii) On 16 August 1978 the Tribunal declared its decision, hereinafter referred to as the Award, under Section 5(2) read with Section 5(4) of the said Act. Thereafter, references were filed by the Union of India and the concerned States before the Tribunal under Section 5(3) of the said Act. After hearing those references, the Tribunal gave its final Award on 7 December 1979. By that Award, the height of the dam was determined at FRL 455 ft, which height according to the Tribunal was required for the purpose of irrigation and for the generation of power. The Tribunal directed the State of Gujarat to take up and complete the construction of the dam. Clause XI of Chapter IX of the final Award contains directions regarding submergence, compulsory acquisition under the Land Acquisition Act, 1894 of land and buildings which would be affected by submergence, and provisions for the rehabilitation of displaced persons.

(iii) The provisions for rehabilitation lay down that in addition to compensation for the property acquired and the payment of special grants, every oustee family would be entitled to and allotted a house site, i.e. a plot of land measuring 60 ft * 90 ft free of cost, at a rehabilitation site providing civic amenities as mandated and every displaced family from whom more than 25 per cent of its land holding was acquired for being in the area of submergence shall be allotted irrigable land to the extent of the land acquired from it, subject to the prescribed ceiling in the State concerned and to a minimum of 2 hectares per family, that irrigation facilities would be provided by the State in whose territory the allotted land is situated. It is further provided that the said land would be transferred to the oustee family if it agrees to take it. The Award directs that Gujarat in the first instance shall offer to rehabilitate the oustees in its own territory but Sub-clause IV(6)(i) of Clause XI of Chapter IX of the Award lays down that in the event of Gujarat being unable to resettle the oustees or the oustees being unwilling to occupy the area offered by Gujarat, the oustees shall be resettled by the home State in the same manner as provided for rehabilitation of the oustee families in Gujarat. Sub-clause IV(6)(ii) of clause XI directs that no submergence of any area shall take place unless all payment of compensation as laid down is made and arrangements for the rehabilitation of the oustees are made and intimated to them. The Tribunal also directed constitution of an inter-State administrative authority, to be known as the NCA, for the purpose of securing compliance with the implementation of the decision and directions of the Tribunal and constitution of a review committee to review the decision of the NCA.

3. The NBA, an anti-dam organisation which is in the forefront of agitation against the construction of the Sardar Sarovar Dam has been in existence since the year

1986. After having chosen different paths to oppose the construction of the dam, in 1994 NBA filed a petition, registered as Writ Petition (C) No. 319 of 1994, before the Hon'ble Supreme Court under Article 32 of the Constitution of India, against the Union of India and others, praying *inter alia* that the Union of India and other respondents be restrained from proceeding with the construction of the dam which was required to be constructed up to 138 metres. The aforesaid Writ Petition filed by NBA was disposed of by the Supreme Court in its judgment dated 18 October 2000. The following observations made by the Supreme Court in paragraph 79 of the judgment dated 18 October 2000 are pertinent:

> The petitioner has been agitating against the construction of the dam since 1986, before environmental clearance was given and construction started. It has, over the years, chosen different paths to oppose the dam. At its instance the FMG was constituted, but its report could not result in the stoppage of construction *pari passu* with relief and rehabilitation measures. Having failed in its attempt to stall the project the petitioner has resorted to court proceedings by filing this writ petition long after the environmental clearance was given and construction started (paragraph 79).

The Hon'ble Court further observed that it had entertained the petition

> with a view to satisfy itself that there is proper implementation of the relief and rehabilitation measures at least to the extent they have been ordered by the Tribunal's Award (…) and the petition in regard to the other issues raised is highly belated (paragraph 80).

4. During the course of hearing it was noted by the Hon'ble Supreme Court that the State of Gujarat, which was the main beneficiary of the project in view of the irrigation facilities which would accrue to it, had liberalised its rehabilitation package by making provision for the allotment of agricultural land to a major son of an oustee as well. It had also constituted a Grievance Redressal Authority to look into the grievances of the project affected families which had settled and would be resettled in Gujarat. Some time in March 2000 the Supreme Court directed the State of Madhya Pradesh to constitute a GRA in the State of MP, as was done by the State of Gujarat, to hear and decide the grievances of the project affected families settled or to be settled in Madhya Pradesh. In pursuance of that direction, this Authority was constituted on 30 March 2000 as an independent autonomous authority whose decisions relating to the rehabilitation of the oustees are made binding on the oustees and the State government.

5. On 18 October 2000 the Supreme Court disposed of petition No. 319 of 1994 by directing that every endeavour would be made to see that the project was completed as expeditiously as possible. The Supreme Court issued *inter alia* the following directions:

i) Construction of the dam will continue as per the Award of the Tribunal.

ii) As the R&R Sub-group has cleared the construction up to 90 metres, the same can be undertaken immediately. Further raising of the height will be only *pari passu* with the implementation of the relief and rehabilitation and on the clearance by the R&R Sub-group. The R&R Sub-group will give clearance of further construction after consulting the three Grievances Redressal Authorities.

6. Thereafter the height of the dam was raised to 90 metres and then it was proposed to raise the height up to 95 metres after complying with the directions given by the Hon'ble Supreme Court. Aggrieved by that decision, the petitioner, NBA, filed another petition before the Hon'ble Supreme Court which was registered as Writ Petition (C) No. 328 of 2002, raising a number of grounds assailing further construction of the dam. That petition however was disposed of by the Supreme Court on 9 September 2002. While disposing of the petition it has been observed by the Supreme Court as follows:

> The GRA having been put in place, there is no reason for this court to interfere. As far as the dispute raised in this petition is concerned, that is over and finalised with the earlier decision of this court. In case an oustee or a person affected by the project has any grievance, it is open to him to approach the GRA.
>
> It is also contended that land for land has not been given. If there is any person so aggrieved or who has a justifiable grievance, it is open to that person to approach the Grievance Redressal Authority, failing which this court. (…)
>
> This Writ Petition is disposed of in the aforesaid terms.

7. By the aforesaid order one should have normally thought that the said order enabled an oustee whose grievance was not redressed by the GRA to approach the Supreme Court for relief by filing a petition enclosing a copy of the complaint filed by him, a copy of the reply, if any, filed by the State, a copy of the decision of the GRA and a statement of the grounds on the basis of which the order of the GRA was being assailed. Such a course would have been not only in conformity with the aforesaid order passed by the Supreme Court in Writ Petition (C) No. 328 of 2002 but would have also enabled this GRA to rectify at the appropriate stage any erroneous view taken by it to avoid repetition of that mistake in future while disposing of the complaints filed by oustees before the GRA. To illustrate, in Case No. 473 decided by this Authority on 25 July 2002, it was observed as follows:

> It is therefore directed that as the complainant is an oustee who has to be rehabilitated at this stage in the light of the judgment of the Hon'ble Supreme Court delivered on 18 October 2000 in Writ Petition (C) No. 319 of 1994 and as now it has been stated on behalf of the complainant on 25 June 2002 before the Authority that he is unwilling to be resettled in Gujarat, the complainant shall be rehabilitated in Madhya Pradesh according to the provisions of the Award and the R&R Policy of the State government. As he is entitled to be allotted irrigable agricultural land, he shall be allotted such land to the extent of his entitlement, provided the Director Agriculture of MP certifies that the land is arable. Irrigation expenses will be borne by the State. If the complainant agrees to take this land it shall be transferred to him as laid down by the provisions of sub-clause IV(7) of clause XI of Chapter IX of the NWDT Award. He shall also be allotted a residential plot in a nearby R&R site where all the civic amenities as mandated by the NWDT Award are available. It is made clear that in case, after transfer of the

> agricultural land, the complainant has a justifiable grievance that the land is not arable, then the Authority shall on receiving a complaint in that behalf get that land examined by experts to ascertain the arability of the land. If according to the opinion of the experts so appointed, the land transferred to the complainant is not arable, then the State shall be directed to allot another piece of land to the complainant according to his entitlement along with a residential plot in case the plot allotted earlier happens to be far away from the agricultural land subsequently transferred to the complainant, and the State shall have to bear such compensation as would be determined by the Authority for all the expenses incurred by him and the inconvenience caused to the complainant.

If the applicants had approached the Hon'ble Supreme Court within reasonable time for setting aside this order and if the order had been set aside, then, as required by judicial discipline, this Authority would not have stuck to the aforesaid procedure for the last 2 years.

8. However, the order dated 9 September 2002, passed by the Hon'ble Supreme Court in Writ Petition (C) No. 328 of 2002, appears to have been constructed by the petitioner NBA, to mean that the Writ Petition (C) No. 328 of 2002 would never stand disposed of and that it had been kept pending to enable any project affected person to file IA in that petition without properly assailing any order passed by GRA. Hence without assailing the procedure followed in Case No. 473, the applicants in IA Nos 4 & 7 filed before the Hon'ble Supreme Court, stated that a valid offer of an allotment of agricultural land was not made to them because the land at Musapura which was offered was not acceptable. The complainants thereafter gave a number of suggestions for acquiring land for allotment to the complainants. These suggestions were not accepted by the State government. Hence, the applicants in IA Nos 4 & 7 were directed by the Supreme Court to approach the GRA, MP. That is how this matter has come up for consideration before the GRA, MP vide the letter dated 24 July 2004 sent by Shri S. Muralidhar and letter dated 28 July 2004 sent by Shri Sanjay Parikh, Advocates, Supreme Court.

9. The matter was fixed for hearing on 7 August 2004. Before the commencement of hearing this Authority had made it very clear at the outset that this Authority will not take into consideration any controversy about the arability of land at Musapura offered to the applicants but will confine its attention to ascertain whether any other arable land or land as suggested by them could be offered to them for allotment with a view to rehabilitate them. The Authority however stated that it would very much appreciate if the matter was amicably settled.

10. During the course of hearing the learned counsel Shri Muralidhar and officials of the State agreed to explore the possibility of amicable settlement of the problems raised by the applicants before GRA. The officials of the State government also stated that they would be able to ascertain the availability of the land of the agriculture farm from the State situated at the village of Khajuri for allotment to oustees to enable the possibility of amicable settlement of the

problems raised. The case, with the consent of the parties, was fixed for hearing on 24 August 2004.

11. On 24 August 2004, when the matter came up for hearing, it was stated before the Authority on behalf of the State that there was considerable progress as regards the availability of farm land situated at Khajuri for allotment to oustees, and it was assured that as soon as the formalities were completed for effecting transfer of the government land to the NVDA as provided in the Revenue Book Circular, and that the said land was transferred to NVDA, it would be added to the corpus of the land bank created by the NVDA for the purpose of offering agricultural land for allotment to eligible oustees and that action would be taken to allot the same to the eligible applicants. It was observed by the Authority that as the applicants and the State as well were exhibiting their positive approach for arriving at an amicable settlement of the grievances of the applicants relating to their rehabilitation, 10 days time was granted as desired, to enable them to do the needful. Therefore, with the consent of the parties the case was fixed for hearing on 4 September 2004.

12. On 4 September 2004 it was stated on behalf of the State that once the NVDA would be in possession of Khajuri land, it would be made available for allotment to oustees according to their entitlement. Written submission dated 4 September 2004 was also filed before the Authority on behalf of the NVDA. It was stated on behalf of the applicants that response if necessary would be filed by 6 September 2004. That response on behalf of the applicants was filed on 6 September 2004. Similarly, further submission on behalf of the State was made on 6 September 2004 stating that with a view to reciprocating the positive attitude shown by the applicants, as a special case, the State was willing to allot agricultural land of Khajuri farm in District Jhabua to the eligible applicants of villages Jalsindhi and Picchodi and as such it was possible to accommodate all the eligible oustees among the applicants in Khajuri farm land, but it was made clear that this may not be treated as a precedent as it would not always be practicable for obvious reasons to offer an allotment of agricultural land as per individual choice of the oustees.

13. These statements submitted on behalf of the parties were placed before the Authority on 5 September 2004, 7 September 2004 being a holiday in the State. Thereafter, after taking into consideration all the facts as brought on record, this order has been passed by the Authority on 11 September 2004.

14. It was stated on behalf of the applicants of village Picchodi by Shri Sanjay Parikh, learned Advocate of the Supreme Court, that those applicants had not approved the Khajuri land and that, as directed by the Supreme Court, the land to be allotted to the oustees should be of their choice. Reliance was placed on the following paragraph of the judgment dated 18 October 2000 passed by the Hon'ble Supreme Court in Writ Petition (C) No. 319 of 1994 (paragraph 180):

> The Award provides that every displaced family, whose more than 25 per cent of agricultural land holding is acquired, shall be entitled to and be allotted irrigable

> land of its choice to the extent of land acquired subject to the prescribed ceiling of the State concerned with a minimum of two hectares land.

It has to be borne in mind while interpreting the aforesaid observation, the Hon'ble Supreme Court has referred to Sub-clause IV(7) of Clause XI of Chapter IX of the NWDT Award which reads as under:

> IV(7): Allotment of Agricultural Lands: Every displaced family, from whom more than 25 per cent of its land holding is acquired, shall be entitled to and be allotted irrigable land to the extent of land acquired from it subject to the prescribed ceiling in the State concerned and a minimum of 2 hectares (5 acres) per family, the irrigation facilities being provided by the State in whose territory the allotted land is situated. This land shall be transferred to the oustee family if it agrees to take it. The price charged for it would be as mutually agreed between Gujarat and the concerned State. Of the price to be paid for the land a sum equal to 50 per cent of the compensation payable to the oustee family for the land acquired from it will be set off as an initial instalment of payment. The balance cost of the allotted land shall be recovered from the allottee in 20 yearly instalments free of interest. Where land is allotted in Madhya Pradesh or Maharashtra, Gujarat having paid for it vide Clause IV(6)(i) supra, all recoveries for the allotted land shall be credited to Gujarat.

It has also to be noted that the choice of the State where an oustee should settle rests with him. It is reasonable to hold that while reproducing the aforesaid clause, the Supreme Court did not want to modify that clause in view of the provisions of Section 11 of the Inter-State Water Disputes Act, 1956 which reads as under:

> Bar of jurisdiction of Supreme Court and other courts: Notwithstanding anything contained in any other law, neither the Supreme Court nor any other Court shall have or exercise jurisdiction in respect of any water dispute which may be referred to a Tribunal under this Act.

The Supreme Court had also noted that the aforesaid Sub-clause IV(7) of Clause XI of Chapter IX of the Award gives choice to the oustees to accept the offer. Therefore, the use of the word 'choice' in paragraph 193 of its judgment dated 18 October 2000 cannot be construed to mean that the land which is required be offered to an oustee should be of his choice. The land to be offered should no doubt be arable and irrigable.

15. It is further made clear that in view of the fact that the matter is being amicably settled, it is not necessary in the opinion of the Authority to deal with the validity of reasons given by the State for not accepting the suggestions made on behalf of the applicants to purchase private land as indicated by them.

16. As the farm land at Khajuri has been offered by the NVDA for rehabilitation of eligible oustees, and the land having been inspected and found suitable by the applicants of Jalsindhi, it is directed that the NVDA shall proceed to rehabilitate the applicants at the appropriate stage in the light of the judgment dated 18 October 2000 passed by the Hon'ble Supreme Court in Writ Petition (C) No. 319 of 1994, by allotting agricultural land to eligible applicants from out of the farm land at

Khajuri, according to their entitlement along with house sites at an R&R site nearby providing all civic amenities as mandated by the Award and other reliefs due to them according to the provisions of the Award and the R&R policy of the State. It is made clear that in the event of failure to do so, the applicants shall have liberty to approach this Authority afresh for redressal of any specific grievance.

17. The case is accordingly disposed of.

18. The Secretary of the Authority is directed to send a copy of the order to the Assistant Registrar (PIL Cell), Supreme Court for placing it on record of IA Nos 4 & 7 in Writ Petition (Civil) No. 328 of 2002. Copies of the order will be also sent to Shri Sanjay Parikh and to Shri Muralidhar, learned counsel for the counsel as desired by Shri Sanjay Parikh, Advocate and to the Vice Chairman and Commissioner (Reh.), NVDA.

Narmada Bachao Andolan v. Union of India (Petition 328 of 2002) Supreme Court Judgment, March 2005

Source: *Narmada Bachao Andolan v. Union of India and Others*, Supreme Court, IA No. 10 in IA No. 4 and IA No. 11 in IA No. 7, Writ Petition (Civil) No. 328 of 2002, Judgment of 15 March 2005 (S.B. Sinha, J.).

Introductory remarks

1. The SSP is one of the most ambitious multipurpose projects, which on completion is expected to produce 1,450 MW of power and supply water for irrigation and drinking purposes to areas not only in the riparian States including Kachchh in the State of Gujarat but even in areas belonging to non-riparian State like Rajasthan.

2. The multiple project by way of construction of a dam over the river Narmada began its journey in 1961. A large number of residents of the States of Madhya Pradesh, Maharashtra and Gujarat are affected by the said construction.

3. The Government of India in exercise of its power conferred upon it under Section 4 of the Inter-State Water Disputes Act, 1956, constituted a Tribunal and made the following reference to it:

> In exercise of the powers conferred by sub-section (1) of Section 5 of the Inter-State Water Disputes Act, 1956 (33 of 1956), the Central Government hereby refers to the Narmada Water Disputes Tribunal for adjudication of the water dispute regarding the inter-State river Narmada, and the river valley thereof, emerging from Letter No. MIP-5565/C-10527-K dated 6 July 1968, from the Government of Gujarat.

4. Another reference by the Government of India was made on 16 October 1969.

5. The state of Gujarat before the Tribunal admittedly made an offer that the oustees can be resettled and rehabilitated in the State of Gujarat wherefor a rehabilitation package would be granted if they opt, therefore, and in the event the oustees opt to stay back in their home State, the entire expenses for the purpose of rehabilitation shall be borne by the State of Gujarat.

6. An Award was made by the said Tribunal in terms of Section 5(2) read with Section 5(4) of the Inter-State Water Disputes Act, 1956 on 16 August 1978. Several references thereafter were filed by the concerned States. As regard relief and rehabilitation, the Award *inter alia* contained mandatory provisions containing Clause XI Sub-clause (IV)(6)(ii) stating that no submergence of any area would take place unless the oustees are rehabilitated. In terms of its Award, the Tribunal directed constitution of an inter-State administrative authority known as the NCA for the purpose of securing compliance with and implementation of the decision and directions of the Tribunal. The NCA in its turn constituted one or more sub-committees including one relating to resettlement and rehabilitation.

Writ Petition

7. The NBA, a non-governmental organisation, which has been in the forefront of the agitation against the construction of the Sardar Sarovar Dam, filed a writ petition before this court raising several issues including relief and rehabilitation.

Before this court a grievance was raised as regard the attitude on the part of the State of Madhya Pradesh as it made an attempt to wriggle out of its responsibilities to provide rehabilitation facilities to the oustees by offering them cash compensation. A contention was further raised that since offers to oustees affected at the 90 metres of the height of the dam to be settled in the State of Madhya Pradesh had not been made, further construction should not be permitted until one year after the resettlement of these project-affected families (PAFs) at 90 metres.

Decision of this Court

9. A three-judge bench of this Court by a judgment and order dated 18 October 2000 in *Narmada Bachao Andolan v. Union of India and others*,[1] disposed of the said writ petition upon issuing various directions. The court *inter alia* opined that:

i. displacement of the tribals and other persons would not per se result in violation of their fundamental or other rights;

ii. on their rehabilitation at new locations they would be better off than what they were;

iii. at the rehabilitation sites they will have more and better amenities than those they enjoyed in their tribal hamlets; and

---

1 Document reproduced above at page 138.

iv. the gradual assimilation in the mainstream of the society would lead to betterment and progress.

10. This Court in its judgment noticed that the Award provided that every displaced family whose more than 25 per cent of agricultural landholding is acquired, would be entitled to be allotted irrigable land of its choice to the extent of land acquired subject to the prescribed ceiling of the State concerned with a minimum of two hectares land. Furthermore, the PAFs will be allotted a house/plot free of cost. The Court noticed that the State governments have liberalised the policy with regard to resettlement and have offered packages more than what was provided for in the Award of the Tribunal. Such liberalised policy included those PAFs who were even encroachers, landless/displaced persons, joint-holders, tapu-land (island) holders and major sons (18 years old). The court noticed various measures taken by the States of Madhya Pradesh, Maharashtra and Gujarat for sustainable development as regard preserving the socio-cultural environment of the displaced persons in these States. This Court noticed that although in terms of the Award those sons of the oustees who had become major one year prior to the issuance of the notification for land acquisition were entitled to be allotted land; the State of Gujarat made a relaxation thereto so as to cover all those who became major up to 1 January 1987. Before us it is contended that the State of Madhya Pradesh also extended the cut-off date to the date of issuance of notification. The Court noticed that the R&R Group and the GRA having been established, a system had come into force for ensuring satisfactory resettlement and rehabilitation of the oustees. The Court furthermore noticed that at the instance of GRA, PAFs were being issued sanads for the lands allotted to them which will ensure provisions of a proper legal document in their favour. The Court also noticed that the sites had been identified by the State of Madhya Pradesh with a view to arrange the resettlement of PAFs, and out of the 92 sites for resettlement of PAFs which were required to be established, 18 were stated to be fully developed, development in 23 sites was in progress; 18 sites where the location and identification of land was complete but development work had not started, and 33 sites were such where location of land for the development was to be decided by the task force constituted for the said purpose. Noticing the variance between the rehabilitation package offered by the States of Madhya Pradesh and Gujarat this court opined (paragraph 247):

> The impression which one gets after reading the affidavit on behalf of the State of Madhya Pradesh clearly is that the main effort of the said State is to try and convince the PAFs that they should go to Gujarat whose rehabilitation package and effort is far superior to that of the State of Madhya Pradesh. It is, therefore, not surprising that the vast majority of the PAFs of Madhya Pradesh have opted to be resettled in Gujarat but that does not by itself absolve the State of Madhya Pradesh of its responsibility to take prompt steps so as to comply at least with the provisions of the Tribunal's Award relating to relief and rehabilitation. The State of Madhya Pradesh has been contending that the height of the dam should be lowered to 436 ft so that a lesser number of people are dislocated, but we find that

> even with regard to the rehabilitation of the oustees at 436 ft the R&R programme of the State is nowhere implemented. The State is under an obligation to effectively resettle those oustees whose choice is not to go to Gujarat. Appropriate directions may, therefore, have to be given to ensure that the speed in implementing the R&R picks up. Even the interim report of Mr Justice Soni, the GRA for the State of Madhya Pradesh, indicates a lack of commitment on the State's part in looking to the welfare of its own people who are going to be under the threat of ouster and who have to be rehabilitated. Perhaps the lack of urgency could be because of a lack of resources, but then the rehabilitation even in Madhya Pradesh is to be at the expense of Gujarat. A more likely reason could be that, apart from electricity, the main benefit of the construction of the dam is to be to Gujarat and to a lesser extent to Maharashtra and Rajasthan. In a federal set up like India, whenever any such inter-State project is approved and work undertaken, the States involved have a responsibility to cooperate with each other. There is a method of settling the differences which may arise amongst them like, for example, in the case of an inter-State water dispute the reference of the same to a tribunal. The award of the tribunal being binding, the States concerned are duty bound to comply with the terms thereof.

The Court issued *inter alia*, the following directions:

> 2. As the Relief and Rehabilitation Sub-group has cleared the construction up to 90 metres, the same can be undertaken immediately. Further raising of the height will be only *pari passu* with the implementation of the relief and rehabilitation and on the clearance by the Relief and Rehabilitation Sub-group. The Relief and Rehabilitation Sub-group will give clearance of further construction after consulting the three Grievances Redressal Authorities.
>
> 5. The reports of the Grievances Redressal Authorities, and of Madhya Pradesh in particular, shows that there is a considerable slackness in the work of the identification of land, acquisition of suitable land and the consequent steps necessary to be taken to rehabilitate the project oustees. We direct the States of Madhya Pradesh, Maharashtra and Gujarat to implement the Award and give relief and rehabilitation to the oustees in terms of the packages offered by them and these States shall comply with any direction in this regard which is given either by the NCA or the Review Committee or the Grievances Redressal Authorities.
>
> 7. The NCA will within four weeks from today draw up an action plan in relation to further construction and the relief and rehabilitation work to be undertaken. Such an Action Plan will fix a time frame so as to ensure relief and rehabilitation *pari passu* with the increase in the height of the dam. Each State shall abide by the terms of the action plan so prepared by the NCA and in the event of any dispute or difficulty arising, representation may be made to the Review Committee. However each State shall be bound to comply with the directions of the NCA with regard to the acquisition of land for the purpose of relief and rehabilitation to the extent and within the period specified by the NCA.

The Present Proceedings

11. As the directions of this court were not implemented in letter and spirit, applications were filed by the petitioners herein for directing the respondents to rehabilitate each of them in accordance with the NWDTA and the orders of this

court, as also for a direction that the orders passed by the GRA be set aside and not acted upon.

12. The petitioners in IA No. 4 of 2004, who are 23 in number, are residents of village Picchodi and the petitioners in IA No. 11 of 2004, who are 14 in number, are residents of village Jalsindhi. In these applications, the petitioners had prayed for a direction upon the respondents not to proceed with further construction by raising the height of the dam until all affected people at the height of 110 metres are rehabilitated in all respects.

13. As the GRA had been constituted by the State of Madhya Pradesh, this court without going into the merit of the matter by orders dated 16 April 2004 and 23 July 2004, directed the parties to agitate their grievances at the first instance before it.

Contentions

14. The contention of the applicants herein is that having regard to the fact that they are Project Affected Families (PAFs) and, thus, being oustees within the meaning of the Award made by Narmada Water Dispute Tribunal (NWDT), each one of them was entitled to the benefits of the rehabilitation package envisaged therein. Such entitlement, according to the applicants, must be extended to:

i. all major sons of the landholders;

ii. those who had also been temporarily affected; and

iii. the heirs of landholders who died prior to the date of notification.

15. It was further contended that in the event, those who had been temporarily affected as also the major sons of the original land holders are held entitled to the benefits of the rehabilitation package, the State of Madhya Pradesh be directed to allot suitable cultivable lands in their favour as the lands situated at Khajuri and measuring 13.40 hectares only would not be sufficient for that purpose.

Proceedings Before the GRA

16. The State contended that every oustee is offered land out of the land bank developed by it as per norms set out in the NWDT Award, and in the event any oustee does not intend to avail the same and finds the Special Rehabilitation Package (SRP) more attractive, he may do so. It was urged that the government had adopted a uniform policy for all the oustees and, thus, the claim for individual preferences cannot be acceded to. It was urged that it was not possible to allot or procure land for allotment as per choice of the applicants as the same is not required to be done under the NWDT Award. It was submitted that it is not possible for the State to procure the land suggested by the oustees and as such either they should accept the land allotted to them or avail the benefit of the SRP.

17. Before the GRA, an owner of land in question, viz. Shri Mahesh Tiwari, appeared and stated that he and his brothers were ready and willing to sell their landholdings measuring 116 acres situated at the village of Devla, at a market value which may be determined by the NVDA according to the procedure laid down in the Land Acquisition Act.

18. Before the GRA, the parties appeared. A piece of land measuring about 13.40 hectares situated at the village of Khajuri was proposed to be allotted by the State. The petitioners of IA No. 11 in IA No. 7 consented thereto.

19. The GRA, however, by reason of an order dated 11 September 2004, having regard to the availability of farm land at Khajuri, which was offered by NVDA for rehabilitation of eligible oustees, directed the State, having regard to the settlement arrived at by and between the parties, to proceed to rehabilitate the applicants at the appropriate stage in the light of the judgment dated 18 October 2000 passed by this court by allotting agricultural lands to the eligible applicants from out of the farm land at Khajuri, according to their entitlement along with house sites at an R&R site nearby and providing the civil amenities as mandated by the Award and other reliefs due to them according to the provisions of the Award and the R&R policy of the State. The State of Madhya Pradesh, however, allotted only 5 land pattas and 7 house plots out of 23 applicants of the village of Picchodi and 5 land pattas and 14 house plots pattas to the 14 oustees of the village of Jalsindhi.

20. The applicants of both the interlocutory applications are, thus, before us.

Admitted Fact

21. It is neither in doubt nor in dispute that applicants herein are PAFs within the meaning of the Award of the Tribunal. It is also not in dispute that acquisition of the land took place, so far as village Jalsindhi is concerned, in terms of the provisions of the Land Acquisition Act in the year 1991 whereas in respect of village Picchodi, it took place in 2000. It is furthermore not in dispute that the applicants belonging to both villages, Picchodi and Jalsindhi, come within the purview of the PAFs, at the height of 95 metres to 100 metres of construction of the dam. It also stands admitted that the present height of the dam is 110 metres.

22. Indisputably, although the State intended to make a distinction between the temporary and permanent oustees, in its affidavit dated 6 May 1999 filed before this Court no such distinction was made and in fact it was emphasised that even temporary submergence even for a short period can affect the oustees badly and, thus, no distinction should be made between temporary and permanent PAFs.

23. Clause XI of the Award indisputably pertains to the directions regarding submergence, land acquisition and resettlement and rehabilitation of displaced persons which would include both permanently and temporarily affected persons.

Relevant Clauses of the Award

24. Sub-clauses II(1), II(2), IV(2)(i), IV(2)(ii), IV(2)(iv), IV(6)(ii), IV(7) and V(3)(iii) of Clause XI of the Award read as under:

> II(1): Madhya Pradesh and Maharashtra shall acquire for the SSP under the provisions of the Land Acquisition Act, 1894, all lands of private ownership situated below the FRL +138.68 m (455 ft) of Sardar Sarovar and all interests therein not belonging to the respective States. If on the basis of the aforesaid, 75 per cent or more land of a contiguous holding of any person is required to be compulsorily acquired, such person shall have the option to compel compulsory acquisition of the entire contiguous holding.

> II(2): Madhya Pradesh and Maharashtra shall also acquire for the SSP, under the provisions of the Land Acquisition Act, 1894, all buildings with their appurtenant land situated between FRL +138.68 m (455 ft) and MWL +141.21 m (460 ft) as also those affected by the backwater effect resulting from MWL +141.21 m (460 ft).

> IV(2)(i): According to the present estimates the number of oustee families below RL 106.68 metres (RL 350 ft) would be 30 spread over 20 villages in Madhya Pradesh and 250 families spread over 20 villages in Maharashtra. Within six months of the publication of the decision of the Tribunal in the Official Gazette, Gujarat, Madhya Pradesh and Maharashtra shall determine by mutual consultation the location of one or two rehabilitation villages in Gujarat to rehabilitate oustees from areas below RL 106.68 metres (RL 350 ft). Gujarat shall acquire necessary lands for the rehabilitation villages and make available the same within two years of the decision of the Tribunal. Within six months of the decision of the location of the rehabilitation villages in Gujarat, Madhya Pradesh and Maharashtra shall intimate to Gujarat the number of oustee families from areas below RL 106.68 metres (RL 350 ft) willing to migrate to Gujarat. For the remaining oustee families, Madhya Pradesh and Maharashtra shall arrange to acquire lands for rehabilitation within the respective States.

> IV(2)(ii): Madhya Pradesh and Maharashtra shall set up adequate establishments for land acquisition and rehabilitation of oustee families. Gujarat shall deposit within three months of the decision of the Tribunal Rupees 10 lakhs each with Madhya Pradesh and Maharashtra in advance towards the cost of establishment and rehabilitation in these States to be adjusted after actual costs are determined. Madhya Pradesh and Maharashtra shall start land acquisition proceedings for areas below RL 106.68 metres (RL +350 ft) within six months of the decision of the Tribunal and convey the lands to Gujarat for project purposes within three years of the decision of the Tribunal. Within 18 months of the decision of the Tribunal, Gujarat shall make an advance payment of Rs 70 lakhs to Madhya Pradesh and Rs 100 lakhs to Maharashtra towards the compensation of land, to be adjusted after actual costs are determined.

> IV(2)(iv): Gujarat shall acquire and make available a year in advance of the submergence before each successive stage, irrigable lands and house sites for rehabilitation of the oustee families from Madhya Pradesh and Maharashtra who are willing to migrate to Gujarat. Gujarat shall in the first instance offer to rehabilitate the oustees in its own territory.

> IV(6)(ii): In no event shall any areas in Madhya Pradesh and Maharashtra be submerged under the Sardar Sarovar unless all payment of compensation, expenses and costs as aforesaid is made for the acquisition of land and properties and arrangements are made for the rehabilitation of the oustees therefrom in accordance with these directions and intimated to the oustees.
>
> IV(7): Allotment of Agricultural Lands: Every displaced family from whom more than 25 per cent of its land holding is acquired shall be entitled to and be allotted irrigable land to the extent of land acquired from it subject to the prescribed ceiling in the State concerned and a minimum of 2 hectares (5 acres) per family, the irrigation facilities being provided by the State in whose territory the allotted land is situated. This land shall be transferred to the oustee family if it agrees to take it. The price charged for it would be as mutually agreed between Gujarat and the concerned State. Of the price to be paid for the land a sum equal to 50 per cent of the compensation payable to the oustee family for the land acquired from it will be set off as an initial instalment of payment. The balance cost of the allotted land shall be recovered from the allottee in 20 yearly instalments free of interest. Where land is allotted in Madhya Pradesh or Maharashtra, Gujarat having paid for it vide Clause IV(6)(i) supra, all recoveries for the allotted land shall be credited to Gujarat.
>
> V(3)(iii): Gujarat shall at each successive stage of submergence intimate to Madhya Pradesh and Maharashtra the area coming under submergence at least 18 months in advance. The inhabitants of the area coming under the respective stages of submergence will be entitled to occupy or use their properties without being required to pay anything for such occupation and use until a date to be notified by the State concerned which shall not be less than six months before submergence. They must vacate the area by the notified date.

25. The provisions of the Award are required to be read along with the definitions of 'oustee' and 'family' contained in Sub-clauses 1(1) and 1(3) thereof which read as under:

> 'Oustee' – An 'oustee' shall mean any person who since at least one year prior to the date of publication of the notification under Section 4 of the Act, has been ordinarily residing or cultivating land or carrying on any trade, occupation, or calling or working for gain in the area likely to be submerged permanently or temporarily.
>
> 'Family' – (i) A family shall include husband, wife and minor children and other persons dependent on the head of the family, e.g. widowed mother. (ii) Every major son will be treated as a separate family.

Submissions

26. The learned counsel appearing on behalf of the applicants submitted that for the purpose of the grant of benefit of the rehabilitation package, no distinction can be made between temporarily and permanently affected people and in this connection our attention has been drawn to the stand taken by the respondent State in the earlier proceedings as also the Award. It was submitted that the major sons of the PAFs being included in the definition of 'family' and treated to be a separate

family, they are entitled to allotment of a separate unit in terms of the Award as also the judgment of this court.

27. The learned counsel would further contend that those applicants who were adults on the cut-off date and whose fathers have passed away are also entitled to the benefit of the rehabilitation package. It was contended that the applicants must be given a choice as regard the site of the irrigable and cultivable lands.

28. The submission of Mr C.S. Vaidyanathan, learned senior counsel appearing on behalf of the respondents, on the other hand, is that the Award contemplates the grant of benefits of the rehabilitation package only to such persons who were affected by reason of the raising of the height of the dam and, thus, all the PAFs are not entitled to the grant of land for land. Further, the contention of Mr Vaidyanathan is that the entire family has to be treated as a unit and the adult sons of a landholder are not entitled to a separate unit unless they were themselves land holders. This question, according to Mr Vaidyanathan, had not so far been specifically considered by this court.

29. According to the learned counsel, Sub-clause IV(7) of Clause XI of the Award clearly specifies the persons who would be entitled to the grant of alternative land. The Award, Mr Vaidyanathan would argue, makes a distinction between permanently affected persons and temporarily affected persons.

Points for Consideration

i. Whether there exists a distinction between temporarily and permanently affected persons in the NWDT Award as well as the judgment of this court?

ii. Whether adult sons are entitled to a minimum of 2 hectares of land as per the NWDT Award and judgment of this court?

iii. Whether those adult sons who became landholders since their fathers passed away, are entitled to the benefit of alternate lands, in place of the acquired lands standing in the names of their deceased fathers?

Determination

Permanently and temporarily affected families

30. Sub-clause IV(6)(ii) of Clause XI makes it imperative that submergence would not be allowed to take place until the complete settlement and rehabilitation of oustees is done, which in view of the definition of 'oustees' would mean both permanently and temporarily affected persons.

31. It has been the consistent stand on the State of Madhya Pradesh that temporarily affected persons would come within the purview of the expression PAFs and there exists no distinction between permanently affected and temporarily affected persons.

32. We may, at this juncture, notice the pattern of rehabilitation of affected families in the SSP from the following chart relied on by the Applicants:

Rehabilitation of Sardar Sarovar Project Affected Families

A Game of Numbers: MP's Diminishing PAF List

Status of R&R at Dam Height EL 95 m of MP PAFs								
Date	Total No. of PAFs	Claimed as Resettled			Balance	Option of Balance		Source of Information
		In MP	In Gujarat	Total		MP	Gujarat	
29/8/01	5,397	1,182	2,385	3,567	1,830	1,378	452	Agenda of 50th meeting of R&R Sub-group
11/11/01	5,379	1,394	2,381	3,775	1,603	782	821	RCNCA (CMS) meeting
8/12/01	5,397	1,399	2,418	3,817	1,580	1,217	363	Agenda of 51st meeting of R&R Sub-group
7/1/02	5,397	1,466	2,691	4,157	1,240	1,150	90	Minutes of 41st meeting of R&R Sub-group
8/2/02	5,397	1,466	2,691	4,157	1,240	1,150	90	Agenda of 52nd meeting of R&R Sub-group
14/5/02	1,883			1,873	10			Minutes of 53rd meeting of R&R Sub-group
31/5/02	1,883*	967	916	1,883	0	0	0	Quarterly status report, NCA
31/12/02	1,883*	967	916	1,883	0	0	0	Half-yearly status report, NCA

* The GoMP has resettled only those PAFs (i) whose agricultural land is coming under permanent submergence and (ii) whose habitation is coming under permanent or temporary submergence due to a 1 in 100-year flood (end notes are taken directly from NCA documents).

Status of R&R at Dam Height EL 100 m of MP PAFs

Date	Total No. of PAFs	Claimed as Resettled			Balance	Option of Balance		Source of information
		In MP	In Guj	Total		MP	Gujarat	
29/8/01	7,913	1,327	2,584	3,911	4,002	2,554	1448	Agenda of 50th Meeting of R&R Sub-group
11/11/01	7,913	1,587	2,684	4,271	3,570	1,902	1668	RCNCA (CMs) meeting
7/1/02	7,913	1,670	3,360	5,030	2,883	2,693	190	Minutes of 51st meeting of R&R Sub-group
8.2.02	7,913	1,670	3,360	5,030	2,883	2693	190	Agenda of 52nd meeting of R&R Sub-group
30/6/02	3,071*	1,990	1,036	3,026	45	45	0	Quarterly Status Report, NCA
14/11/02	3,710*	2,443	1,198	3,641	69	69	0	Minutes of 54th meeting of R&R Sub-group
31/12/02	3,710*	2,443	1,243	3,686	24	24	0	Half yearly Status Report, NCA
13/5/03	3,692*	2,434	1,258	3,692	0	0	0	Minutes of 55th Meeting of R&R Sub-group
30/6/03	3,692*	2,434	1,256	3,692	0	0	0	Half-yearly status report, NCA

* PAFs whose land are temporarily under submergence due to 1 in 100-year flood have not been considered for R&R.

Status of R&R at Dam Height EL 110 m of MP PAFs								
Date	Total No. of PAFs	Claimed as Resettled			Balance	Option of Balance		Source of information
		In MP	In Gujarat	Total		MP	Gujarat	
29/8/01	12,681	1,809	2,802	4,611	8,070	5,489	2,581	Agenda of 50th Meeting of R&R Sub-group
11/11/01	12,681	2,005	2,896	4,901	7,708	5,288	2,420	RCNCA (CMs) meeting
8/2/02	12,681	2,079	3,653	5,732	6,949	5,219	1,730	Agenda of 52nd meeting of R&R Sub-group
14/11/02	12,681*	2,175	3,628	5,803	6,878	5,425	1,453	Minutes of 54th Meeting of R&R Sub-group
13/5/03	5,607**							Minutes of 55th Meeting of R&R Sub-group
30/6/03	8,406***	5,893	2,016	7,909	497	291	206	Half-yearly Status Report, NCA

* This number may change after the declaration of LAQ awards. PAFs whose lands are temporarily submerged due to 1 in 100-year flood have not been considered for R&R.

** Tentative.

*** This number may change due to the addition of genuine PAFs likely to be included after the declaration by the GRA and the passing of the land acquisition award.

33. The contents of the aforementioned chart, are not denied or disputed. They are said to be supported by documents.

34. It is also relevant to notice the gazette dated 31 December 2001 issued by the State of Madhya Pradesh which is as under:

> No.4-73-27.2.2001-1414 – It is informed that because of the water level in SSP for the monsoon of 2002 the villages shown in list 1 will be affected and the oustees shown in list 2 will be affected with respect to their lands, houses and other property. These oustees will be able to make use of submergence affected property until the 31 of December 2001. After that they will have to relinquish this property; all families included in earlier notifications are also included in this notification.

Name of village	No. of PAFs including adult sons	Total effect due to submergence of Sardar Sarovar Project		Effect of submergence in monsoon of 2002	
		No. of houses	Agricultural land (in ha)	No. of houses	Agricultural land (in ha)
Picchodi	428	104	123.497	104	123.497

35. The names of all the 23 applicants of the village of Picchodi find place in the gazette published by the State, the details whereof are as under:

S.N.	Name of PAF and father's name	Landholder/ adult son	Total effect due to submergence of Sardar Sarovar Project		Effect of submergence in monsoon of 2002		Details
			No. of houses	Agricul-tural land (in ha)	No. of houses	Agricul-tural land (in ha)	
12	Mangilal s/o Madia	Adult Son	--	--	--		
34	Ramesh s/o Kalu	LH	--	3.569	1	3.569	
36	Badrilal s/o Kalya	LH	--	--	--	--	Co-sharer of 34
37	Jagan s/o Kalya	LH	--	--	--	--	Co-sharer of 34
38	Sagar w/o Kalya	LH	--	--	--	--	Co-sharer of 34
39	Vediya s/o Dariyav	LH	1	--	1	--	Co-sharer of 34
54	Shankar Rukhadiya	LH	1	1.154	1	1.154	--
55	Sonibai Rukhadiya	LH	--	--	--	--	Co-sharer of 54
56	Shambu Motia	LH	--	0.664	--	0.664	--
216	Pratap Tersingh	LH	--	1.056	--	1.056	--
278	Pokhar Girwar	LH	1	3.152	1	3.152	--
279	Punya Girwar	LH	1	--	--	--	--
281	Buda Banga	LH	1	0.615	1	0.615	--
282	Babu Banga	LH	1	--	1	--	Co-sharer of 281
283	Dhamibai Banga	LH	--	--	--	--	Co-sharer of 281
284	Ratansingh Ranchod	LH	1	4.078	1	4.078	--
285	Radheshyam Ratan	Adult son	--	--	--	--	--
286	Sitaram Ratan	Adult son	--	--	--	--	--
287	Govind Ramsingh	LH	1	1.13	1	1.13	--
288	Sitaram Govind	Adult son	--	--	--	--	--
364	Lanka Pokhar	LH	--	0.243	--	0.243	--

36. The names of the applicants of the village of Picchodi, thus, except Rajaram Pratap, who is an adult son of Pratap Tersingh are contained in the gazette. Similar is the position of the applicants of the village of Jalsindhi whose names also appear in the gazette issued by the State of MP wherein it was categorically stated that they would be affected by submergence in the monsoon of 2002 when the dam height was raised to 95 m. Their names also appear in the Action Taken Report of the State of Madhya Pradesh and the NVDA as was submitted to the Narmada Control Authority with a view to obtaining permission for raising the height of the dam from 90 m to 95 m and then from 95 m to 100 m. In fact, the State had claimed that most of the applicants had already been rehabilitated.

37. It is difficult to accept the contention of Mr Vaidyanathan that the residents of Picchodi village had not been affected at the dam height of 110.64 metres or the house of Pratap Tersingh is not affected. We have noticed hereinbefore that the lands of Picchodi village stood affected at 95–100 m. No material has been placed before us that the oustees of the said village were not affected due to permanent or temporary submergence at the dam height of 110.64 m. No such contention has been raised even before the GRA. Furthermore, it has not been explained that as to how five of them were given the benefit of land for land and house plots.

38. R&R Status of the PAFs at Sardar Sarovar Dam Height EL 95 m as on 31 December 2001 is as under:

	No. of villages affected	Total PAFs	PAFs resettled/allotted agricultural land/paid cash compensation				Balance PAFs to be resettled		
			In Gujarat	In MH	In MP	Total	In Gujarat	In home State	Total
MP	70	5,397	2,691*	0	1,466	4,157	90	1,150	1,240

* Includes ex parte allotment to 253 MP PAFs at EL 95 m.

39. Status of Land Acquisition Awards in the State of Madhya Pradesh at EL 95 m is as under:

(i) For agricultural land

SN	Tehsil	No. of Villages	Awards declared	No. of villages balance for Awards	Notification issued under		
					S.4	S.6	S.9
3	Badwani	20	16	4	4	4	4

(ii) For abadi land

SN	Tehsil	No. of Villages	Awards declared	No. of villages balance for Awards	Notification issued under		
					S.4	S.6	S.9
3	Badwani	16	12	4	4	4	3

40. Despite the same, the State now contends:

> 14.1 That the allegations in the application (IA 4) is that the Government of MP is arbitrarily drawing distinction between temporary and permanent submergence and is not effecting rehabilitation as mandated in the NWDTA, and the directions given in the judgment of this Hon'ble Court. According to sub-clause II(1) (Chapter IX, Clause XI of NWDTA), only such lands of private ownership have to be acquired which fall below FRL (138.68 m). Agricultural lands affected by backwater (afflux) are not to be acquired. As per Sub-clause II(2), ibid., only buildings with their appurtenant land between FRL (138.68 m) and MWL (141.21 m) shall be acquired.

41. The contention of the State of Madhya Pradesh, however, is based on Sub-clause II(1) of Clause XI of Chapter IX of the NWDT Award in terms whereof allegedly only such lands of private ownership have to be acquired which fall below FRL 138.68 m and agricultural lands affected by backwater (afflux) are not to be acquired.

42. It was further contended that in terms of the judgment dated 18 October 2000 of this Court, rehabilitation has to be done *pari passu* with the construction of the dam.

43. It is also relevant to mention that the stand of the State of Madhya Pradesh in terms of the Award was that PAFs should be resettled as a village unit as per the

stipulation of the NWDT Award as far as possible and upon taking practical aspects of the matter into consideration.

44. In terms of the NWDT Award, the irrigable lands and house sites were required to be made available to the PAFs one year in advance of the submergence and requisite amenities were also to be provided. Further, the notices for vacation of the lands are to be given after completion of the R&R of the PAFs on or before 31 December, i.e. 6 months before actual submergence (likely on the 1 of July of the next year). In terms of these stipulations, raising of the dam which would cause submergence would not be permitted unless the rehabilitation programme is carried out. Even in the stipulations of the NWDT decision, which has been accepted by the State of Madhya Pradesh, no distinction was made between permanently affected and temporarily affected families.

45. The Award does not make any distinction between permanently affected families and temporarily affected families. Had it been so, the definition of the 'oustees' would not have been so worded.

46. It is evident that in the Award of the Tribunal no distinction was made between permanently affected and temporarily affected oustees. The State, as noticed hereinbefore, in its affidavit filed before this court in the writ petition, not only failed and/or neglected to raise such a contention but, as pointed out in the rejoinder affidavit filed by the petitioners to the affidavit filed by the State, in fact the State in its affidavits filed before this court had taken a firm stand that permanent oustees and temporary oustees stand on the same footing. The State in support of the aforementioned contention, had also relied upon documents including the views of several committees and their reports. Furthermore, the State had adopted a policy of rehabilitation of oustees, in terms whereof contentions had been raised and a judgment has been obtained and in that view of the matter it is now not open to it to raise a contention which would run counter thereto or inconsistent therewith. The submission of Mr Vaidyanathan to the effect that some of the applicants herein had been granted only house sites, as they were not affected by permanent submergence, cannot, therefore, be accepted. It may be true that the Award makes a distinction between those whose agricultural land had been taken over and those who were in the fringe area and who would face the problem of residence only. However, the applicants herein do not fall in the said category.

47. The Award, as noticed hereinbefore, contained two sub-clauses relating to the directions on the State government for compulsory acquisition of the land by the State of Madhya Pradesh and Maharashtra under the provisions of the Land Acquisition Act. This obligation on the part of the State to acquire land is, thus, neither in doubt nor in dispute. The additional directions are that those persons whose 75 per cent or more land of a continuous holding is required to be compulsorily acquired, will have an option to compel compulsory acquisition of the entire contiguous holding; and acquisition of buildings with their appurtenant land situated between FRL +138.68 metres (455 ft) and MWL +141.21 (460 ft) as

also those affected by the backwater effect resulting from MWL +141.21 metres. The submergence due to maximum water level and backwater would take place only after it reaches full height.

48. In the Action Taken Reports (ATRs) of 90–95 m and 95–100 m, the applicants have been shown as PAFs having been rehabilitated in Gujarat purported to be on the basis of allotment of land made behind their back. The ATR, being a document pursuant whereto or in furtherance whereof permission for increasing the height of the dam was given, cannot be ignored and, thus, the State cannot be permitted to turn round and contend that the applicants are not entitled to be rehabilitated at this stage. It is evident that the State took a different stand at the earlier stage of the proceedings on the assumption that these oustees would go to Gujarat and as such there entitlements were acknowledged, but as soon as they made it clear that they will prefer rehabilitation in the State, their rights are being denied. This attitude on the part of the State, as has been observed in the main judgment, cannot but be deprecated.

49. Sub-clause IV (6)(ii) of Clause XI of the Award states that no kind of submergence in the States of Madhya Pradesh and Maharashtra shall be permitted unless arrangements are made for rehabilitation of the oustees in terms of the directions contained therein. Thus, complete resettlement and rehabilitation of oustees was a condition precedent for submergence.

50. From the following excerpts of the report of the Narmada Control Authority (NCA) which is the highest authority in the matter of implementation of the Award, it is clear that no such distinction can be made:

> Further, it was decided, as per decision in the last meeting of the Sub-group, all possible arrangements for R&R should be made by the concerned State governments. For completing the same in all respect both in regard to oustees affected by the permanent as well as temporary submergence six months ahead from submergence. Actual allotment of land, house plot and payment of compensation etc., and not merely offer of such facilities as per the R&R package, should be made in respect of all PAFs (both categories of affected by permanent and temporary submergence) except in the case, of hardcore PAFs who refuse to accept the package and are unwilling to shift.
>
> Temporary submergence even for a short period can affect the oustees badly and that it is desirable to keep this in mind while rehabilitating the oustees.
>
> In the light of the earlier decision by the NCA on this subject, there should not be any distinction between temporary and permanent PAFs and will be a pre-requisite for the purpose of further raising of the dam.

51. The submission of Mr Vaidyanathan on interpretation of Sub-clauses II(1) and II(2) of Clause XI of the NWDT Award, that such a distinction is implied, is for the foregoing reasons rejected. The said clause applies only to the matter relating to land acquisition at the full height of the dam, i.e. 138.68 metres. This Court did not say in the main judgment that *pari passu* principle applies only to permanently

affected families. If the lands of the applicants are acquired, they are entitled to rehabilitation.

52. This Court in its judgment in *Narmada Bachao Andolan* (supra) permitted construction of the dam up to 90 metres and opined that further raising of the height would be only *pari passu* with the implementation of the relief and rehabilitation measures.

53. In *Black's Law Dictionary*, 5th ed., the term *pari passu* has been defined to mean: 'By an equal progress; equably, ratably; without preference'.

54. The expression *pari passu*, therefore, has a direct nexus with raising of the height vis-à-vis implementation of relief and rehabilitation progress both of which must proceed 'equably' or 'ratably', which would mean that relief and rehabilitation measures must be undertaken as and when the height of the dam is further raised. The said expression should be construed in a meaningful manner.

55. The applicants herein became affected with the raising of the dam at 90 metres and remained affected by the further raising thereof up to 100 metres and, thus, in terms of the directions contained in the Award as also the judgment of this court, it is beyond any cavil that the applicants herein, irrespective of the fact as to whether they are permanently affected or temporarily affected, were entitled to the benefit of the rehabilitation package. We are not oblivious of the fact that the river valley of Narmada is shaped like an inverted cone and the area of submergence increases exponentially for each metre of height raised. We are also not unmindful of the fact that before this court it was contended by the original writ petitioners that whole land up to 138 metres should be acquired, people immediately be resettled, and all requisite studies be done up to that level before permitting the dam height to be raised. It is only in that context this court used the expression *pari passu*.

56. We may notice that an observation has been made by the Chairman of the R&R Sub-group in the meeting held on 11 April 1994 that temporary submergence even for a short period can affect the oustees badly and it is desirable to keep this in mind while rehabilitating the oustees. In the meeting held on 18 December 1998, it was observed:

> In the light of the earlier decision by the NCA on this subject, there should not be any distinction between temporary and permanent PAFs and will be a pre-requisite for the purpose of further raising of the dam.

57. Our attention has been drawn to various orders of the GRA to the effect that a distinction has been made between the temporary affectees and permanent affectees. We do not subscribe to the said view.

58. We are of the opinion that all the applicants who were both permanently and temporarily affected by submergence by reason of raising of the height of the dam to the present height would be entitled to the benefit of the rehabilitation package.

Major Sons

59. The definition of family indisputably includes major sons. A plain reading of the said definition clearly shows that even where a major son of the land-holder did not possess land separately, he would be entitled to the grant of a separate holding. The State of Gujarat, it is trite to notice, has extended this facility even to unmarried daughters.

60. The definition of 'family' has to be read along with that of the 'oustee'. We may notice that 'oustee family' and 'displaced family' have interchangeably been used in the Award. They, thus, carry the same meaning.

61. In paragraph 180 of the main judgment, this court noticed that every affected family must be allotted land, a house plot and other amenities. In paragraph 203 thereof, it was noticed:

> According to the Tribunal's Award, the sons who had become major one year prior to the issuance of the notification for land acquisition were entitled to be allotted land.

62. It is now well-settled that when the interpretation clause used an inclusive definition, it would be expansive in nature.

63. In G.P. Singh's *Principles of Statutory Interpretation*, 9th ed. (2004), at page 166, it is stated:

> The word 'includes' is often used in interpretation clauses in order to enlarge the meaning of the words or phrases occurring in the body of the statute. When it is so used these words and phrases must be construed as comprehending not only such thing as they signify according to their nature and import but also those things which the interpretation clause declares that they shall include.

(See also *Godfrey Phillips India Ltd v. State of UP*).[2]

64. Once 'major son' comes within the purview of the expansive definition of family, it would be idle to contend that the scheme of giving 'land for land' would be applicable to only those major sons who were landholders in their own rights. If a person was a landholder, he in his own right would be entitled to the benefit of the rehabilitation scheme and, thus, for the said purpose, an expansive definition of family was not necessarily to be rendered. Furthermore, if such a meaning is attributed as has been suggested by Mr Vaidyanathan, the definition of 'family' would, to an extent, become obscure. As a major son constitutes 'separate family' within the interpretation clause of 'family', no meaning thereto can be given.

65. In IA No. 11 of 2003, there is no dispute as regard the age of the concerned applicants. In that case, two of the landholders Athiya and Khatriya died even prior to the issuance of the notification. This Court, in paragraphs 152 and 176 of the main judgment, specifically referred to the entitlement of the major sons (18 years

---

2 2 (2005) 2 SCC 515. 2005 AIR SCW 613.

old). The major sons, therefore, cannot be denied the said benefit. A half-hearted contention was raised on behalf of the State that those who had not been granted land might not have become major on the date of notification. Such a contention had not been raised before the GRA. We at this stage cannot permit a new plea to be raised and that too without any pleading and supporting material brought on records in that behalf.

66. Each of the 8 applicants were, thus, in reality a landholder in their own right since their fathers Athiya and Khatriya died even prior to issuance of the notification under Section 4 of the Act. They, therefore, could not have been directed to be given only a house plot on the ground that they were adult sons of the landholders. The applicants, Athia Dhoklia and Khatria Peecha, not only had asked for an allotment of land in the State of Madhya Pradesh, they had filed these applications long ago. It is to be noticed that Noorjiya s/o Mahariya had not been given the benefit of an allotment of land although his brother Bunda and his mother Kajli had been recognised as eligible for the allotment of agricultural land to the extent of 2 hectares each. There is, thus, no ground to deny the said benefit to Noorjiya.

67. Several contentions involving factual dispute had, we may notice, not been raised before the GRA. The GRA had been constituted with a purpose namely, that the matters relating to the rehabilitation scheme must be addressed by it at the first instance. This Court cannot entertain applications raising grievances involving factual issues raised by the parties. The GRA, being headed by a former Chief Justice of the High Court, would indisputably be entitled to adjudicate upon such disputes. It is also expected that the parties should ordinarily abide by such decision. This Court may entertain an application only when an extraordinary situation emerges.

Choice of Land

68. In a case of this nature we do not accept the contention raised on behalf of the applicants herein that the oustees are entitled to opt for land of their choice and the State is bound to acquire or purchase lands for the said purpose. The State has constituted a land bank. Normally, those lands which are available from the land bank should be allotted and in relation thereto, the parties may have a choice. But they cannot reject such land only unless it is shown that the land is not irrigable or cultivable or otherwise unsuitable. In view of the dicta of this Court, that the oustees would be better off at the rehabilitated place, they should be offered lands which are really cultivable or irrigable. They are also entitled to the basic civil amenities and benefits specified in the Award. In this view of the matter, if and when necessary, the GRA would be entitled to consider the matter in accordance with law and pass suitable directions.

69. This Court, in the main judgment, did not say that the oustees are to be relocated as a community. The question of rehabilitation inevitably would arise as and when they become entitled thereto.

Extent of Land

70. It is not in dispute that the Award provided that every displaced family, whose 25 per cent or more agricultural landholding has been acquired, shall be entitled to be allotted irrigable land to the extent of land acquired subject to the prescribed ceiling of the State with a minimum of two hectares of land.

71. It is, however, not in dispute that the lands offered by the NVDA, a state forum, have been found acceptable by the applicants belonging to the village of Jalsindhi. We direct the respondents to allot such lands immediately to them. Having regard to the fact that the farm lands available at Khajuri village would be insufficient for allotment to the applicants of IA No. 11, the matter may be considered afresh by the GRA. We agree with the opinion of the GRA that the applicants therein would not be entitled to an allotment of land of their choice but the land offered to them should be irrigable and cultivable in terms of the judgment of this court as well as the Award of the Tribunal. We hope and trust that the parties hereto shall render all cooperation with the GRA for the purpose of finding suitable irrigable and cultivable lands for allotment thereof to the applicants of Picchodi village at an early date and preferably within a period of three months from the date of communication of this order.

Conclusion

72. These applications are disposed of with the aforementioned directions. In the facts and circumstances of the case, there shall be no order as to costs.

Special Rehabilitation Package

Narmada Valley Development Authority, Proposal for Amending the Terms of Resettlement and Rehabilitation under NWDT, 2001

Source: Amendment to the stipulations in Sub-clause IV(7) of Clause XI of NWDT Award, Letter from H.N. Tiwara, Director (TW), Narmada Valley Development Authority, Bhopal to the Executive Member/Member (E&R), Narmada Control Authority, Indore, Document No. 79/Socio/GRA/2000/674, 28 March 2001/17 May 2001.

Dear Sir,

1. About 33,014 families are being affected in MP by the SSP and out of these only 18,890 PAFs, as per Action Plan 1,933, are to be resettled in MP and only 830 PAFs are eligible to be allotted land in Madhya Pradesh in lieu of land acquired from them. Although as per the initial assessment only 2,000 ha of land was required for these 830 PAFs, out of the 14,124 PAFs allocated for resettlement in Gujarat, many of them are changing their option and as a result as per provision under Clause IV(6)(i) of the NWDT Award, the State of MP is required to arrange land for such PAFs also. The number of such PAFs is increasing day by day necessitating the requirement of a larger area of land in MP. Besides, the PAFs are also demanding their resettlement on the principles of the village as a unit as provided in the rehabilitation policy of the State. This would mean that the agricultural land in sufficient measure has to be arranged at the relocation sites or as close to them as possible.

2. Since there is a paucity of cultivable government land in Madhya Pradesh as pointed out at various fora, it is difficult to arrange more land for the oustees in the contingency of accepting change in their option exercised earlier. Although efforts are on to identify and procure as much land as possible for allotment to the oustees, past experience does not raise our hopes very high. As such, there is a need for considering suitable amendments in Sub-clause IV(7) of Clause XI of the NWDT Award and introducing a provision on the option of the oustee to accept compensation in full in one instalment to enable him to purchase land and settle down as per his choice. It may be recalled that the Hon'ble Chief Minister of Madhya Pradesh in the meeting of the Review Committee NCA, held on 10 January 2001, had very explicitly observed that if the PAFs are required to be allotted agricultural land and arrangement of the land is to be made by the State, and the resettlement is to be village as a unit then it will be very difficult to adhere to the time frame for the resettlement of the PAFs as approved by the NCA.

3. In the light of the above, the State of Madhya Pradesh proposes that following Sub-clause IV(7)(a) of Clause XI, a proviso may be added in the NWDT Award below the Clause IV(7):

> Provided that the displaced family shall have the option of obtaining full compensation for settling down and purchasing land in a village of their choice by submitting an application to this effect in writing, to the concerned officer of the State of Madhya Pradesh or Maharashtra. An option once exercised shall be final and no claim for allotment of land can be made afterwards.

4. The Sub-clause VI of the Clause XI of the NWDT Award – 'nothing contained in Clause XI shall prevent the alteration, amendment and modification of all or any of the foregoing clauses by agreement between all party States'. In view of this provision the aforesaid suggested amendment may please be presented in the next ensuing meeting of the NCA for approval by consensus in the interest of the displaced family as it would enable them to settle down and purchase land in a

village of his choice at the earliest. This will also help to complete the project as schedule.

Narmada Control Authority, Consideration of Proposal for Special Rehabilitation Package, 2001

Source: Amendment to Stipulations in Sub-clause IV(7) of Clause XI of NWDT Award Concerning Allotment of Agricultural Lands, Narmada Control Authority, Minutes of the 63rd Meeting, 20 November 2001, Item No. LXIII-4(671).

The representative of GoMP presented a detailed background as contained in the agenda item regarding an amendment to stipulations in Sub-clause IV(7) of Clause XI of the NWDT Award and requested the Authority to approve the proviso as proposed by GoMP.

The representative of GoM informed that they were broadly in agreement with the amendment but cautioned that a fool-proof procedure should be evolved by the States to ensure that PAFs purchasing land by themselves do purchase agricultural land free from all encumbrances since empowering PAFs to purchase land on their own may lead to several problems including the fixing of cost of the land. Legal implications of this issue are also to be examined.

Vice Chairman, SSNNL, GoG informed that the proposal made by the GoMP, in fact, amounts to a liberalisation of the R&R policy of State government over and above the NWDT Award. He further stated that this would help the PAFs to purchase land of their choice and settle. As such, it gives an additional option to the PAFs to purchase land of their own choice and is an additional facility to PAFs losing land. He, therefore, emphasised that this proposal may be agreed. Commissioner (R), SSPA, GoG was of the view that empowerment to PAFs to purchase land on their own needs to be properly monitored through a proper mechanism to ensure that the land purchased by PAFs is free from all encumbrances.

Secretary, Ministry of Social Justice and Empowerment, was of the view that the land for land policy should be strictly followed, as stipulated in the NWDT Award. The amendment/provision should cover the fundamental aspects of NWDT. Regarding the amendment suggested by GoMP, he suggested that all facts should be furnished before the Law Ministry to ascertain whether the amendment shall be acceptable in the court of law before taking any decision in this regard.

The Chairman, NCA was of the view that GoMP should make all efforts to purchase private land for allotment to PAFs.

The Authority, on the basis of the above deliberations, decided to set up a sub-group to deliberate on this issue and frame a suitable draft of the amendment and examine the implications of the proposal of GoMP including the mechanism to ensure that the compensation is utilised to purchase the agricultural land so that the basic provision of land for land in NWDT Award is not altered. The sub-group comprises the following members:

1. Dr S.V. Joshi, Principal Secretary, Government of Maharashtra,
2. Shri V.K. Babbar, Commissioner (R) & CEO, SSPA, Government of Gujarat,
3. Shri N.B. Desai, Director, SSNNL, Government of Gujarat,
4. Shri Rajneesh Vaish, Director (R), Government of MP,
5. Shri H.N. Tiwari, Advisor, Government of MP,
6. Shri N.D. Tiwari, Member (E&R), NCA-Convenor.

The Committee will submit its recommendations to the MoWR immediately, for seeking the necessary legal opinion.

Narmada Valley Development Department, Terms of Special Rehabilitation Package, 2001

Source: Madhya Pradesh Government, Narmada Valley Development Department, Letter to Vice-Chairman, Narmada Valley Development Authority, Narmada Bhavan, Bhopal, No. 12/1/21/27/2/98/1286, 27 November 2001–4 December 2001.

Subject: The disbursement of an additional liberalised rehabilitation package and financial assistance to the oustees of Sardar Sarovar

Reference: Memo of the Department No. 12/1/21/27/2/98 dated 15 May 2000

According to the above-referred memo, the State government has decided to approve/sanction an additional liberalised package to the oustees who are affected by the submergence from the SSP.

(a) Special rehabilitation package:

Every oustee whose agricultural land has been acquired due to submergence due to Sardar Sarovar is to be remitted/disbursed the special rehabilitation package as follows:

1) The value of land will be determined for irrigated/non-irrigated land per acre in the said village by a process of the selection of sale prices (bikri chhaant) for one year prior to the notification under Section 4(1). Similarly, the value of land will be determined for irrigated/non-irrigated land per acre in villages of the same tehsil apart from those affected by submergence by the process of the selection of sale prices. Apart from this, the value of land will also be determined for irrigated lands in the Sakalda command area for a period of one year. The highest of these three figures will be taken. The year 2000–2001 will be taken even for cases where the section 4(1) notification has been done before this date.

The special rehabilitation package will be payable as follows:

Evaluated per acre cost * acquired land + 30 per cent solatium = Award amount

That special rehabilitation package will be payable to only those oustees who

(1) will not demand land-for-land from the government;

(2) will buy irrigated land for themselves from the given amount;

(3) will get ousted voluntarily and hand over the vacant land to the government;

(4) who are landowners before 1 January 2000, meaning those who have bought submergence lands after 1 January 2000 will not be eligible for the special rehabilitation package.

(b) The rates for financial assistance for the rehabilitation grant and buying productive assets will be amended/revised as follows:

Rehabilitation grant	Present rate	Revised rate
1. Landless agricultural labourer/SC/ST farmer/small/marginal farmer	11,000	18,700
2. Other landless	5,500	9,350
3. Other families	5,500	9,350

Financial assistance for purchasing productive assets

Rehabilitation grant	Present rate	Revised rate
1. Landless agricultural labourer/landless SC/ST farmer	29,000	49,300
2. Other landless	19,500	33,150

(c) Transportation grant

Those oustees who opt for individual transport instead of transport provided by the project authorities will be paid a sum of Rs 5,000 instead of Rs 500.

Expenditure for Contingency Plan:

The establishment expenditure for the contingency plan for the oustees affected temporarily every year by the floods caused due to dam construction will be Rs 2 crores, to be borne by the rehabilitation package.

This additional liberalised rehabilitation package will be applicable to all the oustees of the Sardar Sarovar Project. This expenditure should not be borne by the project.

This approval has been notified by the finance department of Bhopal/Gwalior on page No. 1902/P 3127/B-10/4/2001 dated 10 December 2001.

Documents Relating to the Decision to Raise the Level of the Dam from 110.64 to 121.92 Metres, 2006

Resettlement and Rehabilitation Sub-group, Clearance for Construction up to 121.92 Metres, March 2006

Source: Narmada Control Authority, Resettlement and Rehabilitation Sub-group, Minutes of the 63rd Meeting, Item No. LXIII-2 (345), 8 March 2006.

Review of Status of R&R at Dam Height EL 121.92 m of the Sardar Sarovar Project (SSP)

The Chairperson, R&R Sub-group of NCA informed that the permission to raise the dam height up to EL 110.64 m was accorded by NCA following the procedure evolved by the Hon'ble Supreme Court by its judgment dated 18 October 2000, and accordingly the dam height was raised to EL 110.64 m by the end of June 2004. Thereafter, the Hon'ble Supreme Court in a Interlocutory Application (IA) in the Writ Petition (Civil) 328 of 2002 filed by the PAFs of two submergence villages of Madhya Pradesh gave two important directions that major sons of the landed PAFs who are losing more than 25 per cent of their land are also entitled for the land allotment and no distinction should be made between the permanently and temporarily affected PAFs due to the raising of the height of the dam. In view of this judgment, more PAFs of the above category up to EL 110.64 m and beyond had to be resettled before rehabilitation could be considered complete.

The Chairperson mentioned that the R&R Progress was reviewed in the 62nd R&R Sub-group Meeting held on 12 September 2005 and 75th NCA meeting held on 29

November 2005 wherein party States were requested to submit Action Taken Reports (ATRs)/progress reports for completion of the R&R works up to the dam height EL 121.92 m by 31 December 2005. She stated that the States have submitted ATRs/progress reports and the same has been monitored/verified on a sample basis in the field and consultations with the GRAs of three State governments is over.

Thereafter, the Chairperson requested the Director (IA&R), NCA to present the status of R&R at dam height EL 121.92 m based on the Action Taken Reports (ATRs)/progress report furnished by the States and consultation with the GRAs.

The Director (IA&R), NCA has made a presentation highlighting the procedure of clearance for raising the Sardar Sarovar Dam by the R&R Sub-group of the NCA, the status of R&R between dam height EL 110.64 m and EL 121.92 m as reported by the States, the details about the verification/monitoring findings of ATRs, and the 'opinion' of the GRAs based on the consultations the committee of the R&R Sub-group of NCA had with them.

Thereafter, the Chairperson, R&R Sub-group of NCA requested representatives of State governments to inform on the progress of R&R in their States corresponding to the dam height EL 121.92 m and their views thereon. The R&R status as reported by the representative of the State governments are as follows:

A. Gujarat

A.1. The representative of the Government of Gujarat (GoG) informed the Sub-group that they have submitted the ATRs/clarifications/progress reports on completion of the R&R works related to the PAFs of Madhya Pradesh, Maharashtra and their home State to the NCA Secretariat and GRA. The ATR/progress reports present a clear picture of resettlement of 3,432 PAFs (1,148 PAFs of Gujarat; 46 PAFs of Maharashtra and 2,238 PAFs of Madhya Pradesh) between dam height EL 110.64 m and EL 121.92 m. The PAFs in between the dam height EL 110.64 m and EL 121.92 m have been resettled at 75 R&R sites (Gujarat PAFs in Gujarat), 97 R&R sites (Madhya Pradesh PAFs in Gujarat) and 12 R&R Sites (Maharashtra PAFs in Gujarat) having all civic amenities as mandated by the NWDT Award and the policy of the State government. The detailed account of R&R facilities/packages, viz. allotment of agricultural land, house plots, civic amenities, etc., extended to the PAFs as per the NWDT Award stipulations and liberalised policy of the States have also been presented in the ATR.

A.2. The consultation with GRA Gujarat were held on 08 February 2005 at Ahmedabad. The 'opinion' of the GRA is as follows:

> After a judicious evaluation of the relevant information generated by the detailed and comprehensive site survey, GRA is of the considered opinion that in the resettlement and rehabilitation of the PAFs affected/likely to be affected between EL (>) 110.64 m and 121.92 m in Gujarat, there has been substantial compliance of the provision contained in the NWDT Award and relevant Government of

Gujarat Resolutions on R&R. GRA is also satisfied with the provision of basic civic amenities and social infrastructure for the PAFs resettled in the R&R sites surveyed. On this basis, therefore, it can be reasonably presumed that the PAFs affected/likely to be affected between EL (>) 110.64 m and 121.92 m have been properly and adequately resettled and rehabilitated.

B. Madhya Pradesh

B.1. The representative of the Government of Madhya Pradesh (GoMP) informed that they have submitted the ATRs/clarifications on completion of the R&R works related to the PAFs of MP in December 2005 and January 2006 to the NCA and the GRA. According to the GoMP representative, 16,156 PAFs including backlog after the judgment of the Hon'ble Supreme Court dated 15 March 2005, are likely to be affected in between the dam height EL 110.64 m and EL 121.92 m. Out of this 16,156 PAFs, 2,238 PAFs have been resettled in Gujarat and 13,918 PAFs in the home State at 89 R&R sites.

B.2. The Chairperson requested Director (IA&R), NCA to present the outcome of the consultation meetings the Committee of the R&R Sub-group of NCA had with the GRA MP.

B.2.1. Director (IA&R), NCA submitted that the GoMP has submitted ATRs/clarifications on 29/30 December 2005 and 9/12/16 January 2006 to the NCA Secretariat. The sample field verification/monitoring by NCA officials has been carried out by undertaking field visits to submergence villages/R&R sites on 9–11, 19–21 January 2006 and 1–3 February 2006 to assess the status of the R&R package extended to the PAFs and civic amenities at the R&R sites established for the aforesaid affected PAFs. The sample monitoring/verification of ATRs for assessment of status of the R&R package covers 28 villages, out of 177 submergence villages. 89 R&R sites have been planned for dam height up to EL 121.92 m; 86 R&R sites had been visited earlier, i.e. during 2004 and 2005 by NCA officers and it was noticed that the developmental works were still in progress at ten old R&R sites. Therefore, this time these ten R&R sites and six new R&R sites were verified which includes three sites up to FRL also. The participation of the Sarpanch, Deputy Sarpanch and Panch have been ensured wherever they were available.

B.2.2. Director (IA&R), NCA further informed that the sample verification/monitoring of ATRs indicate that out of seven tehsils, in one tehsil, i.e. Badwani, the R&R progress was slow. Besides, civic amenities at six R&R sites, namely Nimbola, Khujawan, Khajuri, Morgari, Dharampuri and Jalkoti, were under various stages of development. According to the monitoring report, out of 13,918 PAFs resettled in the home State, 4,262 PAFs (includes 1,722 major sons of landed PAFs) are entitled for land. Out of these 4,262 PAFs, 3,834 PAFs have opted for the Special Rehabilitation Package (SRP) and 428 PAFs have been allotted government land.

B.2.3. Director (IA&R), NCA submitted that initially the ATRs were examined and clarifications sought on different points from the GoMP vide NCA's letter dated 25 January 2006. The Commissioner (R/Field), NVDA vide letter dated 03 February 2006 has submitted compliance/clarifications. The monitoring/verifications reports were forwarded to Commissioner (R), NVDA with a copy to GRA MP vide letters dated 20 and 21 February 2006 by NCA for obtaining the necessary compliance on the observations from the Government of Madhya Pradesh. The Commissioner(R), NVDA vide letter dated 25 February 2006 has submitted compliance/clarifications/assurance to the NCA Secretariat.

B.2.4. The Director (IA&R), NCA stated that the first consultation meeting with the GRA MP was held by the Committee of R&R Sub-group of NCA on 14 March 2005 but due to the judgment of the Hon'ble Supreme Court dated 15 March 2005, the GRA in his opinion delivered on 22 March 2005 has held that a revised ATR is required to be submitted by GoMP in the light of the aforesaid judgment of the Hon'ble Supreme Court, the NCA need to verify the correctness of the entries in ATRs and then the Committee may have another meeting with the GRA.

B.2.5. The Committee of the R&R Sub-group of NCA had the second consultation meeting with the GRA on 08 June 2005 for an interim dam height EL 112.0 m wherein the GRA opined that the permission to increase the dam height further cannot be granted until the backlog is cleared.

B.2.6. The Committee of the R&R Sub-group of NCA held the third consultation meeting with the GRA on 23 February 2006 regarding completion of the R&R works up to EL 121.92 m. The GRA has held views as follows:

- Hon'ble Chairman, GRA appreciated the systematic approach and the sample verification done by NCA officials.
- The Hon'ble Chairman, GRA with regard to the implementation of SRP, opined that the introduction of the SRP is to secure compliance with that provision of the Award which lays down the allotment of agricultural land to eligible oustees. It is merely a step taken to secure compliance with the provisions of the Award and therefore the SRP does not violate any provisions of the Award or any direction of the Supreme Court. According to him, the SRP has been in force for quite some time but no grievance has been made before the GRA by the PAFs against the use of the SRP.
- The GoMP should keep a contingency plan in place to avoid any submergence during the forthcoming monsoon of 2006.
- The GoMP should evolve a mechanism to ensure that the SRP is translated to purchase agricultural land only and submit the progress report on a monthly basis to the NCA Secretariat/GRA.

The GRA in his detailed opinion mentioned that in view of the material on record, clearance can be given by the R&R Sub-group to raise the height of the dam up to EL 121.92 m if:

(a) It is satisfied on the basis of verification by the NCA officials that in the amended ATRs which were submitted on behalf of NVDA, MP, and that the work at all the R&R sites would be substantially completed by June 2006, in accordance with the provisions of the NWDT Award relating to rehabilitation as regards PAFs whose rehabilitation would become necessary by virtue of increasing the height of the dam up to EL 121.92 m;

(b) That these PAFs have received all payments as required by the provisions of Sub-clause IV(6)(ii) of Clause XI of Chapter IX of the NWDT Award and;

(c) That arrangements have been made for rehabilitation of these PAFs in accordance with the directions given in the NWDT Award and intimated to these PAFs.

The representative of GoMP informed that the compliance assurance/clarifications on the above points have been submitted to the NCA Secretariat vide letters dated 3 February 2006 and 25 February 2006.

B.2.7. To a query from the Chairperson regarding payment of the second instalment to the PAFs who opted for the SRP, the work related to the purchase of land by PAFs from the SRP and compliance on the points contained in the 'opinion' of the GRA, the Secretary to the Government of MP has given commitment in writing which is as follows:

> In view of the concern expressed by the R&R Sub-group of the NCA over the completion of payment of the second instalment to the PAFs who opted for the SRP, the Government of MP categorically assured that the work related to the purchase of land by PAFs and payment of the second instalment will be completed by 15 June 2006, i.e. before the onset of monsoon 2006. The Government of MP will ensure that all payments will be made under sub-Clause IV(6)(ii) of Clause XI of the NWDT Award to the PAFs by 15 June 2006. Within a week, the Government of MP will prepare a monthly action plan. The Government of MP also agrees that the action plan will be reviewed by the sub-committee of the Sub-group on a monthly basis.

B.2.8. The Vice-Chairman, NVDA also informed that more than 75 per cent of the PAFs who had been paid the first instalment had already obtained exemption from the registration fees for the purchase of land.

B.3. The Sub-group, on the basis of the 'opinion' of GRA MP and the commitment/assurance of GoMP on the 'opinion' and monitoring/verification reports of NCA officials, observed that R&R requirements have substantially been complied by GoMP and further directed GoMP as follows:

1. The work in progress at six R&R sites should be completed expeditiously.

2. All the PAFs resettled in MP are provided R&R packages for which they were entitled but still residing in the submergence villages should be shifted well before monsoon.

3. A mechanism should be evolved to ensure that the SRP is translated to purchase of agricultural land only. A progress report to this effect should be submitted on a monthly basis to the NCA Secretariat/GRA.

C. Maharashtra

C.1. The Government of Maharashtra representative informed that 789 PAFs are likely to be affected between the dam height EL 110.64 m and EL 121.92 m. Out of this 789 PAFs, 46 PAFs have been allocated for R&R in Gujarat and the balance 743 PAFs in the home State. All 789 PAFs had already been resettled, i.e. 46 PAFs in Gujarat and 743 PAFs in Maharashtra at nine R&R sites equipped with all civic amenities as mandated by the NWDT Award and State Policy.

C.2. The Chairperson requested the Director (IA&R), NCA to present the outcome of the consultation meeting the Committee of the R&R Sub-group of NCA had with the GRA Maharashtra.

C.2.1. Director (IA&R), NCA submitted that the Government of Maharashtra has submitted ATR for 595 PAFs on 24 February 2005 and the progress reports for the 148 additional PAFs on 31 January 2006 and 28 February 2006 to the NCA and GRA. The field verification by NCA officials has been carried out on 24 February 2005, 31 December 2005 and 1 January 2006 and the findings have been reported during the consultation meetings and three follow-up meetings with the GRA. He informed that the consultation meeting with the GRA Maharashtra was held by the Committee of the R&R Sub-group of NCA on 16 March 2005 and thereafter three follow-up meetings were convened by the GRA on 10 June 2005, 13 July 2005 and 31 January 2006. The Hon'ble Chairman GRA, in the consultation meeting convened on 16 March 2005, has held that the GoM should assess the implications of the judgment of the Hon'ble Supreme Court dated 15 March 2005 in Maharashtra and report the status of R&R including those PAFs declared by him up to EL 121.92 m. In view of this, GRA has not delivered an 'opinion'.

C.2.2. The Director (IA&R), NCA further informed that the GRA Maharashtra has convened first a follow-up meeting on 10 June 2005 of the consultation meeting and sought compliance on the following points from GoM:

- PAFs as declared by the GRA and their R&R status,
- A progress report on civic amenities,
- The status of Savariya Digar village and nearby villages which are likely to become marooned due to backwater of the Udya river, a tributary of the Narmada,

- The implication of the Hon'ble Supreme Court's judgment dated 15 March 2005 in Maharashtra.

The GRA, after taking into account the submission and assurance of GoM officials, expressed satisfaction on the substantial compliance made by GoM regarding R&R works. Regarding the allotment of land to major sons of landed PAFs, he opined that matter could be discussed further at the government level. No 'opinion' was delivered.

C.2.3. The GRA convened a second follow-up meeting of the consultation committee on 13 July 2005 and reviewed the status of compliance on the points that emerged during the first follow-up meeting. The GoM have submitted the status report on R&R, the stand of GoM that the judgment of the Hon'ble Supreme Court dated 15 March 2005 is not applicable to GoM and the findings of a Task Committee of GoM that Savariya Digar and other nearby villages will not come under the category of Tapu land (island) and hence only access to these villages is required. The interim arrangements made for the access to these villages, like barges, country boats etc., have also been reported.

C.2.4. The GRA convened a third and final follow-up meeting on 31 January 2006 of the consultation committee wherein GoM officials submitted a detailed status of R&R up to EL 121.92 m, information on the proposal for the construction of flyover bridges for proper access to Savariya Digar village and other nearby villages, schemes for irrigation, and the plan for the economic development for the resettled PAFs etc. Thereafter, the GRA had opined as follows:

- The GoM has made substantial compliance on the points outlined in the consultation meeting and follow-up meetings.
- The Chairman, GRA Maharashtra, after taking into account submissions and assurance of GoM officials, expressed satisfaction on the substantial compliance made by GoM regarding the completion of R&R works up to dam height EL 121.92 m.

C.3. The Secretary (Relief & Rehabilitation), Government of Maharashtra has confirmed the above position as reported by Director (IA&R), NCA.

C.4. The Chairperson stated that she has received several representations from various NGOs and social activists claiming that a large number of PAFs are still not resettled at dam heights EL 90 m, EL 100 m and EL 110.64 m. She desired to know the status in the light of the claims made by the above groups.

The representatives of the State governments, namely MP, Maharashtra and Gujarat, had categorically submitted that R&R works up to dam height EL 110.64 m is complete. However, they unanimously were of the opinion that the mechanism to redress the grievances of the PAFs is already in place in all three State governments where aggrieved persons may approach for the redressal of his/her grievances.

The Sub-group on the basis of the above deliberations decided as follows:

- The Sub-group has examined and assessed the progress of R&R as reported by the officers of the three State governments during the meeting and also the reports related to sample monitoring/verification of the ATRs/progress reports for the completion of R&R works pertaining to PAFs of dam height EL 121.92 m in the States by NCA officials. It has also gone through the records/opinions of the consultations with the three GRAs that a committee of the Sub-group has undertaken. The consultations with all three GRAs were satisfactory.
- Keeping in mind the reports submitted by the States, the progress on the ground and opinion of GRAs, the R&R Sub-group were of the view that raising of the dam height to EL 121.92 m would be in order while keeping in mind the principle that there will be no submergence without rehabilitation.
- In view of the concern expressed by the R&R Sub-group of NCA over the completion of payment of second instalment to the PAFs who opted for the SRP, the Government of Madhya Pradesh categorically assured that the work related to the purchase of land by PAFs and the payment of the second instalment will be completed by 15 June 2006, i.e. before the onset of monsoon 2006. Within a week, GoMP will prepare a monthly action plan.
- The Government of Madhya Pradesh also agrees that the action plan will be reviewed by the sub-committee of the Sub-group on a monthly basis. The Government of Madhya Pradesh will ensure that all payments will be made under Sub-clause IV(6)(ii) of Clause XI of the NWDT Award to the PAFs by 15 June 2006.
- The Sub-group reiterated the need for constant monitoring of the rehabilitation works and asked the State governments to prepare a contingency plan to take care of any unforeseen event which could affect the PAFs during the monsoon of 2006.
- With these conditions, the R&R Sub-group convey its clearance to NCA to consider giving permission for raising of the Sardar Sarovar dam height from existing EL 110.64 m to EL 121.92 m.

Narmada Control Authority, NCA Decision, March 2006

Narmada Control Authority, Minutes of the Seventy-Sixth (Emergency) Meeting, Item No. LXXVI (Emer.)-l (756), 8 March 2006.

Review of Preparedness for Raising the Spillway Portion of the Sardar Sarovar Dam (SSD) to EL 121.92 m

1. The Chairman, NCA welcomed all the members of the Authority and officers of the Central and State governments to this emergency meeting. The Chairman pointed out that the Hon'ble Supreme Court has required that any clearance for raising the spillway of the Sardar Sarovar Dam from its present level of 110.64 m can only be considered if the relief and rehabilitation and related directions of the Supreme Court have been complied with. The Chairman pointed out that as the R&R Sub-group had decided to meet today in the forenoon to consider whether rehabilitation has been satisfactorily completed in accordance with the directions of the Supreme Court, this emergency meeting was scheduled this evening in order that there would be no delay in considering the matter of according permission to further raise the height of the spillway. The Chairman also informed that the mechanical run of the 5th unit of the River Bed Power House (RBPH) has been done and efforts are also being made to commission the last unit of the RBPH at the earliest.

2. Thereafter the Chairman, NCA requested the Chairperson, R&R Sub-group of NCA to apprise the NCA of their deliberations.

3. The Chairperson, R&R Sub-group read out the decision of the Sub-group, which is reproduced below.*

4. The Executive Member, NCA informed that the Environment Sub-group (ESG) of NCA in its 41st meeting held on 6 January 2005 recommended to NCA to permit raising the height of the dam to the next stage of construction, i.e. up to EL 121.92 m from the environmental angle. He then read the following extract from the minutes of the 41st meeting of the ESG:

> 4.1 It was observed by the members of the Sub-group that there was substantial compliance on most of the areas, but still there are gaps, which are required to be filled up by concrete actions on the ground by the party States, and considering that the Governments of Gujarat, Maharashtra and Madhya Pradesh have conveyed their assurance for completion of all the remaining activities by the end of March 2005, the Sub-group gave clearance for raising the dam height from EL 110.64 m up to EL 121.92 m and at the same time asked the party States to implement all the environmental safeguard measures, as discussed, in the letter and spirit. The Sub-group recommended to the NCA to permit raising the height of the dam to the next stage of construction, i.e. up to EL 121.92 m from the environmental angle.

5. The Chairman, NCA observed that the ESG has recommended raising of the height of the dam to EL 121.92 m, keeping in view the assurance given by the Governments of Gujarat, Maharashtra and MP to complete all the remaining

* The text of the Sub-group decision is omitted here. It is reproduced above at page 311.

activities by the end of March 2005. He desired to know the extent to which these assurances have been fulfilled.

6. The Executive Member, NCA informed that the status of the environmental related activities of SSP were again reviewed in the 42nd meeting of ESG held on 7 April 2005.

7. MD, SSNNL informed that the specific activities which were required to be completed by the end of March 2005 have been completed and those activities, which are of a continuing nature would, of course, be continued. This view was confirmed by the representatives of the Governments of Madhya Pradesh, Maharashtra and Gujarat. With regard to continuing activities, the party-States of Gujarat, Maharashtra and Madhya Pradesh assured that all such activities related to environment safeguard measures, which are necessary to be completed before the onset of the monsoon, shall be completed and compliance shall be reported in the next meeting of the ESG.

8. The Chairman, NCA said that as per the Supreme Court decision in WP No. 319/1994, the permission to raise the dam height from 110.64 m to 121.92 m will be given by the NCA from time to time, after it obtains the clearances from the R&R Sub-group and Environment Sub-group. The NCA noted that in accordance with the Orders of the Supreme Court, the R&R Sub-group and the Environment Sub-group, have recommended increasing the height from 110.64 m to 121.92 m.

9. Thereafter, the Chairman, NCA pointed out that earlier after submersion to EL 100 m the State government of Madhya Pradesh had in its orders dated 29 August 2003 adopted a policy according to which all the Gram Panchayats of the concerned areas would be provided necessary information and extended an opportunity to raise any grievances before the GRA.

9.1 The Chairman observed that the above policy of the Government of Madhya Pradesh is a good practice and hence the NCA requested the Government of Madhya Pradesh to continue the good practice by sending the ATRs to the Gram Panchayats and affording them a similar opportunity.

10. The representative of the Government of Madhya Pradesh said that the order of 2003 was issued after the height of the dam was raised to 100 metres at that time and, as this is a good practice, the Government of Madhya Pradesh would send copies of the ATRs to all the concerned Gram Panchayats as was confirmed to the Chairman, NCA by the Chief Secretary of the Government of Madhya Pradesh vide his D.O. letter dated 4 March 2006.

11. Taking into account the recommendations of both the Sub-groups of NCA and the assurance of all the States, the following decisions were taken by the NCA:

> The SSNNL is permitted to raise the height of the spillway of Sardar Sarovar Dam to EL 121.92 m as per the approved design. The project authorities should comply with the conditions laid down by the two Sub-groups of NCA and complete the

above construction by end of June 2006. The project authorities should ensure that the non-overflow blocks are raised to safe elevations or suitable measures are taken so as to avoid over-topping of these blocks. Particular care is to be taken by all the party States in preparing the contingency plans to provide timely safeguards in case of any untoward incident and to ensure that there is no submergence before rehabilitation.

Group of Ministers, Note on the Assessment of Resettlement and Rehabilitation Sites and Submergence of Villages of the Sardar Sarovar Project, April 2006

Source: A Brief Note on the Assessment of Resettlement and Rehabilitation (R&R) Sites and Submergence of Villages of the Sardar Sarovar Project, Note of the Group of Ministers, 9 April 2006.

The Group of Ministers (GoMs) comprising Prof. Saifuddin Soz, Minister of Water Resources, Srimati Meira Kumar, Minister of Social Justice and Empowerment, and Shri Prithviraj Chavan, MOS in the PMO, deputed by the Hon'ble Prime Minister to Madhya Pradesh, arrived in Indore late in the evening on 6 April 2006.

Soon after arrival in Indore, a meeting was held with Madhya Pradesh Chief Minister Shri Shivraj Singh Chouhan and some of his cabinet colleagues and officers.

The Madhya Pradesh Government made a presentation and wanted the GoMs to appreciate that the Madhya Pradesh Government had taken concrete steps to rehabilitate Project Affected Families (PAFs) and that Rehabilitation and Resettlement would be completed by 30 June 2006. In that connection, the GoMs was requested to visit some sites such as Khalghat, Dharampuri, Lakhangaon and Borlai.

When asked as to how many SC/ST families were affected, the government could not provide any information.

Early in the morning of 7 April 2006, the GoMs left for a visit to rehabilitation and submergence sites.

The GoMs visited Khalghat, Dharampuri, Lakhangaon, Borlai 1, 2 and 3, Awalda, Piplud and Nisarpur. The GoMs was stopped at other places including Picchodi where people narrated their tales of woe. The representatives of Narmada Bachao Andolan (NBA) had insisted in Delhi in their memorandum that a visit to Borlai, Awalda, Piplud and Nisarpur would be necessary to find out whether the claim of the Government of Madhya Pradesh that the PAFs had been rehabilitated was correct.

Khalghat

The GoMs visited Khalghat site where the Madhya Pradesh Government had offered land to 407 families. Only two families had accepted the land. The top soil there is black. The people say that they have to dig 10 feet deep to find the cultivable land. The government had not succeeded in persuading the oustees to accept the land. Hundreds of people on the spot complained before the GoMs that the government had not conducted a proper survey and offered the land without consulting the oustees. Shri Mohan Lal Sharma (resident of Gazipur, District Dharampuri), who spoke on behalf of oustees, complained before the members of the NVDA that the Madhya Pradesh Government had acted in haste and allotted the land which was totally uncultivable. The members of the NVDA did not contradict Shri Mohan Lal.

It was for the first time that the GoMs heard from Shri Mohan Lal that the Income Tax Department deducted Rupees one lakh from every unit of 10 lakhs that was paid to the oustees by way of compensation and for purchase of land. It was Shri Mohan Lal again who said that people were pressurised to accept cash. He said that a bribe of Rs 20,000 had to be paid for the receipt of every cheque that was given to the oustees. The crowd present on the spot gave full-throated support to Shri Mohan Lal who they said had represented their grievances correctly.

Dharampuri

From Khalghat, the GoMs went to Dharampuri. It is the largest area selected by the Madhya Pradesh Government for the settlement of oustees and 4,000 PAFs are slated to be settled there (no, they are shown to have been settled there already). Not a single plot of land has been occupied by any PAF.

Approximately 2,000 people, who had gathered there, explained through their representatives (about a dozen people explained their grievances) in detail and said that it was not possible for anybody to settle there as no infrastructure had been built. In official papers, it has been indicated that the PAFs have been settled. The GoMs was amazed that no sanitation, no drinking water, no system of sewage, no roads, much less the facilities like hospital, water reservoir, school, post office etc., have been provided there. There is no hope that such infrastructure will be built there soon. The officials presented a status report in respect of R&R sites at Dharampuri indicating that facilities that need to be provided by way of infrastructure would be provided and certain facilities were in progress, whereas the GoMs saw nothing happening.

In any case, the PAFs, under no circumstances, can be settled there before 1 July 2006 when the monsoon sets in.

Dharampuri had been shown to the GoMs as a success story by the Madhya Pradesh Government and it turned out to be the worst example of not doing anything by way of settlement when there was apparently no difficulty in respect of resources. The people there showed to the GoMs two dry water pumps and a heap of stones that had been dumped there a day before the GoMs' arrival indicating that roads would be built soon.

Lakhangaon and Borlai 1–3

From Dharampuri, the GoMs proceeded to visit R&R sites at Lakhangaon, Borlai 1, 2 and 3. The GoMs saw some stray dwellings without any infrastructure such as drinking water, sewage system, electricity and roads, etc.

As per the status report of the Madhya Pradesh Government, 18,965 PAFs were indicated to have been settled, whereas the GoMs could see, at the most 80 incomplete dwellings in Lakhangaon and Borlai 1–3 (including the dwellings shown to the GoMs at Picchodi). The GoMs found that there were no amenities of life like drinking water, roads and electricity, etc. anywhere. The only water reservoir in Borlai 3 was not operational. In Borlai 1–3, the GoMs could not see any semblance of infrastructural facilities. In Lakhangaon, of course, some structures by way of a hospital and a primary school (not functional) and three hand pumps showed the fact that a determined leadership can settle the oustees. One can see that PAFs (18 of them) here can be settled properly at Lakhangaon within the next six months, provided effective effort is directed towards that goal.

Visit to Piplud, Awalda and Nisarpur

The members of Narmada Bachao Andolan had, among other things, urged the Minister for Water Resources that the GoMs should visit Piplud, Awalda and Nisarpur so that GoMs could ascertain properly whether the oustees from these submergence villages had been settled or not.

Piplud

Nearly 2,000 people had gathered at Piplud to state their tales of woe. They pointed out that not a single PAF had been offered cultivable agricultural land anywhere in Madhya Pradesh as per the Narmada Water Disputes Tribunal (NWDT) Award. At this place, ten representatives explained in detail that the NWDT Award had been flouted and even land acquisition proceedings had not been completed. They narrated that many of the oustees had not been offered compensation for their houses. They also said that many of those who had been pressurised to accept cash compensation, had been given amounts of money with which they could not purchase even 2 acres of land, let alone 5 acres of land to which they were entitled. A tribal named Kailash from Bhikheda explained that more than 300 persons from Manavar Tehsil had been shown in records to have been offered one plot of land measuring 7 hectares in Borgaon and, strangely enough, the land was not cultivable. The representatives of these oustees gave details of how people were pressurised to accept cash compensation and how people had to pay bribes for the receipt of cheques.

Awalda

The GoMs next proceeded to Awalda, a far-flung village inhabited mostly by adivasis and tribals. The GoMs felt that, by any standards, the entire population in the village lived below the poverty line. A group of 1,500 people met the GoMs there and the youth there were restive and narrated their tales of woes in a very emotional manner. It was here that people said that they had not been given land for land and that they had been pressurised to accept cash, and an oustee, Shri Motilal Patidar of village Chhota Barde, said that he had been given Rs 5,53,000 whereas he needed to purchase 5 acres of land for Rs 13 lakhs and that this cash award was not adequate at all but he was forced to accept it. The entire crowd raised slogans against cash compensation and they said it was a fraud and officers insisted on this because they received bribes. It was in this crowd that a woman Sajan Bhai Yadav, who had come from Pipri Gaon, narrated most pathetically how they (the oustees) had lost their home and hearth, their Khaliyan and Shamshan and they were now on the roadside. The crowd said that the Gram Sabha was never consulted and every thing was happening 'zabardasti' (by force). The crowd said that they had no money to travel to Bhopal and seek redressal of their grievances. The people at Piplud and Awalda had categorically stated that no member of the R&R Sub-committee or the Grievance Redressal Authority (GRA) ever visited these villages. It was stated in good detail that the chairman of the redressal authority was headquartered at Bhopal and he once in a while visited Indore and never did Justice Sohani, Chairman of the Grievance Authority visit these villages.

Nisarpur

The GoMs finally came to Nisarpur where a crowd of 4,000 people voiced its grievances through 18 representatives which included four women. The administration had succeeded in telling the crowd that they must allow the representatives to present their case in an orderly manner.

So, M/s Parasmat Kanawat (having been Sarpanch of Nisarpur for 30 years) Pradeep Kumar, Bagirath Rathore, Mangti Bhai and others narrated pathetic tales of extreme high-handedness on the part of the NVDA. They categorically stated that not a single family of the PAFs in Nisarpur had been settled as per the orders of the Supreme Court. They said in one voice that until they were rehabilitated, there was no question of allowing the height of the dam to be raised. They said that the R&R sites at Nisarpur, which are supposed to settle 700 families of oustees, did not have any infrastructure, and the house plots for building their houses were located at depressed land which gets waterlogged and one of the essential needs of the site was drainage, which was missing. The people said that NVDA officials claimed that R&R sites would be ready by 30 June 2006 and it was a rude joke to the oustees. The oustees said that these claims were absolutely false. They claimed that the R&R Sub-group and the GRA had never visited the Narmada Valley during the last six years. They rejected cash payments as a fraudulent practice. The crowd vociferously demanded that their rehabilitation and resettlement should be a precondition for raising the height of the dam.

General Observations

1) The complaints from various quarters that the rehabilitation and resettlement of oustees of Sardar Sarovar Dam has not taken place in consonance with the orders of the Supreme Court have been found to be correct.

2) In spite of the extreme shortcomings in respect of rehabilitation and resettlement of oustees, the Madhya Pradesh Government can organise proper effort for rehabilitation, say, within a year from now. The leadership there has to show its political will to accomplish this gigantic task.

3) The Chief Minister of Gujarat had assured the Water Resources Ministry, in a recent letter, that rehabilitation and resettlement in Madhya Pradesh could be fully ensured in accordance with the orders of the Supreme Court. It will have to be ascertained as to how Gujarat Government could come to the rescue of the Madhya Pradesh Government in this behalf.

4) The outcry against the SRP (cash award) must be responded by stopping this practice as it has bred corruption, and thousands of people the GoMs met in the valley have rejected the same as a practice breeding corruption. Besides, this practice has been adopted by the State government with the approval of the GRA. It is yet to be seen whether the Supreme Court would find this practice to be in line with its specific instruction which said: 'every displaced family whose more than 25 per cent of agricultural landholding was acquired would be entitled to be allotted irrigable land of its choice to the extent of land acquired subject to the prescribed ceiling with a minimum of two hectares land and that project-affected families (PAFs) would be allotted a house/plot free of cost'.

5) The GoMs found that there was no moral and legal justification for deducting Rs 1 lakh by way of income tax for every unit of Rs 10 lakhs that is to be given to an oustee as a settlement under SRP.

6) In due course of time, the government must explore a better system of redressal of grievances than what is sought under the present GRA. It is a fact that 5,000 petitions for redressal of grievance are pending before the GRA. The Chairman of the GRA is headquartered at Bhopal and his visits even to Indore are few and far between, which has meant a great hardship for the oustees.

7) The position of the adivasis (oustees), particularly in the areas like Kakrana and Kharia Bhadal, which the GoMs could not visit, is reportedly very bad. The GoMs will pay a visit there if required, in due course of time.

8) The Ministry of Water Resources had been insisting that Madhya Pradesh Government should adopt the mechanism of referring ATRs to Gram Panchayats according to the previous government's assurance given in the then Chief Minister's letter of 29 August 2003. Since the Madhya Pradesh Chief Minister has finally responded positively on this issue recently, the ATRs will now be sent to Gram Panchayats, as stated by the Chief Minister in his meeting with us on 7 April 2006. That will, however, not have any effect on the problems at hand.

9) The GoMs felt concerned about SCs and STs for whom there doesn't seem to be any special provision in respect of rehabilitation and resettlement.

10) The reports of the R&R Sub-group and the GRA on the basis of which the NCA granted permission for raising the height has been largely paperwork and it has no relevance with the situation on the ground.

Union of India, Status of Resettlement and Rehabilitation, April 2006

Source: Application for Directions on Behalf of the Union of India, *Narmada Bachao Andolan v. Union of India*, IA No. 18-22/2006 in WP (Civil) No. 328 of 2002, 17 April 2006.

Sardar Sarovar Project, Status of Resettlement and Rehabilitation (R&R) at Various Elevations of Sardar Sarovar Dam on 7 April 2006

					PAFs resettled so far in				Balance PAFs likely to be resettled in		
Sl. No.	Elevation (corresponding afflux)	State	No. of Villages affected	Total PAFs	Gujarat	Maha-rashtra	MP	Total PAFs resettled	Gujarat	Home State	Total
1	110.64 m	Gujarat	14	3,578	3,578	0	0	3,578	0	0	0
	(128.16m)	Maharashtra	29	2,663	846	1,817	0	2,663	0	0	0
		MP	104	7,166	2,141	0	5,025	7,166	0	0	0
		Total	147	13,407	6,565	1,817	5,025	13,407	0	0	0
2	121.92 m	Gujarat	19	4,726	4,726	0	0	4,726	0	0	0
	(134.32 m)	Maharashtra	30	3,452	892	2,560	0	3,452	0	0	0
		MP	177	23,322	4,355	0	18,967	23,322	0	0	0
		Total	226	31,500	9,973	2,560	18,967	31,500	0	0	0
3	138.68 m	Gujarat	19	4,728	4,726	0	0	4,726	2	0	2
	(455 ft)	Maharashtra	33	3,698**	946	2,560	0	3,506	77	115	192
		MP	192	43,021***	6,037	0	18,967	25,004	8,087[$$]	9,930[$$]	18,017
		Total	244	51,447	11,709	2,560	18,967	33,236	8,166	10,045	18,211

The figures at a given level are cumulative

* The figures are based on the ATRs/progress reports submitted by the State governments. It includes the backlog of dam height EL 110.64 m in MP.

** This number may increase due to the addition of genuine PAFs to be declared by GRA and the State government.

*** Tentative.

[$$] The allocation for R&R in Gujarat is yet to be assessed.

Status of Resettlement and Rehabilitation, Submission of Shri Shanti Bhushan, April 2006

Source: Annexure 10, Written Submissions of Shri Shanti Bhushan, Senior Advocate on Behalf of the Applicants, in the matter of *Narmada Bachao Andolan v. Union of India*, IA No. 21-22 of 2006 in Writ Petition No. 328 of 2002.

Game of Numbers – Madhya Pradesh

Source	Date	PAFs up to 121.92 m	PAFs between 110.64 m and 121.92 m	PAFs claimed to be resettled		110.64 m backlog to the resettled	Balance to be resettled	Total balance to be resettled	Annexure and page number in the record of the case
				In Gujarat	In MP				
Minutes of the 62nd Meeting of the R&R Sub-Group	12/9/05	30.690	-	4,262	13,026	13,233	13,402*	13,402	Annexure-A/18 at p 184-201
Status Report, Narmada Control Authority	31/12/05	28,742		4,729	12,468	13,233	11,545	24,778	Annexure-A/37 to the Rejoinder at p 122
Press Release issued By the NVDA	18/2/06	24, 421	-	-	-	-	No mention of status	No mention of status	-
Letter from Prof. Soz to Shri L.C. Jain	6/3/06	-	17,255	3,339*	13,916*	No mention of status	?**	-	Annexure-A24 to additional affidavit of 11/3/06, pp 235–7
Minutes of the 63rd Meeting of the R&R Sub-group	8/3/06	-	16,156	2,238*	13,918*	No mention of status	Nil	Nil	Annexure-A30 to additional affidavit of 30/3/06, pp 272–90B
Status report placed as Annexure 2 in UoI's Application	7/4/06	23,322	-	4,355	18,967	No mention of status	Nil	Nil	
Press release issued by the NVDA	17/4/06	24,421	-	5,456	18,965	No mention of status	Nil	Nil	

* These figures correspond to PAFs between 110.64 and 121.92 m.

** This letter mentions that 13,916 PAFs are to be resettled in MP but doesn't specify whether they are actually resettled or yet to be resettled.

Supreme Court Order of 17 April 2006

Source: *Narmada Bachao Andolan v. Union of India and Others*, Supreme Court of India (Hon'ble the Chief Justice, Hon'ble Mr Justice K.G. Balakrishnan, Hon'ble Mr Justice S.B. Sinha), Record of Proceedings, IA No.18-22 in Writ Petition (Civil) No. 328 of 2002, Order of 17 April 2006.

Order

These applications have been filed complaining about the relief and rehabilitation measures to be provided to the project affected families in terms of the Award made by the Narmada Water Disputes Tribunal and the judgments of this court reported in 2000 (10) SCC 664, 2005 (4) SCC 32 and other orders passed by this court from time to time. In terms of the Award and decisions of this court, the further raising of the height of the dam has to be only *pari passu* with the implementation of the relief and rehabilitation measures and on clearance by the R&R Sub-group. The R&R Sub-group is required to give clearance for further construction after consulting the Grievance Redressal Authority. When the judgment was pronounced by this court in the year 2000, the height of the dam was 90 metres. Subsequently, it has been increased from 90 metres to 95 metres, from 95 metres to 100 metres and from 100 metres to 110 metres. The challenge to the relief and rehabilitation measures that had been provided when the heights were earlier raised led to the judgment reported in 2005 (4) SCC 32.

There is no manner of doubt that all relief and rehabilitation measures have to be provided to the oustees in letter and spirit of the Award and decisions of this court. The State governments have seriously disputed the applicant's assertion that the relief and rehabilitation measures have not been provided in terms of the Award and the judgments of this court. According to the State of Madhya Pradesh, well before issue of notice to the oustees, the said measures had been provided for. Narmada Control Authority, on 8 March 2006, accorded permission for raising the height of the dam from 110 metres to 121.92 metres. The work of raising the height is in progress. An application has also been filed by the Union of India referring to the visit by a Group of Ministers to the site and a brief note of the assessment of the said Group of Ministers has been placed on record. According to the said application, the matter requires examination. Prayer in that application is to permit the Central Government to establish a suitable mechanism to undertake a review of the measures taken so far by the State of Madhya Pradesh and ensure rehabilitation of all project affected families in the State of Madhya Pradesh to be completed within a period of about three months. The period of three months has been sought for in the application as, according to the learned Additional Solicitor General, the villages as a result of the increase of the height of the dam are expected to be submerged in water on the onslaught of the monsoon towards the end of August 2006. This application has been presented in court today. The State of Madhya Pradesh disputes the note prepared by the Group of Ministers.

We have also been informed that in the Review Committee, there was a sharp, rather equal division among the members. We have heard for some time the learned counsel appearing for either side. We have also appealed to them that the matter deserves to be examined in a fair and reasonable manner in a cool and calm atmosphere and not in the charged emotive atmosphere. We hope that they will be able to persuade the concerned stakeholders about this requirement as the creation of a cool and calm atmosphere will facilitate the early disposal of the matter.

Having heard the learned counsel, we feel that before we consider in depth the prayer strenuously made today for stopping the on-going work, it is necessary to give opportunities to the State governments and others to file replies to all the applications having regard also to the fact that, as of today, we have on record the permission granted on 8 March 2006, and at the same time, the matter does not brook any delay lest it may become fait accompli. It has been made clear to the State governments that if the relief and rehabilitation to the unfortunate oustees are not granted in letter and spirit, this court will have no option but to stop the on-going construction. Under these circumstances, we direct that replies by all concerned shall be filed within a period of one week. Rejoinder affidavit thereto can be filed within one week thereafter. The matter shall be listed on 1 May 2006.

We clarify that pendency of these matters would not preclude the Union of India to take permissible steps and initiative with a view to resolve the controversy in accordance with the judgments, above-referred. Mr Gopal Subramaniam, learned Additional Solicitor General, states that a reference has been made by the Review Committee to the Prime Minister. We take this fact on record.

Supreme Court Order of 8 May 2006

Source: *Narmada Bachao Andolan v. Union of India and Others*, Supreme Court of India (Hon'ble the Chief Justice, Hon'ble Mr Justice K.G. Balakrishnan, Hon'ble Mr Justice S.B. Sinha), Record of Proceedings, IA No. 18-22 in Writ Petition (Civil) No. 328 of 2002, Order of 8 May 2006.

Order

We have heard learned counsel for the parties for a considerable time. The dispute is about the rehabilitation of the oustees in terms of the Award of the Narmada Water Disputes Tribunal and decisions of this court in *Narmada Bachao Andolan v. Union of India* reported in (2000) 10 SCC 664 and (2005) 4 SCC 32.

One of the issues is whether the oustees are to be rehabilitated in the State of Madhya Pradesh *pari passu* with the completion of the dam or a year prior thereto.

Yet another issue is about the validity and the operation of the Special Rehabilitation Package (SRP). There are certain other issues as well. During the course of hearing, Mr Gopal Subramaniam, learned Additional Solicitor General, has brought to our notice an Office Order dated 24 April 2006 issued by the Government of India constituting the Sardar Sarovar Project Relief & Rehabilitation Oversight Group comprising of Shri V.K. Shunglu, retired CAG, Government of India, as Chairman and Prof. G.K. Chadha, Former Vice Chancellor, Jawaharlal Nehru University, and Dr Jaiprakash Narayan, Convenor, Loksatta, as its members. This Oversight Group has to report on the status of rehabilitation in the State of Madhya Pradesh of the Project Affected Families (PAFs) to the Prime Minister through the Minister of Water Resources. The specific terms of reference of the said Group are as under:

> (a) Based on the reports of the State government ascertain the number of project affected families affected due to submergence caused by the raising of the height of the dam.
>
> (b) Estimate, on the basis of sample checks, the number of project affected families who may not have received so far, in full measure, the resettlement and rehabilitation package as per the norms laid down by the Award and the orders of the Supreme Court.
>
> (c) Ascertain, on the basis of sample checks, if offers of alternate land to eligible oustees were made in a fair and transparent manner.
>
> (d) Ascertain, on the basis of sample checks, if such offers were voluntarily refused by such oustees who preferred to accept the special rehabilitation package of the Government of Madhya Pradesh.
>
> (e) Ascertain when all measures of rehabilitation, resettlement and civic amenities as mandated by the Award (paragraph 184 of Supreme Court Judgment of October 2000) will be in existence.
>
> (f) Recommend a system to ensure that all families affected by an increase in the height of the dam to EL 121.92 metres receive, within the next three months, the benefit of the relief and rehabilitation package as per the norms laid down by the Award, the relevant orders of the Supreme Court and of the Grievance Redressal Authority of Madhya Pradesh.

The said Group had its first meeting on 27–28 April 2006. The summary record of discussions of the said meeting has been placed before us. The Group desired to have a detailed account of PAFs between the elevation levels EL 110.64 metres and EL 121.92 metres in the State of Madhya Pradesh and decided to request the Narmada Valley Development Authority to provide the same. The following decisions were taken in that meeting:

(i) NCA to (a) provide an updated status of relief given to PAFs; (b) provide the submergence area for different levels of dam; (c) design a draft questionnaire to be posed to the PAFs so that the NSSO can finalise the same by 9 May 2006; and (d) set up a control cell at Indore for processing the data.

(ii) NSSO to (a) constitute 40 teams consisting of two members for conducting the field work and simultaneously processing the data commencing from 19 May 2006 to complete the same within one month; and (b) enumerate particulars of non-listed persons who claim to be oustees.

(iii) Each team of the NSSO to be accompanied by a revenue official of the NVDA.

(iv) A separate mechanism to scrutinise the claims of the non-listed persons.

(v) Ministry of Water Resources to constitute four teams to visit R&R sites to report on its development.

(vi) OSG to visit some villages after 15 June 2006.

According to the aforesaid decision, 40 teams are to be constituted; each team consisting of two members for conducting the field work and simultaneously processing the data commencing from 19 May 2006 to complete the same within one month and also to enumerate particulars of non-listed persons who claim to be oustees. The Group has decided to visit some villages by 15 June 2006. The next meeting of the Group is fixed for tomorrow, i.e. 9 May 2006.

It was submitted by the learned Additional Solicitor General that the Group will submit its report by the end of June 2006 to the Prime Minister through the Minister of Water Resources in terms of the Office Order, and that the Prime Minister would be in a better position and would be better informed of facts on receipt of the report from the said Group. We hope and expect that the Prime Minister will be able to take a decision on or before 3 July 2006. We also hope and trust that all concerned would fully cooperate with the teams and the work assigned to the Oversight Group and would not cause any obstruction in the project assigned to the Group. The decision taken shall be placed on record along with the requisite material by 6 July 2006. We would also be in a better position of facts which are highly disputed on receipt of such report.

As at present, we are not inclined to restrain the continuing work of raising the height of the dam.

List these matters for further hearing on 10 July 2006.

*Shunglu Committee Report, July 2006**

Source: Report of the Sardar Sarovar Project Relief and Rehabilitation Oversight Group on the Status of Rehabilitation of Project Affected Families in Madhya

* The full text of this document can be downloaded at http://www.ielrc.org/content/c0608.pdf.

Pradesh (Shri V.K. Shunglu Chairman, Prof. G.K. Chadha Member, Dr Jaiprakash Narayan Member), New Delhi, 3 July 2006 (extracts).

Chapter 4: Conclusions and Recommendations

Conclusions

4.4 No substantial deviation was discovered in the number of PAFs. Claims, now made during survey could not be substantiated. However, while enumerating PAFs for the next stage (FRL), NVDA should also make use of this enumeration and judge each case on its merit. Even after that process, the opportunity to approach the GRA will remain available.

4.5 Over 4,000 cases have been referred by the GRA to the NVDA. Progress in disposal of these cases requires to be expedited, and in any event all cases need to be disposed of during the next three months. NCA should monitor the progress of disposal.

4.6 NSSO verification with regard to payments made and entitlements, as recorded in the ATR, revealed that by and large the information in the ATR corresponds to the ground reality.

4.7 The offer of government land in place of land acquired did not meet the requirements of PAFs and was the most significant reason for their opting for the SRP.

4.8 SRP, in the opinion of the GRA, is a legitimate substitute for providing land for land as stipulated in the NWDT Award. The progress of disbursement of the second instalment and consequential acquisition of land under the SRP has picked up significantly in recent weeks.

4.9 Facilities created at R&R sites throw up a mixed picture, good at 37 sites, average at 25 sites and poor at 24 sites. Deficiencies at most sites can be removed by developing uneven plots and by proper maintenance and repair of roads and buildings. The deficiencies can be removed during the present financial year. For the present, the number of developed plots exceeds the number of PAFs who need to be rehabilitated.

4.10 Poor progress in shifting PAFs to R&R sites is due to the provisions of the NWDT Award (please refer to the discussion in paragraphs 3.8 to 3.12) and the absence of the threat of submergence, amongst the PAFs.

Recommendations

4.11 The present system would require some changes. During 2006–07, the height of the dam would not increase significantly and may not thus require GoMP to apply and secure further permission, in terms of the procedure outlined in paragraph 1.9.4. And yet a great deal needs to be done including:

(a) An action plan should be prepared for 37 sites classified as good. This plan should be based on 100 per cent survey of PAFs assigned to these sites and should spell out dates by which current habitation will be vacated and new sites occupied. Based on the provision of services, health and education, electricity connections, etc, the NCA may monitor the shifting of PAFs to resettlement sites. The NCA would need to be considerably strengthened to do so.

(b) A programme of up-gradation of all sites classified as average or poor and of seven sites reserved for PAFs between dam height 121.92 m and 138 m needs to be drawn up. An overall review to accommodate all PAFs i.e. (a) about 19,000 PAFs up to 122 m + (b) 15,000 beyond 122 m, say 34,000 PAFs in all, in relation to facilities available at existing sites would require to be undertaken. Since the acquisition of agriculture land up to FRL has already taken place, and because the SRP seems to be working, the major issue now is largely about relocation to new homesteads. A connected issue relates to 3,879 PAFs who were entitled to land for land, who opted for the SRP and from amongst whom nearly 1,650 have purchased land of their choice. Land purchases at places of their choice may well preclude many of them from settling at the R&R sites assigned to them. An increase in number of PAFs accepting cash instead of resettlement supports this and should be taken into account.

(c) In July 2005 about 4,000 claims were forwarded by the GRA to the NVDA for verification and settlement. While some of these have already been disposed of, the number is not very large. The NVDA under the superintendence of the GRA would therefore be required to dispose of all claims before arriving at a final list of PAFs.

(d) The Supreme Court had directed that all major sons would also be entitled to land where the PAFs are entitled to land. The NVDA should confirm to the NCA that this exercise has been completed even in respect of major sons whose cases were rejected earlier in view of a different interpretation.

(e) All functions in relation to the SSP in MP are to date performed by NVDA. The district administration has played at best a supportive role. Considering the kind of fieldwork required and the time available it may be desirable for the district administration to be more directly involved in planning and implementation of the several actions required for securing compliance with the NWDT Award and Supreme Court directions. The modus vivendi is within the competence of GoMP.

(f) GoMP should, within two months, prepare a plan of action, assign milestones, secure, approval of the R&R Sub-group, and commence action for the resettlement of PAFs up to FRL.

(g) Monitoring to secure compliance should remain the responsibility of the NCA. The NCA would certainly need considerable support. The kind of support which became available to the OSG was clearly predicated on an extraordinary situation which required the NSSO to achieve deadlines by diverting staff from other projects. The Ministry of Water Resources needs to evolve mechanisms to provide adequate professional support to the NCA to discharge its obligations.

Supreme Court Order of 10 July 2006

Source: *Narmada Bachao Andolan v. Union of India and Others*, Supreme Court of India (Hon'ble the Chief Justice, Hon'ble Mr Justice K.G. Balakrishnan, Hon'ble Mr Justice S.B. Sinha), Record of Proceedings, IA Nos 18-22 and IA Nos 23, 24 in Writ Petition (Civil) No. 328 of 2002, Order of 10 July 2006.

In terms of the order dated 8 May 2006, Mr Gopal Subramaniam, learned Additional Solicitor General, has filed in Court the report of the Sardar Sarovar Project Relief and Rehabilitation Oversight Group under the chairmanship of Mr V.K. Shunglu on the status of rehabilitation of the project affected families in Madhya Pradesh. An additional affidavit of the Commissioner – Projects, Ministry of Water Resources, Government of India, has also been filed. Learned Additional Solicitor General has also placed before us (in a sealed cover) the decision dated 8 July 2006 taken by the Prime Minister on consideration of the said report which, *inter alia*, states that the said Oversight Group seems to have given a fairly accurate picture of the facts and circumstances as they obtained on the ground. It further states that while relief and rehabilitation measures are of primary importance, considering the fact that work on construction of the dam is going to be stopped by the State of Gujarat from 10 July on account of the onset of the monsoon, and also the fact that the shortcomings in relief and rehabilitation work brought out in the report can be remedied by an accelerated pace of implementation. It would not be appropriate, in the light of the material and observations contained in the report, to pass any direction or orders at this stage stopping the construction of the dam which is designed to serve a larger public interest. The decision further states that the Government of Madhya Pradesh, however, shall be directed that relief and rehabilitation work must be accelerated during the period when there would be stoppage of work so that all the relief and rehabilitation sites can be brought at par with each other and that the same should be attempted to be completed before work on construction of the dam is resumed.

Mr Ashok Desai, learned counsel appearing for the State of Gujarat, has placed before us the position of construction of the project as on 10 July 2006 and submits that the construction has reached the height of 119 metres except in some spillway blocks. Learned counsel states that, except completing the remaining construction in spillway blocks so as to make it level at 119 metres, further construction at this stage beyond 119 metres will not be carried out on account of the onset of the monsoon. Though Mr Shanti Bhushan, learned counsel, submits that the construction in spillway blocks from block No. 28 to block No. 36 may not be permitted to be carried out, we find no reason to restrict the construction at this stage particularly when it is pointed out that non-construction in the remaining spillway blocks may endanger the safety of the construction already made.

Parties are permitted to file their responses to the report within three weeks and replies thereto may be filed within three weeks thereafter.

List these matters in the month of September 2006.

Chapter 7

Project Assessments

The Morse Report and Official Responses

Bradford Morse & Thomas R. Berger, Sardar Sarovar – Report of the Independent Review, 1992

Source: Bradford Morse & Thomas R. Berger, *Sardar Sarovar – Report of the Independent Review* (Ottawa: Resource Futures International, 1992), pp. xi–xxv and 349–58.

Letter from Bradford Morse (Chairman) and Thomas R. Berger (Deputy Chairman) to Lewis T. Preston (President, The World Bank) of 18 June 1992

Dear Mr President:

On 1 September 1991 we began our Independent Review of the Sardar Sarovar dam and irrigation projects in India. Since then we have spent much time in India; we conferred with ministers and officials of the Government of India and of the Governments of Gujarat, Maharashtra and Madhya Pradesh; we met with non-government organisations and concerned citizens; we received hundreds of submissions. We travelled throughout the Narmada valley, to villages and relocation sites, to the dam site, the upstream area, the command area, and downstream. We also visited Kachchh and other drought-prone areas of Gujarat.

We have talked to whomever we thought could help us in the task assigned to us, that is, to conduct an assessment of the measures being taken to resettle and rehabilitate the population displaced or otherwise affected by the Sardar Sarovar Projects, and of the measures being taken to ameliorate the environmental impact of the Projects.

The World Bank has made an important contribution to the advancement of human and environmental concerns by developing policies for the resettlement and rehabilitation of people displaced or otherwise affected by Bank supported projects and for the mitigation of the environmental effects of such projects. Similarly the Government of India has developed a comprehensive environmental regime to reduce the environmental impact of public works projects. In spite of these positive factors, however, we believe that the situation is very serious. We have discovered fundamental failures in the implementation of the Sardar Sarovar Projects.

We think the Sardar Sarovar Projects as they stand are flawed, that resettlement and rehabilitation of all those displaced by the Projects is not possible under prevailing circumstances, and that the environmental impacts of the Projects have not been properly considered or adequately addressed. Moreover, we believe that the Bank shares responsibility with the borrower for the situation that has developed.

The Sardar Sarovar Projects are intended to bring drinking water to Kachchh and other drought-prone regions of Gujarat, and to irrigate a vast area of that state as well as two districts of Rajasthan. This requires a large reservoir on the Narmada river and an extensive canal and irrigation system. The Sardar Sarovar dam, under construction on the Narmada river, at Navagam and Kevadia, will impound water to a full reservoir level of 455 feet. It will submerge 37,000 hectares of land in three states: Gujarat, Maharashtra and Madhya Pradesh.

The Sardar Sarovar dam, along with planned developments farther upstream, is designed to divert 9.5 MAF of water from the Narmada river into a canal and irrigation system. The canal itself is the biggest in the world in terms of its capacity, and will extend 450 kilometres to the border with Rajasthan. The main canal is 250 metres wide at its head and 100 metres wide at the Rajasthan border. The aggregate length of the distribution network is 75,000 kilometres. It will require approximately 80,000 hectares of land, more than twice as much land as the submergence area.

The Sardar Sarovar Projects constitute one of the largest water resources projects ever undertaken; their impact extends over an immense area and affects a very large number of people, especially tribals. At least 100,000 people, in 245 villages, live in the area affected by submergence. In Gujarat and Maharashtra almost all are tribals. A great many of them are encroachers, that is, they have no formal title to their land. There are thousands of tribal people in the submergence area of Madhya Pradesh as well, many of whom are encroachers. In Madhya Pradesh there are also many caste villages where the inhabitants are engaged in conventional agriculture.

In addition to the 100,000 people living in the villages in the submergence area, there are likely to be 140,000 farmers who will he affected by the canal and irrigation system. Finally, there are the people living downstream, below the dam, numbering thousands more, whose lives will be significantly affected.

In 1985 the Bank entered into credit and loan agreements with the Government of India and the Governments of Gujarat, Madhya Pradesh and Maharashtra relating to the construction of the dam and the canal. Under these agreements the Bank has treated only the people whose villages will be affected by submergence as 'project-affected' persons entitled to be resettled and rehabilitated. Our first task has been to consider the measures being taken for the resettlement and rehabilitation of these people. But our Terms of Reference refer to persons 'displaced/affected by the reservoir and infrastructure'. We were also asked by President Conable to consider, under our Terms of Reference, the status of resettlement and compensation for 'canal-affected persons'.

On the environmental side, our Terms of Reference require us to consider measures being taken to ameliorate the impact of 'all aspects of the Projects'. To do this we have reviewed the extent to which there has been compliance with the Bank's and India's requirements for the Projects. We have also considered hydrology and water management issues and their relationship to environmental impact upstream, downstream, and in the command area. Without an understanding of these matters it is impossible to appreciate what the environmental impact of the Projects may be, and thus to determine what ameliorative measures are appropriate.

The idea of damming the Narmada goes back many years, but its realisation has been complicated by the fact that the river passes through three States, which could not agree upon division of project costs and benefits. In 1969 the dispute was referred to the Narmada Water Disputes Tribunal, established under India's *Interstate Water Disputes Act, 1956*. In 1979 the Tribunal handed down its Award. The Tribunal, by agreement of the States, for the purpose of distribution of benefits accepted the figure of 28 MAF as the flow of the Narmada. It went on to apportion 9 MAF of water to Gujarat – the water to be diverted into the canal for use in that State (another 5 MAF was to be delivered to Rajasthan). The hydroelectric benefits were divided among the three riparian States. The assumptions upon which the Tribunal's Award were based included a second dam project, Narmada Sagar, which was to be built, concurrently with Sardar Sarovar, upstream in Madhya Pradesh, as part of a basin-wide storage system.

In 1985 the World Bank made credits and loans, totalling US $450 million, to India and the States of Gujarat, Maharashtra and Madhya Pradesh, to help finance the construction of the Sardar Sarovar dam and canal. A second application has been made for US $350 million to complete the canal. And there is now before the Bank an application for an additional US $90 million for an associated project, the Narmada Basin Development Project.

The Narmada Water Disputes Tribunal laid down conditions regarding resettlement and rehabilitation of those who would be displaced by submergence in Madhya Pradesh and Maharashtra, described in the award as the 'oustees'. The Government of India imposed certain environmental conditions. The Bank's credit and loan agreements contain requirements relating to both. There is a dispute within India and worldwide over the question of whether India and the States have lived up to these conditions and requirements.

Our Terms of Reference require us, in making our assessment, to consider all of the Bank's existing operational directives and guidelines, bearing in mind that the credit and loan agreements were approved in 1985. Under Bank policy at that time resettlement and rehabilitation and environmental impact had to be appraised at the threshold of a project. Yet there was no proper appraisal made of the Sardar Sarovar Projects; no adequate appraisals of resettlement and rehabilitation, or of environmental impact, were made prior to approval. The Projects proceeded on the basis of an extremely limited understanding of both human and environmental impact, with inadequate plans in place and inadequate mitigative measures under way.

It is noteworthy that the Bank has seen fit to establish our review. The Bank has provided us with all necessary documents, has engaged in the frankest discussions with us, and has given us the latitude we needed to do our job. We think it unlikely that any other international aid organisation has ever established a review with a mandate as sweeping as ours in connection with a project, no matter how controversial. The Bank's willingness to do so is a tribute to its determination to understand what has gone wrong with the Projects. Similarly, we have had the cooperation of the Government of India, of the Governments of Gujarat, Maharashtra and Madhya Pradesh, of NGOs, and of people affected by the Projects.

In the past, when high dams have been built, people living in the submergence area have often been evicted without proper compensation, often without due process. This has happened in developed countries as well as developing countries. Compensation, in such cases, did not usually include anything more than cash, and the cash was more often than not inadequate, rarely if ever sufficient to buy replacement land.

Since the Second World War, developed and developing countries have built high dams in rural, forest and frontier regions. Usually, this has resulted in incursions on the lands of indigenous and tribal people. It was the special situation of these people that first gave rise to measures to protect persons subject to involuntary resettlement.

The earliest international recognition came in 1957. In that year the International Labour Organization (ILO) passed Convention 107; it required that indigenous or tribal oustee families be 'provided with lands of quality at least equal to that of the lands previously occupied by them, suitable to provide for their present needs and future development'. India ratified Convention 107 on 29 September 1958.

In 1979 in India itself, the Narmada Water Disputes Tribunal stipulated that landed oustees in Maharashtra and Madhya Pradesh were to receive land for land, in fact, a minimum of two hectares (five acres) of land. Moreover, the Tribunal held that major sons of landed families (i.e. those aged 18 and over) were to be treated as separate families. Landless oustees, under the Award, were to receive only a house lot.

India and the three States, Gujarat, Maharashtra and Madhya Pradesh, take the position that under the Tribunal Award, land for land was only intended for the benefit of landowners holding formal title, i.e. having what is known in India as revenue land. They say that the Tribunal did not intend that tribal people cultivating encroached land in the forest, to which they have no formal title, should receive any land on resettlement.

In 1980 the World Bank, for the first time, adopted a general resettlement policy. Indeed, the Bank made clear that there must be not only resettlement but also rehabilitation. It therefore provided that, on resettlement, displaced persons should 'regain at least their previous standard of living'. Such persons were to include those displaced by dams and canals. Moreover, in 1982 the Bank developed a policy specifically designed for tribal people. It provided that their customary usage of land should be respected, and required that they should only be displaced when the borrower can implement 'measures that will effectively safeguard the integrity and well-being of the tribal people'.

In 1985, when the credit and loan agreements were signed between the Bank and the three States, no one knew the scale of the displacement that would result from the Sardar Sarovar Projects, nor did anyone have anything like a true picture of the peoples who were to be displaced, nor had the people themselves been consulted. In fact, resettlement policies for Sardar Sarovar, both those of the States and to a great extent those of the Bank, have been based on the measures set out in the 1979 Tribunal Award. But the award sought mainly to adjudicate an interstate dispute. It did not, and should not have been expected to, design policies that would meet the needs of the affected people of the Projects as a whole. It did not even mention the Gujarat oustees, nor did it concern itself with the people potentially affected by the canal and irrigation system, nor did it take into account the cultural attributes of the oustee population; in the Award there is no discussion of tribal peoples, or of encroachers, or of the real meaning of 'landlessness'.

In 1985, when the credit and loan agreements were signed, no basis for designing, implementing and assessing resettlement and rehabilitation was in place. The numbers of people to be affected were not known; the range of likely impacts had never been considered; the canal had been overlooked. Nor had there been any consultation with those at risk. Nor were there benchmark data with which to assess success or failure. As a result, there was no adequate resettlement plan with the result that human costs could not be included as part of the equation. Policies to mitigate those costs could not be designed in accord with people's actual needs.

When the Bank signed the agreements for the Sardar Sarovar dam, it adopted the definition of landed oustee as set out in the Tribunal's Award, which did not include encroachers. Moreover, it did not address the question whether major sons were to receive land. This created the possibility of dispossession for the majority of tribal oustees.

At the same time the Bank, in the agreements for the canal, made no separate provision for those persons displaced or otherwise affected by the canal. Oustees were defined as those affected by submergence and project infrastructure, even though, in 1980, the Bank had acknowledged that resettlement is necessary in the case of persons displaced by canals and irrigation systems.

In 1990 the Bank announced a comprehensive resettlement policy applying to oustees generally, and in 1991 a specific resettlement policy relating to tribals. These policy statements reiterated and elaborated the principles laid down a decade earlier.

These Bank policies reflect the global adoption of new concepts of human rights. They constitute a recognition that large-scale projects, especially in rural, forested and frontier areas, may displace people just as do war and natural calamities. They focus on people who are being displaced by the advance of development, and require that in any project the human rights of the oustees must be respected. According to ILO 107, these are rights not to be impaired on grounds of national sovereignty or national interest. These considerations may justify undertaking a project but, according to ILO 107, they do not justify the nullification of these human rights if a project goes ahead. The governments of the three States claim that they are prepared to implement the Award of the Tribunal and to live by the Bank credit and loan agreements. There is disagreement, however, over interpretation. Gujarat, which has 4,700 oustee families, adopted a policy in 1988 which offers two hectares of land to all landed oustees. It also offers two hectares of land to those designated as landless; tribals and others who may be cultivating encroached land therefore receive two hectares of land. Under Gujarat's policy, in keeping with the Tribunal's Award, major sons also receive two hectares.

The Government of Gujarat and the Governments of Madhya Pradesh and Maharashtra contend that Gujarat's policy goes beyond the requirements set out in the Tribunal Award and the Bank agreements. Maharashtra, which may have as many as 3,000 families to be resettled, and Madhya Pradesh, with as many as 23,000 families to be resettled, are prepared to offer two hectares of land to landed oustees. But they are not willing to provide two hectares for major sons. Neither Madhya Pradesh nor Maharashtra acknowledges any rights of encroachers to adequate land on resettlement.

This disparity in State policies has resulted in a dispute over the meaning of the Tribunal Award and the requirements of the Bank credit and loan agreements. The dispute may seem technical but upon its outcome depends the chances of thousands of oustees to land on resettlement.

The first aspect of the dispute relates to major sons. It is said that Madhya Pradesh and Maharashtra are obliged under the Tribunal Award to provide two hectares for major sons of families displaced from revenue lands. The States say they are not. Yet the direction by the Tribunal that every major son be treated as a separate family stands without qualification, express or implied. What other purpose would this provision serve except to enable each major son to claim the same entitlement as the family to which he belongs? In our view the failure of Madhya Pradesh and Maharashtra to provide a minimum of two hectares of land to each major son in any landed family constitutes non-compliance with the Tribunal Award.

Of course, even if the Tribunal's Award were to be adopted, as regards major sons, by Madhya Pradesh and Maharashtra, it would still benefit only the major sons of landed families, for the Tribunal did not acknowledge any right in encroachers to be treated as landed.

This brings us to the second aspect of the disagreement, relating to encroachers. As noted above, Madhya Pradesh and Maharashtra say that encroachers must be treated as landless oustees with no entitlement to adequate land for cultivation on resettlement. The dispute here is whether tribal people, holding their land by customary usage, are entitled to be treated as landed oustees. Madhya Pradesh and Maharashtra say they are not, that they are illegal occupiers.

The result is that, in Madhya Pradesh and Maharashtra, thousands of tribal families, who are classified as landless but who are in fact cultivating land, may not receive any or adequate land on resettlement. Both States have provided that encroachers who can prove that they were cultivating encroached land prior to a certain date (in Maharashtra, 1978; in Madhya Pradesh, 1987) will be entitled to have their interests recorded. But these arrangements depend on documented proof which does not often exist. We estimate that, under the States' view, at least 60 per cent of tribal oustees engaged in cultivating land in Madhya Pradesh and Maharashtra will not receive adequate land on resettlement.

There are more than 60 million tribal people in India, many of them dependent on land they and their forebears have cultivated for generations. In 1987 the United Nations World Commission on Environment and Development (the Brundtland Commission) addressed the need for respect for indigenous and tribal land and resource rights. It said:

> The starting point for a just and humane policy for such groups is the recognition and protection of their traditional rights to land and the other resources that sustain their way of life – rights they may define in terms that do not fit into standard legal systems.

Central to the Bank's credit and loan agreements with India and the three States is the objective requiring that *all* oustees, including those described as landless, be enabled as a result of resettlement and rehabilitation measures taken on their behalf, to '*improve or at least regain* the standard of living they were enjoying prior to their displacement' (emphasis added). How can this be guaranteed in the case of oustees for whom cultivation is their one skill and at the heart of their social, economic and cultural lives, except by providing them, on resettlement, with land to cultivate? In 1984 the Narmada Control Authority, established to oversee the Projects, declared: 'For tribals, there is no rehabilitation more effective than providing land as the source of livelihood'. We have concluded that it is in fact the only way to ensure that they improve or at least regain their standard of living. The result of classifying encroachers as landless oustees means that people who are in fact cultivating land they regard as their own will become landless labourers. This is not rehabilitation. It does not leave them at least as well off as before.

The tribal people in Madhya Pradesh and Maharashtra are aware of the issue, and what it will mean for them if they are resettled as landless labourers. When we visited Bamni, a tribal village in Maharashtra, the people told us, 'We are farmers, not labourers'. In our view Maharashtra and Madhya Pradesh, in failing to provide adequate land on resettlement for rehabilitation of encroachers, have not complied with the Bank credit and loan agreements.

The States point out that under the Award and World Bank agreements all oustees have the right to resettle in Gujarat, where landed and landless oustees alike are to receive two hectares of irrigable land. Madhya Pradesh and Maharashtra contemplate that a very large number of oustees will therefore resettle in Gujarat. In fact, under Madhya Pradesh's plan for resettlement, its resettlement sites are to provide only 10 per cent of the land needed for its oustees.

But many oustees do not wish to go to Gujarat, for reasons which have to do with language, culture and other ties to their region. It would, for many of them, be a long cultural journey. Under both the Tribunal Award and under the Bank credit and loan agreements, oustees have the right in be resettled in their own State. It is true that in the last eighteen months Gujarat has achieved a measure of success in implementing resettlement, but Gujarat has thus far resettled something like 3,000 families. To resettle and then rehabilitate so many more oustees – perhaps 15,000 families – from Maharashtra and Madhya Pradesh would be an enormous task for Gujarat, and would impose a severe strain on its resources, which are not unlimited. Moreover, it is not just a question of resettlement; it also entails rehabilitation. The States may be able to effect the physical removal of thousands of families to land in Gujarat, but we do not think that it will be possible for Gujarat to rehabilitate them all.

It is important, we think, not to leave the matter there. The fact is that in Madhya Pradesh virtually no steps have been taken towards resettlement and rehabilitation. Even if Madhya Pradesh were to alter its policies, would it be reasonable to expect that Madhya Pradesh could implement a policy conforming to the Tribunal Award and Bank agreements, if it were prepared to adopt one, within the time remaining before inundation? Could the implementation of resettlement and rehabilitation be done in time? We have reluctantly concluded that the answer must be 'No'.

In Maharashtra there are 33 submergence villages divided between two talukas or districts – Akkalkuwa and Akrani. Because of cultural links between Akkalkuwa and adjacent districts of Gujarat, a Maharashtra-to-Gujarat migration has always been part of the resettlement and rehabilitation plan for some Akkalkuwa villages. So far, some 400 Akkalkuwa families have relocated in Gujarat.

Maharashtra oustees, however, have the right to relocate within their own State. To this end, after its release was agreed by the Ministry of Environment and Forests in 1990, 2,700 hectares of forest land near Taloda were made available for resettlement.

Resettlement and rehabilitation in Maharashtra are beset by serious difficulties. As noted earlier, Maharashtra's policy fails to provide adequate land to encroachers (and major sons). The significance of this failure is revealed by the fact that none of the 24 Akrani villages are deemed by the Maharashtra plan to have any revenue land. The villagers of all 24 communities can only qualify for encroacher status, with one-acre land benefits.

Also, the number of oustee families in Maharashtra is much larger than originally anticipated; the Tribunal estimated 450 families; by 1988 the figure had grown to 2,000; it is now judged to be approximately 3,000. The Taloda forest land is not large enough to provide the land to which this number of oustees is entitled, even if the tribal people in the 24 Akrani villages are treated as landless. Additional forest land is unlikely to be released – the Taloda case is seen as unique.

This raises questions about the right of choice provided for in the Tribunal Award and by the Bank agreements. That right ensures that displaced families, though obliged to leave their homes, ought not to be compelled to leave their home State. It is true that the bare right of choice remains. But the disparity in benefits means that they must choose between migrating to Gujarat or giving up their standard of living.

The only resettlement policy applicable to all three States is the Bank's. But Bank policy has not been respected. The Projects were not appraised in accordance with Bank requirements, basic information had not been gathered and adequate plans for resettlement and rehabilitation were not in place.

Notwithstanding Gujarat's success in providing land for submergence oustees, it has not provided land on resettlement for those oustees displaced in 1960–61, when the lands of six villages of Kevadia were expropriated to establish the construction site for the dam. To be sure, some of these villagers have received a measure of cash compensation. But since 1985 these people have been covered by the Bank agreements. Their entitlement to land should have been acknowledged seven years ago, yet the Bank has failed to secure an acknowledgment by Gujarat of their entitlement under the Bank agreements, let alone conveyance of appropriate lands.

Indeed, it is only recently that the Bank has urged – though it has never insisted – that India and the States comply with the 1979 Tribunal Award regarding major sons, and develop policies to match the overarching objective of the Bank agreements in order to ensure land for encroachers.

Nor is it only that the Bank has failed to enforce the Award and agreements. It has, in the case of the canal, failed to obtain a covenant in its agreement with Gujarat to require compliance with Bank policy. What about those villagers living in the path of the canal? Construction of the canal and irrigation system will affect as many as 140,000 families, of whom perhaps 13,000 – no one knows how many – will lose much or all of their land. People losing land to the canal and irrigation system are offered compensation under the Land Acquisition Act of 1984. The number of such persons is a matter of competing estimates. But this much is clear: acquisition of land under the Land Acquisition Act has often meant that farmers losing land have been compensated at rates substantially lower than replacement costs.

The responsibility in this regard appears to us to rest with the Bank. It did not include resettlement benefits for canal oustees in the 1985 credit and loan agreements, even though such had been a part of Bank policy for five years.

Evolving respect for human rights has established new norms for resettlement and rehabilitation. The Bank's policies have been influential in establishing these norms, and India has adopted many of them. It ratified ILO 107 in1958. India and the three riparian States signed the 1985 credit and loan agreements with the Bank. At the end of the day, however, the failure of India and the States to enforce the relevant provisions of the Tribunal Award and the Bank agreements, and the Bank's failure to enshrine its policies in the agreements, mean that involuntary resettlement resulting from the Sardar Sarovar Projects offends recognised norms of human rights – human rights that India and the Bank have been in the forefront to secure.

In 1972, after the Stockholm Conference, a new consciousness of environmental issues emerged. In India, as elsewhere, in the 1970s and 1980s this was reflected in new environmental laws, guidelines and practices. We have already noted the absence in India of a national policy in the field of resettlement and rehabilitation (the matter is regarded as a State responsibility). In the environmental field, however, the Government of India has developed a comprehensive structure of policies for environmental protection and assessment of environmental impact.

In 1983 environmental clearance for the Sardar Sarovar Projects was not forthcoming from India's Ministry of Environment and Forests because of a lack of information on environmental impact. In 1985 the Bank approved the credit and loan for the Projects. An appropriate environmental assessment was not made. In the Bank's 1985 Staff Appraisal Report no mention is made of the controversy that was holding up environmental clearance in India. The Bank required an environmental work plan by December 1985. It was not done. The date was extended to 1989. The workplan is still not available.

It was not until 1987 that a conditional environmental clearance for the Projects was given by India's Ministry of Environment and Forests. It was provided in the clearance that, instead of environmental impact studies being done before approval of the Projects, they were to be done *pari passu*, that is, concurrently with construction – an approach that we believe undermines the very basis for environmental planning. There was, however, an explicit schedule providing for the completion of the environmental impact studies by 1989. Most of the studies were not completed by 1989. Many have still not been completed. Without proper data and studies, proper assessments of environmental impact cannot be made and effective ameliorative measures cannot be developed.

The history of the environmental aspects of Sardar Sarovar is a history of non-compliance. There is no comprehensive impact statement. The nature and magnitude of environmental problems and solutions remain elusive. This feeds the controversy surrounding the Projects. As with the resettlement and rehabilitation issues, this has placed our review in a difficult position. To complete our work, we have had to assemble basic ecological information to establish the likely effects of the Projects upstream, downstream and in the command area. This work should have been done by others before the Projects were approved.

The design and operation of a multi-purpose project like the Sardar Sarovar Projects depends on the hydrology of the river. Understanding impacts, therefore, begins with an understanding of the hydrology and the nature of the changes that will be caused by the engineering works.

During the proceedings before the Narmada Water Disputes Tribunal, the States agreed on a figure of 28 MAF as the average annual stream flow to be expected three years out of four. The Tribunal accepted this figure as a basis for the apportionment of the benefits of the Projects. It also provided a benchmark for design of the dam and canal.

We found discrepancies in basic hydrological information related to these works. We therefore examined the streamflow data and did our own analysis. We found that there is good reason to believe that the Projects will not perform as planned. The problems relate to the sequence and timing of streamflows and the capacity of the dam and canal to store and divert water. The effects of Sardar Sarovar upstream, downstream and in the command area, therefore, will be different from what has been assumed to date whether or not the upstream NSP is built as planned. A realistic operational analysis upon which to base an environmental assessment is lacking. This alarmed us and it should alarm others, especially for a megaproject with such far-reaching implications as Sardar Sarovar.

For the area upstream of the dam there are piecemeal studies that suggest that the impact on biodiversity will be minimal. But there has been no attempt properly to assess the cumulative effects of the impacts arising from the NSP. Although the NSP is not within our terms of reference, the resulting cumulative impacts will almost certainly be serious. The Bank has placed itself in a difficult position by agreeing to proceed with the SSP before the environmental implications of directly related projects upstream are understood.

Programmes in the upstream region for compensatory afforestation and catchment area treatment are under way. We believe that these programmes, however successful in the short term, are likely to fail because of the lack of participation by local people. It is our view that achieving the necessary cooperation is not likely to be possible within the construction schedule imposed by Sardar Sarovar.

The backwater effect of sedimentation upstream of the dam is also an issue which has been ignored. Our analysis indicates this effect could mean a rapid, continuing, and cumulative rise in water level in the river above the reservoir. This can cause flooding to extensive areas of densely populated farmland. The human and environmental impacts could well be severe.

The construction of a dam on a free-flowing river has obvious implications for the downstream ecosystem, all the more so when proposed developments upstream will divert most of the river flows. But we found that no assessment of downstream impact has been done. Some of the basic information is only now being gathered. The implications of the Sardar Sarovar Projects for the geomorphology of the lower reaches of the river and its estuary and for the fishery and the people living in the region are unknown. We were able to assemble enough information to indicate that the impacts will be serious. It is likely, for example, that the hilsa fishery, the largest on the west coast, on which thousands of people depend, will suffer severe losses or be eliminated completely. The mitigative measures currently proposed are inadequate.

The shortcomings we have found in environmental assessment also extend into the command area. Although properly integrated studies are lacking, we have found that there are likely to be serious problems with waterlogging and salinity. Assumptions used in design of the canal and irrigation network, and on the development of mitigative measures, are questionable. We can only conclude that, when taken together, the problems that will arise in the command area will be quite similar to those identified by the Bank in many other irrigation projects in its 1991 India Irrigation Sector Review.

The priority water use is domestic consumption. We were surprised therefore to find that the plans for the delivery of water to the people in the villages and other centres in the drought-prone regions of Gujarat were only in the earliest stages of development. Apart from guidelines and intentions, we had little to review. We could not make any proper assessment as required by our Terms of Reference.

We have been conscious throughout our review of the close connection between the Projects' engineering design and the human and environmental impacts. This can be most clearly observed in the field of public health.

Large-scale irrigation projects such as the Sardar Sarovar Projects are known to carry health risks. From the first phases of construction, through creation of canals and ponds, to establishment of the reservoir itself, there are inevitable dangers, of a large-scale increase in water-borne diseases. These have been documented since the 1930s, and World Bank-assisted projects have witnessed some of the problems that can occur.

Yet, as recently as January 1992, we find that the Bank's consultant says that the Sardar Sarovar Projects appear to have been 'planned, designed and executed without incorporation of Health Safeguards'. He describes various parts of the Projects as 'death traps' and as 'taking Malaria to the doorsteps of the villagers' and as creating 'ideal breeding sites' for malarial mosquitoes. He reported a total collapse of vector control measures. The incidence of malaria has risen sharply in villages near the dam; local clinics have recorded deaths from malaria. The failure to anticipate and prevent malarial hazards is a part of the failure to implement measures to mitigate the impacts of the Projects.

The Bank is now proposing a Narmada Basin Development Project, and is considering providing a US $90 million credit for this purpose. The connections between this project and Sardar Sarovar are many. Although the Basin Development Project appears to address many of the problems raised during our review, and we recognise that some parts have merit, we have concluded that it will not succeed in meeting the stated objective as 'a comprehensive programme to tackle the growth and sustainability needs of the basin'. Furthermore, the staff appraisal report for the proposed Basin project fails to acknowledge the linkages that also exist with the NSP. The Bank may be moving incrementally towards involvement in another major development project without prior consideration of the possible social and environmental consequences.

In spite of non-compliance with Bank resettlement and environmental requirements, the Sardar Sarovar Projects are proceeding – in the words of Chief Minister Patel of Gujarat – as 'an article of faith'. It seems clear that engineering and economic imperatives have driven the Projects to the exclusion of human and environmental concerns. Social and environmental tradeoffs have been made that seem insupportable today.

The Bank has followed what it describes as an incremental strategy, in an attempt to secure compliance with its resettlement policies. India has done much the same in its adoption of the *pari passu* principle with regard to environmental issues. These approaches, however, have failed to achieve their objectives. Moreover, they signify that these crucial matters – resettlement and environment – are of only secondary importance.

We are well aware of the scale of the development task facing India, of the importance India places on irrigation in increasing production in the agricultural sector, and of the longstanding partnership between India and the Bank in this endeavour. But our Terms of Reference are specific. They require us to consider the Bank's policies, India's environmental regime, and the credit and loan agreements. These emerge from the context of Bank–India relations just as surely as does the longstanding partnership in the enhancement of agricultural production between the Bank and India. If there was no intention of following Bank policy or India's regulatory regime, it would have been appropriate to acknowledge this. In any event, the incremental strategy has been counter-productive.

The Bank, in crafting our Terms of Reference, invited specific recommendations which 'should include, as appropriate, any recommendations for improvement of project implementation (...)'. If essential data were available, if impacts were known, if basic steps had been taken, it would be possible to know what recommendations to make. But we cannot put together a list of recommendations to improve resettlement and rehabilitation or to ameliorate environmental impact, when in so many areas no adequate measures are being taken on the ground or are even under consideration.

Important assumptions upon which the Projects are based are now questionable or are known to be unfounded. Environmental and social trade-offs have been made, and continue to be made, without a full understanding of the consequences. As a result, benefits tend to be overstated, while social and environmental costs are frequently understated. Assertions have been substituted for analysis.

Every decision as to the Sardar Sarovar Projects has always been, and will continue to be, a decision for India and the States involved. Together, they have spent a great deal of money. The foundations of the dam are in, the dam wall is going up, the turbines have been ordered and the canal is completed to the Mahi River. No one wants to see this money wasted. But we caution that it may be more wasteful to proceed without full knowledge of the human and environmental costs.

We have decided that it would be irresponsible for us to try to patch together a series of recommendations on implementation when the flaws in the Projects are as obvious as they appear to us. As a result, we think that the wisest course would be for the Bank to step back from the Projects and consider them afresh. The failure of the Bank's incremental strategy should be acknowledged.

Whatever decisions the Bank makes about its role in the Projects, it must bear in mind the critical importance of consultation with the people of the valley and along the route of the canal. Such consultation would be in accord with the Brundtland Report, which said that in the case of tribal people, 'they must be given a decisive voice in the formulation of resource policy in their areas'. The same must he achieved for non-tribals as well. As Prime Minister Rajiv Gandhi said to the United Nations on the adoption by the General Assembly of the Brundtland Report, 'The search for the right answers must go on relentlessly. It is a worldwide endeavour to which India pledges its unstinting support'.

Our job has been to make an assessment. We have done so. We have, in the course of our work, made many friends in India. We wish to assure them that our sole desire has been to find the truth and report it. We hope that our assessment may advance the search for constructive and creative solutions.

Chapter 17: Findings and Recommendations

We have completed an assessment of resettlement and environmental aspects of the Sardar Sarovar Projects. In this chapter we draw together the findings of our review, already explained in the preceding chapters, and set forth the recommendations which, in our judgment, are appropriate to these findings.

The Findings

Resettlement and Rehabilitation

- The Bank and India both failed to carry out adequate assessments of human impacts of the Sardar Sarovar Projects. Many of the difficulties that have beset implementation of the Projects have their origin in this failure.
- There was virtually no basis, in 1985, on which to determine what the impacts were that would have to be ameliorated. This led to an inadequate understanding of the nature and scale of resettlement.
- This inadequate understanding was compounded by a failure to consult the people potentially to be affected.
- Failure to consult the people has resulted in opposition to the Projects, on the part of potentially affected people, supported by activists. This opposition has created great obstacles to successful implementation.
- In drafting the terms and conditions of the 1985 credit and loan agreements, the Bank failed to take adequate account of the fact that a large proportion of those at risk from the development of the Sardar Sarovar Projects are tribal people. This meant that insufficient account was taken of the principles enshrined in the Bank's 1982 Operational Manual Statement outlining its policies regarding tribal people.
- In these and other ways, the Bank failed to follow the principles and policies it set out in 1980 and 1982. In addition, the Bank's overarching principle embodied in the 1985 credit and loan agreements by which resettlement and rehabilitation were to be judged, namely that oustees improve or at least regain their standard of living as quickly as possible, was not consistently advanced or insisted upon with sufficient force or commitment.
- The Bank failed to consider the effects of the Projects on people living downstream of the dam. We recommend that the Bank develop a policy to deal

with the plight of persons affected downstream. They may not come within the rubric of resettlement, but their situation should be addressed.

- As a result of both the inadequate database and the failure to incorporate provisions of the Bank's policies in the 1985 credit and loan agreements, the provisions for resettlement and rehabilitation do not adequately address the real needs of those to be affected.
- In particular, the agreements allowed a distinction between 'landed' and 'landless' oustees which failed to recognise the realities of life in the submergence villages.
- Similarly, the rights of encroachers were not acknowledged. The only way of implementing resettlement policy, at least in the case of the Sardar Sarovar Projects, in a way that restores oustees' previous standard of living is by provision of adequate land. This is of special relevance to the oustees of Maharashtra and Madhya Pradesh.
- The people of the six villages affected by construction and development of Kevadia Colony were not appropriately and adequately compensated. The Bank failed to ensure that this be done as required by the 1985 agreements. We recommend that the Bank require India to provide land for the families of the six villages, with an adjustment for cash compensation received in the interim, as appropriate.
- Relocation and resettlement of the people of the rock-filled dyke villages was implemented in a way that meant that the Bank's overarching principle of resettlement and rehabilitation, i.e. that no one should suffer a fall in standard of living, was not likely to be achieved.
- The Bank failed to ensure that those affected by construction of the canal and irrigation system would be entitled to resettlement benefits.
- We recommend that the Bank should use its good offices to ensure that Gujarat provides resettlement benefits to canal-affected persons, especially those farmers who are rendered marginal or landless.
- The policies of the riparian States failed to anticipate the needs of major sons, and adopted what we regard as an unduly restrictive interpretation of the Tribunal Award's provision for major sons. Maharashtra and Madhya Pradesh continue to maintain this interpretation and provide inadequate benefits to major sons of landed families.
- In 1987–88 the Government of Gujarat expanded its resettlement and rehabilitation policies to provide two hectares of irrigable land to all oustees, including the landless, encroachers and major sons. This represented a policy package that came nearer than any thus far set out anywhere in India to establishing a basis for successful resettlement.

- Despite Gujarat's improved policy, Maharashtra and Madhya Pradesh continued to limit the provision of two hectares of land to 'landed' oustees. This means encroachers and major sons (including the major sons of landed oustees) are not entitled to benefits in their own States that meet the Bank's overarching principle of resettlement and rehabilitation. The proportion of oustees thus vulnerable to a reduced standard of living is at least 60 per cent.
- The disparity between Gujarat's policy and the policies of Maharashtra and Madhya Pradesh has meant that oustees' right to choose between relocation in Gujarat and their own State has been rendered meaningless.
- Implementation of resettlement in Maharashtra has been limited by both policy deficiencies and availability of irrigable land.
- Implementation of resettlement in Madhya Pradesh has been limited by policy deficiencies, inadequate institutional commitment, continuing failure of consultation and limited availability of suitable resettlement land.
- This state of affairs in Madhya Pradesh has produced a situation in which, even if Madhya Pradesh were to adopt a policy with benefits equal to Gujarat's, such a policy could not now be implemented, given the time necessary to meet the requirements of the Sardar Sarovar Projects.
- Resettlement of oustees in Gujarat has entailed a scattering of families and villages among many different sites. This is in part a result of choices made by oustees. It is also a result of inadequate land at resettlement sites to accommodate all oustees who wish to have land there. This has contributed to some separation of families, especially in the case of oustees from the rock-filled dyke villages.
- Gujarat is unlikely to be able to resettle a large proportion of oustees from Maharashtra and Madhya Pradesh. Even if land were available for relocation sites, resettlement and rehabilitation at these sites presents major problems.
- The record of resettlement and rehabilitation in India, which has been unsatisfactory in virtually every project with a large resettlement component, should reasonably have prompted the Bank to adopt a less flexible standard for resettlement and rehabilitation of project-affected people. In this context, the Bank's incremental strategy to obtain compliance, made explicit in 1989, greatly undermines prospects for achieving successful resettlement and rehabilitation.

<u>Environment</u>

- Measures to anticipate and mitigate environmental impact were not properly considered in the design of the Projects because of a lack of basic data and consultation with the affected people.

- The Bank's appraisal took no account of the fact that environmental clearance in India was not forthcoming in 1983 from the Ministry of Environment and Forests because of insufficient information.
- Under the 1985 credit and loan agreements, the Bank required an environmental workplan to be developed by the end of 1985, later extended to 1989. It is still not available, resulting in a disjointed, piecemeal approach to environmental planning that is both inefficient and ineffective.
- In 1987 India's environmental clearance for the Projects was given, despite the fact that the information required prior to the Projects' clearance was unavailable. In order to overcome this deficiency, studies were to be conducted *pari passu* with construction. The clearance was conditional on completion of these basic studies by 1989. Most remain to be completed. We believe that the *pari passu* policy greatly undermines the prospects for achieving environmental protection.
- Significant discrepancies in the hydrological data and analyses indicate that the Sardar Sarovar Projects will not perform as planned either with or without the upstream NSP. A realistic operational analysis of the Projects upon which to base an impact assessment has not been done.
- The cumulative impacts of the SSP together with the related upstream developments, especially the NSP, are very likely to be far reaching, yet they have not been studied.
- The afforestation and catchment area treatment programmes proposed upstream are unlikely to succeed within the timetable of the Projects because of the lack of consultation with, and participation of, villagers in the affected areas.
- The compensatory afforestation approach being taken by Gujarat in Kachchh, if continued, will lead to a steady decline in the quality of forests. The practice of replanting marginal forest lands in substitution for better lands that will be submerged, means that the forests will be diminished in value.
- The impact associated with the backwater effect of sedimentation in the upper reaches of the reservoir has not been considered. Our assessment has concluded that it will be significant.
- The downstream ecological implications of dam construction have not been considered. Important but limited data have only recently begun to be collected. The downstream impacts are likely to be significant, including severe losses to, if not the elimination of, the last important hilsa fishery in western India.
- There has been no comprehensive environmental assessment of the canal and water delivery system in the command area. Information we have gathered

leads us to believe that there will be serious problems with waterlogging and salinity. We also found that many of the assumptions used in project design and for the development of mitigative measures are suspect.

- Despite the stated priority of delivery of drinking water, there were no plans available for review.
- The existing threat from malaria within the command area is serious. The Projects have been designed and executed without appropriate safeguards. The failure to adopt measures to reduce the likelihood of the spread of malaria illustrates the breakdown between assurances offered by the Bank and India and the reality on the ground. We recommend that the Bank use its good offices to ensure that preventive measures are taken as a matter of urgency to address the public health problems posed by water-borne diseases in the Projects area.
- The newly proposed Narmada Basin Development Project, although it appears to address some of the problems highlighted in our review, fails to address key issues, many of which are the same as have caused problems with the Sardar Sarovar Projects. Although some specific elements have merit, the Basin Development Project adopts a piecemeal approach, falling far short of the work that the Bank's own missions have said is needed for proper basin development. The implications of Narmada Sagar for basin development are overlooked.
- Bank requirements that the Basin Development Project not entail forced relocation and proceed on the basis of a participatory approach to forest management and catchment area treatment, as proposed, are laudable but unrealistic, given the hostility towards the Projects in the region and the time frames envisaged by the Projects.

The Bank

We have made findings that reveal a failure to incorporate Bank policies into the 1985 credit and loan agreements and subsequent failure to require adherence to enforceable provisions of these agreements. Much of what has gone wrong with Sardar Sarovar Projects is the result of such failures over a range of resettlement and rehabilitation and environmental matters.

How did this happen?

It is apparent that there has been, and continues to be, deep concern among Bank officers and staff that India should have the means to enhance agricultural production. The Sardar Sarovar Projects were seen as offering enormous benefits, especially in terms of delivery of drinking water and irrigation.

There developed an eagerness on the part of the Bank and India to get on with the job. Both, it seems, were prepared to ease, or even disregard, Bank policy and

India's regulations and procedures dealing with resettlement and environmental protection in the hope of achieving the much-needed benefits.

Experience worldwide, in developed as well as developing countries, has shown that by factoring in and allowing for human and environmental considerations at the outset, projects can be substantially improved. To be effective, resettlement and environmental planning must be integrated into the design of projects; otherwise they become costly and burdensome add-ons.

These considerations lead to an examination of issues that focus on the Bank itself. Our work in conducting the independent review has encouraged us to make a number of observations which may be of value.

Embedded in the World Bank's operational directives is a resolve to establish *ex ante* project assessment. This requires an investment by the Bank of time and money and personnel with appropriate expertise, with on-the-ground studies and consultation as part of the planning of a project.

There should be a review of Bank procedures to ensure that the full reach of the Bank's policies is being implemented. The Bank should establish whether the problems we have found in the case of Sardar Sarovar are at issue in other projects in India and elsewhere. Our findings on this project may well indicate a need on the part of the Bank to strengthen quality control.

The Projects

The Terms of Reference provided that our assessment should include, as appropriate, recommendations for improvement of implementation. The absence of proper impact assessments and the paucity of undisputed data have limited our ability confidently to make project-specific recommendations of the kind that were contemplated. We have limited ourselves to recommendations with respect to the Kevadia villagers, the canal oustees, downstream policy, and the protection of public health, that should be carried through regardless of the fate of the Projects.

Our findings indicate that the Sardar Sarovar Projects are beset by profound difficulties. These difficulties have their genesis in the earliest phase of the Bank's involvement in the Projects, for they turn on the absence of an adequate database and failure to consult with the people whose lives and environment were and continue to be affected.

Lack of data meant that the Bank was not able, in the early 1980s, to appraise the Projects properly. No one is sure about the impacts of the reservoir and the canal on either people or the land. Without knowing what impacts were likely to be, we found it difficult to the point of impossibility to assess measures by which they might be mitigated; much of our work has therefore been devoted to gathering our own limited information base.

People who live in the villages and depend on the resources of the valley should have played a central part in determining the Projects' impact. Both their knowledge and their vulnerabilities are integral to any understanding of what is at issue. At the same time, failure to consult has fuelled intense opposition to the Projects which, as we have pointed out, has itself become a serious obstacle to design and implementation of mitigative measures.

These factors – absence of adequate data, failure of consultation, and hostility towards the Projects in the Narmada Valley – bear on every aspect of implementation. Our Terms of Reference invite us to recommend measures to improve implementation. It seems to us that the essential condition, the very starting point of any such recommendation, requires that these underlying difficulties be addressed.

But the underlying difficulties – the failures that reach back to the origin of the Projects – cannot be overcome by a patchwork of studies. The limited information base which we constructed is inadequate for the purpose. Nor is it a question of applying more intense pressure to Maharashtra and Madhya Pradesh in order to secure improved resettlement policies. As we say, the difficulties are profound. The Bank's incremental strategy and India's *pari passu* policy, adopted to deal with resettlement and environmental problems, have for the most part failed. A further application of the same strategy, albeit in a more determined or aggressive form, would also fail. As long as implementation continues in these ways, problems will be compounded rather than mitigated.

Absence of human and environmental assessment *ab initio* creates the impression that the demands of engineering carry far more weight in the Bank than the needs of the people to be affected or of the environment. The Bank's incremental strategy (and the Bank's concurrence in India's *pari passu* policy) strengthen this impression. Readiness to bear with non-compliance thereafter confirms it.

Decisions as to the future of the Sardar Sarovar Projects and the Bank's participation in them are within the exclusive domains of India and the Bank. But implementation of the Projects requires measures that go to the heart of the problems in which the resettlement and environmental components of the Sardar Sarovar Projects have become mired. We have been at pains in the section of this chapter summarising our findings to demonstrate how those problems of human and environmental impact encompass all aspects of the Projects, including the uncertainties of hydrology, the upstream questions, the impact downstream, the command area issues, the health risks, the deficiencies in resettlement policy and implementation in each of the three States as well as the canal. None of these issues can he ignored.

It seems to us that the matters we have raised are fundamental. It would be prudent if the necessary studies were done and the data made available for informed decision-making before further construction takes place. Implementation requires that the Bank take a step back. Otherwise, the possibility of making sound decisions will be further compromised.

Little can be achieved while construction continues. What would a step back achieve? First, it would afford an opportunity to design the kinds of human and environmental impact studies that are still needed. Second, it would permit the assessment of the results of such studies, to see whether modifications of the Projects might be in order. Third, it would provide a chance to consider what resettlement and rehabilitation policies might meet the needs of the oustees, and how these could be implemented in a way that is consistent with the Bank's policies and principles as set out in its Operational Manuals and Directives.

Even though proponents describe Sardar Sarovar as the most studied and least implemented project in India, we do not agree. The Projects may well be the most talked about in India, but the fact is that their human and environmental consequences have not been studied, and their engineering, design and operation would profit from further analysis.

There is a need to consider Sardar Sarovar in the social and environmental context of the Narmada valley as a whole, to consult, inform and involve the people affected by the Projects throughout the Narmada valley, those affected in the command area, and those living downstream. The opposition, especially in the submergence area, has ripened into hostility. So long as this hostility endures, progress will be impossible except as a result of unacceptable means.

A way must be found to rebuild confidence, to demonstrate goodwill, and to send out an unambiguous message that the Bank continues to be committed to its principles and its policies.

Reflections

In the case of the Sardar Sarovar Projects, India has bound itself to meet standards for resettlement and rehabilitation more exacting than any it had agreed to in the past.

We do not expect perfect justice; in an imperfect world it cannot be obtained. There is no doubt that in the national interest, people can be required to resettle. However, India, in conformity with the development of international standards of human rights, has subscribed to certain minimum conditions that must be observed even when the national interest is involved. They reflect the inalienable human rights of the oustees. We believe that these norms must be adhered to.

Nor do we insist upon an unattainable standard in environmental impact assessment and mitigation. However, to construct the Sardar Sarovar Projects, India has availed itself of world-class engineering technology. Should it settle for less than adequate standards in the application of social and environmental science?

We are aware of the statement in the eleventh principle of the Rio Declaration presented to the 1992 United Nations Conference on Environment and Development:

> Standards applied by some countries may be inappropriate and of unwarranted economic and social cost to other countries, in particular developing countries.

But the environmental standards for the Sardar Sarovar Projects were established by India itself. On the resettlement side, standards were determined by the Narmada Water Disputes Tribunal and agreed to by India and the States in the credit and loan agreements.

We have felt obliged to illuminate what we think are flaws in the Sardar Sarovar Projects. It should not be thought that these would only be found in India or confined to the Sardar Sarovar Projects. The fragile assumptions which have supported this project can be found elsewhere. Failure to consider the human rights of the displaced and failure to consider environmental impacts occur in the development of megaprojects in both developed and developing countries.

If the human rights obligations identified by ILO Convention 107 and in Bank policy are acknowledged and respected, if the commitment to the environment is real, and if these are properly integrated into project design at the outset, more effective and equitable development will ensue. Some believe that these requirements make it more difficult, often more costly, to build megaprojects like Sardar Sarovar. This implies that human and environmental costs are to be heavily discounted in project planning and execution. But hard lessons from the past have taught us that this is unacceptable. In some cases it may be that alternatives to projects that cause compulsory relocation on a large scale or severe environmental impact may have to be sought.

We have found it difficult to separate our assessment of resettlement and rehabilitation and environmental protection from a consideration of the Sardar Sarovar Projects as a whole. The issues of human and environmental impact bear on virtually every aspect of large-scale development projects. Ecological realities must be acknowledged, and unless a project can be carried out in accordance with existing norms of human rights – norms espoused and endorsed by the Bank and many borrower countries – the project ought not to proceed.

The Bank must ensure that in projects it decides to support the principles giving priority to resettlement and environmental protection are faithfully observed. This is the only basis for truly sustainable development.

Letter from Mr Chitale, Secretary, Ministry of Water Resources to Mr Vergin, World Bank Concerning the Morse Report, 1992

Source: Secretary to the Government of India, Ministry of Water Resources, 7 August 1992, D. O. No. 21/1/92-PP (extracts).

Dear Mr Vergin,

Reg: (i) Narmada River Development (Gujarat) – Sardar Sarovar Dam & Power Project (Credit No. 1552-IN Loan No. 2497-IN) and (ii) Narmada River Development (Gujarat) – Water Delivery & Drainage Project (Credit No. 1553-IN) – Report of the Independent Review.

1. The report of the Independent Review (IR) of the Sardar Sarovar Dam was presented to the President of the World Bank on 18 June 1992. Thereafter, the Bank deputed three separate teams for visit to the Project areas and they had extensive discussions with the State and central organisations on technical aspects of the project's planning and also the issues associated with the Rehabilitation and Resettlement (R&R) and environment in the light of the comments contained in the report. The World Bank mission, comprising three teams, during their visits to the three States from 13 to 22 July 1992 had also the opportunity of inspecting the engineering works, the R&R villages in Gujarat, Maharashtra and Madhya Pradesh, and also the downstream and command area environment. The various aspects of planning and implementation of the project as referred by the IR have been discussed and clarified to a large extent during the mission's work in India. These issues were further discussed in the Narmada Control Authority. The position is in brief as below.

2. At the outset, I must mention that the IR have clearly overstepped their Terms of Reference and have commented on several issues which were not within their jurisdiction. Their comments on the award of the Tribunal indicate a scant respect for the well laid out statutory processes in the country. They have also dilated on questions of tribal culture and their cultural journey, their place in the Hindu society and their role in the independence movement without a proper understanding of the sociological processes in India and these peoples' urge for development after Independence. The classification of tribals as 'true tribals', 'genuine tribals', etc. is misleading as such classification does not exist. The Constitution of India has a comprehensive list of tribals listed in the schedule to the Constitution. Such comments in the IR do not deserve any consideration. Further, the team had the full opportunity for discussions with the highest level of people including the Chief Ministers but it was unfortunate that the information conveyed to the team through these discussions had not been properly utilised, and distorted versions of environmental and rehabilitation scenarios have appeared in the report.

3. IR makes a mention about the concept of human rights in the process of resettlement. The observations of the IR on this account are erroneous, illogical and prejudiced. The traditional rights of tribals have been appreciated and honoured by the NWDT as well as the concerned State governments. Relocation of tribals is justified in the interest of public purpose and national economic development. Concerned tribals have been fully compensated in accordance with

the laws of land and in accordance with the directives contained in NWDT, further liberalised by the concerned governments. The basic objective of the R&R package is to improve the standard of living and the quality of life of PAPs. Therefore, in the process of implementation of R&R policy, there is no violation of human rights whatsoever so far as Sardar Sarovar Project is concerned.

4. The basic right of human individuals is the one of development and that is what is being ensured through different development programmes including the water resources development programmes such as the Narmada Development Projects. There is, therefore, no question of any violation of the human rights when such projects are undertaken or implemented.

Government of Gujarat, Comment on the Report of the Independent Review Mission on Sardar Sarovar Project, 1992

Source: Government of Gujarat, Comments on the Report of the Independent Review Session on Sardar Sarovar Project, August 1992 (extracts).*

1. Introduction+

The Sardar Sarovar Project consists of a dam and power project and an irrigation and water supply scheme. It includes a 139 m high concrete dam, a 1,200 megawatt power house, transmission lines, a water conveyance system including the Narmada Main Canal (460 kilometres and 40,000 cusecs) to the Rajasthan border, and 31 branch and a large number of distributory canals.

The project will irrigate about 2 million hectares of land in Gujarat and another 20,000 hectares in Rajasthan; provide drinking water for at least 30 million people and over 8,000 villages in dry and semi-arid and highly uncertain rainfall conditions and supply over 5 billion units of electric power.

The Project will accelerate Gujarat's agricultural growth to over 5 per cent compound annual and will provide a major land and water sustainable development plan for each one of its 13 agro-climatic regions. It will have a decisive impact on poverty and unemployment in the State.

The blueprint of Sardar Sarovar was honed in the early 1980s through planning methodologies which included a number of innovations. The command area was sub-divided into 13 agro-climatic regions based on climate (temperature, average and distribution and rainfall, average and distribution), soil conditions,

---

* The full report is available online at http://www.ielrc.org/content/c9201.pdf.

\+ This response of the Government of Gujarat is mainly based on the report of the High Level Group headed by Prof. Yoginder K. Alagh of the Sardar Patel Institute of Social and Economic Research, Ahmedabad.

groundwater and surface water regimes, and other characteristics essential for irrigated agriculture. A massive research effort was mounted to study the behavioural responses of the Gujarati farmer to the availability of limited quantities of irrigation water, the purpose being to design the delivery system taking into account expected economic outcomes and the farmers' responses. Every university and research institute in the social sciences in Gujarat was involved in conducting socio-economic benchmark studies in each taluka (sub-region) of the command. Drainage and groundwater models were developed for the first phase of the project and extensive observation wells set up to monitor groundwater. The project was designed to supply limited quantities of surface water and to preclude waterlogging, the farmer was to be encouraged to pump out groundwater but if he did not do so, since ground level water were monitored on a 15-day basis, the water would be pumped out by the Project authorities and conveyed back into the irrigation delivery system which was designed for this purpose. Behavioural studies of the Gujarati farmers cropping responses and the requirement to generate employment for the predominantly scheduled caste and scheduled tribe landless labourers were used to develop econometric and programming models to design the capacities for the branches and distributories. A specialized environmental prognostication study was completed. An independent Planning Group insisted on realistic information being used to work out costs and benefits. For example, thousands of crop cutting experiments of farmers' fields were used to work out the benefits of irrigated agriculture by number of watering, cost of cultivation studies were retabulated to work out the costs and the profitability of irrigated agriculture and computer models used to design the irrigation network and its costs realistically.

Expertise was used within India and the world. The hydrology of the dam was studied through real time with ten daily flows, at different stages of irrigation development, alternatives of stages of development of the number of dams in the entire valley and under different climatic conditions. Dynamic hydraulic studies were used to design technical parameters of the largest irrigation canal in the world and computer controlled irrigation system designed with electronic controls up to the level of each village and measurement designed below that to the field.

A detailed census was done of all project affected persons by one of the country's leading social science research organisations. A policy making set-up was established to oversee the rehabilitation system under a respected and powerful political leader of tribal origin in the State. The plan was published and discussed openly. Criticisms were welcomed and non-governmental organisations involved.

No wonder even the World Bank, a somewhat conservative funding body, declared that 'the Sardar Sarovar Plan represents a break with past approaches to the planning, design and construction and operation of irrigation projects in India', and again 'within a period of three years an impressive array of high quality studies and designs have resulted, including a comprehensive framework plan by the Narmada

Planning Group and production by the Irrigation Department of high quality designs, specifications and procurement documents'. These improved planning procedures are now proposed to be extensively used in India in other projects. Gujarat will strengthen the planning machinery, and maintain its autonomy and primacy in the project machineries.

The World Bank set up an Independent Review under the Chairmanship of Mr Bradford Morse in 1991. With characteristic openness, the Review Missions were welcomed in the State. Some of the members knew outstanding individuals in Gujarat. The elected leadership and the official machinery and non-governmental organisations and the press took keen interest in the Review and there were considerable expectations that the Sardar Sarovar plan, which had been built in the early 1980s, would be subjected to constructive review and particularly of the rehabilitation and environmental studies, where considerable developments have taken place in the last decade, and that the Review would examine progress and give constructive proposals for improvement.

The Independent Review report presented in June 1992 was highly disappointing. The Review ignored the entire planning studies and the vast amount of data, methodological improvements and improved policies which were brought to play in the project. It criticised the hydrology of the Project without once referring to 51 pages and more than 17 tables in the Master Plan which demonstrated conclusively that adequate water was available. It talked of waterlogging in the project by pointing out that delivery losses would be higher than those planned by referring to other projects and ignoring the actual measurement used from delivery of water in structures of the kind proposed to be constructed in the Sardar Sarovar system. It ignored socio-economic structures developed by painstaking studies of the tribal population to be affected, by one of India's leading social science research organisations, and superimposed views of colonial and British anthropologists which treated tribals as noble savages to be protected from the mainstream. The Review made extremely uncultured remarks on India's religions and went out of the way to be sarcastic on the leadership of India's freedom movement. This report is privileged to present a point by point rebuttal of the intellectual bankruptcy parading as arrogance of the so called Independent Review.

Reports of Other Organisations, Committees and Public Hearings

Report of the Five Member Group, 1994

Source: Report of the Five Member Group (Vasant Gowarikar, Ramaswamy R. Iyer, L.C. Jain, V.C. Kulandaiswamy, Jayant Patil), set up by the Ministry of Water

Resources to Discuss Various Issues Relating to the Sardar Sarovar Project, 21/28 April 1994 (extracts).*

Introduction

1.2 Terms of Reference

1.2.1 The terms of reference of the Group were not spelt out in detail in the Office Memorandum of 5 August 1993, as is customary when the Government sets up Commissions or Committees or Groups. The only indication was (as already mentioned) that the Group was 'to continue the review discussions initiated during the end of the June 1993 of all issues related with the Sardar Sarovar Project'. As mentioned earlier, a list of those issues was also made available to the Group. Subsequently, however, the Union Minister of Water Resources and Parliamentary Affairs made a statement in the Lok Sabha on 10 August 1993 in reply to a Calling Attention Motion. (...) The statement observed, *inter alia*, that 'according to Clause 16 of the final order of the Tribunal, the parameters of shares of utilisable waters by the States, the FRL, MWL of the Reservoir and the FSL of Navagam canal are made subject to review at any time after a period of 45 years from the date of publication of the decision of the Tribunal in the official Gazette'. It added that 'according to sub-clause 6 of Clause 11 relating to submergence, land acquisition and rehabilitation, alteration, amendment and modification of any of the provisions of Clause 11 is permitted by agreement between all the party States'. The statement proceeded to refer to the discussions initiated at the end of June 1993 on all issues relating to the Sardar Sarovar Project, the constitution of the Five Member Group through the Office Memorandum of 3 August 1993, and its amendment by the Office Memorandum dated 5 August 1993; and concluded by stating that the Group had already started functioning and was required to give a report within a period of three months.

1.2.2 The Group was not entirely clear as to the bearing that the Minister's Lok Sabha Statement would have on the terms of reference of the Group, and sought a clarification on this point from the Ministry of Water Resources. In his reply, Secretary (Water Resources) observed as follows:

> [A]ccording to the Award of the NWDT, the parameters of the Sardar Sarovar Dam like Height, Full Reservoir Level, Maximum Water Level and Full Supply Level of the canal along with other specified provisions of the award will neither be reviewed nor changed until 2025, that is, 45 years after the notification of the Award.
>
> The discussion on various issues related with the Sardar Sarovar Project (SSP) were initiated in the end of June 1993 in this clear context. The Five Member Group has to continue these discussions in the same context with all concerned opinion groups as per the MoWR O.M. dated 5 August 1993.

---

* The full report is available online at http://www.ielrc.org/content/c9402.pdf.

> The project has been planned, designed, approved and is being implemented by the four party States in accordance with the NWDT award. Discussions being held by the Group will, therefore, possibly focus on such issues which can lead to improvement on implementation and benefits planned for Sardar Sarovar Project.

1.2.3 A statement similar to his Lok Sabha Statement was made by the Minister of Water Resources in the Rajya Sabha on 27 August 1993 (…) in which there was the following additional paragraph:

> It is hoped that some useful suggestions about R & R and environmental concerns may be received which will help speedy implementation of the project. There is no danger anticipated to the project.

1.2.4 Later on, a further letter dated 30 September 1993 was received from Secretary, Water Resources in which he stated: 'it is not, I repeat NOT a REVIEW GROUP – Similarly, the project is not under Review'.

1.2.5 Those six documents, namely, the Office Memorandum of 5 August 1993, the list of the issues which had been raised in the meetings of 29 and 30 June 1993, the Minister's statements in the Lok Sabha and later in the Rajya Sabha, and the two letters from Secretary, Water Resources Dated 19 August and 30 September respectively, together indicated the Government's view of the scope of the Group's work. One further document was the statement regarding the Group made by the Government in the counter-affidavit filed by it in the Gujarat High Court in response to a writ petition regarding the constitution of the Group to which a reference is made later in this report (see paragraph 1. 4). In that statement the Government observed as follows:

> The Government has considered setting up of this Group as a sounding board or a listening post for the Government, as discussions are considered to be an important component of the democratic process for getting representative views on various issues. It is hoped that some useful suggestions on resettlement and rehabilitation and environmental concerns may be received and that the same would help the speedy and effective implementation of the project.

(…)

III. Benefits

3.1 Introductory

3.1.1 The main benefits claimed for the project are the provision of irrigation, for 1.8 million hectares (including areas in Saurashtra and Kachchh), the assurance of drinking water for parts of Saurashtra and Kachchh, and the generation of hydroelectric power, including the balancing of the regional system through peaking facilities. The critics of the project argue that these benefits have been greatly exaggerated and are unlikely to materialise, and that therefore the financial, social, environmental costs of the project, which can be justified only by the benefits expected to accrue from the investment, would lose that justification. As the questioning of the benefits thus calls into question the project itself, it needs to be very carefully considered. Though we have had the benefit of many detailed presentations made and documents provided to us, we propose to discuss the issues as far as possible in broad and non-technical terms (so as to make the report readily intelligible to the general reader), but with some rigour, with a view to arriving at a finding whether a sufficient basis has been established for the doubts expressed, and if so, whether a more detailed and thorough examination is warranted.

3.2 Hydrology

3.2.7 In accordance with our terms of reference, as clarified by the Ministry of Water Resources, we have to take the basic features of the project as given. Questions relating to hydrology, which have a bearing on project design, are thus not within our purview. We have gone into the question of the quantum of flows in the river only from the point of view of the possible impact, if any, that a reduced quantum of, flows may have on the planned benefits. Within the limited time available to us, and on the basis of material presented to us, we are unable to come to a firm conclusion on the question of hydrology, in the face of an evident conflict of views on the part of knowledgeable persons. *As we are anxious to see that the intended scale and spread of benefits is not jeopardised, we recommend that the Government should quickly have this issue examined and resolved once for all. We also recommend that if in fact the dependable quantum of flows is of a lower order than had been assumed earlier, the implications of this for the planned benefits, and the consequent steps that need to be taken (for example, a possible re-ordering of priorities among different uses, and a re-allocation among different regions), should also be examined quickly.*

3.3 Irrigation

3.3.1 At the outset, before going into specific points relating to irrigation, we must dispose of a general issue, which in our view, is a red herring. One of the standard justifications often put forward for any irrigation project (including the SSP) is that it will result in an increase in the production of food grains and will thus help in meeting the food needs of a growing population. On the other hand, a point made in criticism of the SSP is that the cropping patterns in the SSP command may favour cash crops rather than food grains. In our view both the justification and the criticism are somewhat misleading. In the first place, the distinction between 'food crops' and 'non-food crops' (or 'cash crops') is not a well-defined one. Even food grains such as paddy and wheat, if produced essentially on a commercial basis, could be regarded as cash crops; contrariwise, even some cash crops may meet essential human needs. Moreover, some areas in the country may specialise in growing food grains whereas others may go in for oil seeds, pulses, cotton, jute, sugarcane, sunflower, and so on; and this may well be in the larger national interest. If sugarcane is frowned upon in a water-short area, the reason is not that it is a cash crop but that it consumes too much water. Secondly, an irrigation project can only provide water for irrigated agriculture; it cannot determine what the farmers in the command area will grow. Actual cropping patterns rarely conform to the assumptions made in a project. It has now come to be generally accepted that water should be supplied in predetermined quantities on a volumetric basis and charged for at prices that would discourage wasteful use, leaving it to the farmers to decide what they should do with the water. Cropping patterns will then be influenced by market forces and relative prices. A view now prevalent is that the liberalisation of industry should be followed by the liberalisation of agriculture and that even the exports of agricultural produce should be freely allowed. Whether this will result in a shift from food grains to more profitable crops, whether food grains may then have to be imported, and whether such developments (if they occur) should be accepted, are larger issues of agricultural policy which cannot be discussed in the context of a specific irrigation project. What we are concerned with in this section is the provision of irrigation water, and in that context what is relevant is the relative water demands of different crops and not the distinction between food crops and cash crops. Having got that point out of the way, we may now proceed to consider issues relating to the irrigation benefit expected from the SSP.

3.3.2 The approach to irrigation proposed in the project is to provide a low delta of water so as to extend the benefit to a large area; take the water as far as possible through gravity (using lift only where unavoidable); introduce methods of automatic and semi-automatic control such that water is provided when and where needed; improve operational efficiency; supply water in bulk and on a volumetric basis to water users' groups and charge for the water at appropriate rates, so as to promote better water management and conservation, and discourage water-intensive crops; and ensure the conjunctive use of groundwater and surface water. Many of these ideas have been in the air for some time; they have been recommended by several Commissions and Committees which have gone into questions of irrigation management, users' participation and the pricing of irrigation water. The SSP proposes to adopt and operationalise these recommendations as far as possible.

3.3.3 The project is expected to provide irrigation to 1.8 m.ha. This includes 0.8 m.ha in North Gujarat, 0.386 m.ha in Saurashtra and 0.037 m.ha in Kachchh. The extension of irrigation benefits to drought-prone areas in Gujarat (North Gujarat, Saurashtra and Kachchh) has been claimed as one of the major benefits of the project.

3.3.4 The doubts expressed by critics of the project on the planned irrigation benefits are as follows:

- The flows in the river at the project site will be only 23 MAF and not 28 MAF; the delay in the NSP will also affect the availability of water in the SSP; reduced water availability will mean reduced benefits.
- '*En route* flows' are now being taken into account; this had not been envisaged earlier.

- The assumed irrigation efficiency of 60 per cent is unrealistic; seepage and operational losses will be higher than has been assumed.
- The probability of water-intensive crops coming up in the head reaches is high, having regard to past experience and the fact that sugar factories are being licensed in the command area.
- There is a strong possibility of valuable agricultural land in the command area being put out of use because of the emergence of waterlogging and salinity conditions.
- In the area downstream of the dam (including Vadodara, Bharuch, etc.) there will be substantial water needs for municipal and industrial purposes; big industrial complexes are coming up in this area; and if the needs of these areas are fully met, the water available for irrigation elsewhere will get reduced.

On these grounds critics conclude that irrigation coverage in the project cannot be 1.8 m.ha; and further that the water will not reach Saurashtra and Kachchh.

3.3.5 We have already dealt with the question of hydrology and pointed out that opinion is divided on the subject, and have accordingly recommended that the Government of India should have this question examined and settled once for all very quickly (paragraph 3.2.7 above).

3.3.6 The point that the utilisation of *en route* flows had not been envisaged earlier but is now being taken into account, does not seem to us to be a significant criticism. Regardless of whether or not this had been envisaged earlier, it seems to us that it makes sense to utilise such *en route* flows.

3.3.7 In regard to the question of irrigation efficiency, seepage loss, operational loss, etc., we have carefully considered the scepticism expressed by the critics as well as the explanations offered by the project authorities in their publications and by the Ministry of Water Resources in their comments on the points made by the NBA. We have also put the question to some of the experts who appeared before us. What is involved here is a question of judgement. Critics believe that a 60 per cent efficiency is far higher anything than that has been achieved so far; that seepage and operational losses will be higher than the levels assumed; that the sophisticated control methods proposed are novel efforts in this country and may not operate as planned; that to the extent that there is a departure from the kind and extent of canal-lining envisaged earlier, the seepage loss will increase; that taking all this into account irrigation efficiency will be nearer 40 per cent than 60 per cent and that this would mean that the area covered will be less than the projected figure. The supporters of the project argue that the country cannot remain for ever at the efficiency levels of the past; that levels *must be* improved; that irrigation efficiencies of more than 60 per cent have in fact been reached in certain instances in Gujarat in Rabi 1991 (Aji-II 66.91 per cent, Demi-II 68.03 per cent, Uben 71.32 per cent); that losses in the entire distribution system are based on details worked out for the pilot block of Kapurai distributory; that with the innovations proposed in the project irrigation efficiency is bound to increase; that though these are pioneering efforts there is no reason to believe that they will not work; and indeed that they must be made to work, and that the scepticism expressed in this regard is not warranted. Some of these arguments have, in turn, been questioned by the NBA.

3.3.8 *Having given careful consideration to both the criticisms and the answers we have come to the conclusion that an irrigation efficiency of 60 per cent is achievable; but whether it is actually achieved will depend on the nature and quality of the lining of the canals, and on the manner in which the innovative measures of controlled distribution, pricing and water management which are envisaged are operated. We recommend that determined efforts should be made to achieve the irrigation efficiency level of 60 per cent. The doubts expressed in this regard should be given due consideration and should spur the project management to greater efforts.*

3.3.9 *The doubts expressed on the score of the possible emergence of water-intensive cropping patterns in the head reaches also need to be taken very seriously.* Past experience in other projects throughout the country clearly warrants such doubts. As the reservoir fills up and the head reaches of the system get completed, there is abundant availability of water; there is an understandable desire to use that water to maximum profit; water intensive-cropping patterns (paddy, sugarcane, etc.,) develop; eventually when the full canal system gets completed there is an inadequacy of water because much of it has been commandeered by the head reach farmers; it becomes very difficult indeed to reduce water-use by them; and the result is that the tail-enders get little or no water. This has been the pattern of irrigation development in this country. Even in the SSP command it has been pointed out that sugar factories are being licensed in the head reaches. It may be argued that water has not been assured for sugarcane growing but it will undoubtedly be difficult to deny water for this purpose once the cropping pattern gets established. The answer given to these doubts, namely that there will be a sophisticated control system, bulk supply in volumetric terms, proper pricing, the provision of low delta, etc., is not without force, but it would not do to underestimate the difficulty of withdrawing water once a pattern of water-use gets established and vested interests develop; and these interests are usually able to acquire certain degree of political influence. It does not follow that these developments must be accepted as inevitable. *What needs to be done is to devise effective measures to minimise this possibility.*

3.3.10 Several suggestions have been made in this regard. These include the following:

- Starting the supply from the tail end first.

- Operating the water distribution system (low delta, automatic control, bulk and volumetric supply, proper pricing) *strictly*, without allowing it to be compromised.
- Creating a vested right for the tail-enders by declaring the water-shares of different areas in advance and giving it a good deal of publicity, thus allowing tail-enders also to acquire a certain degree of political strength.
- Facilitating the eventual reduction of water to the head reaches by putting them on notice in advance and by providing a statutory backing through legislation for such withdrawal.

There is some merit in all these suggestions though the extent of their practicability may vary. *We would commend all of them for careful consideration by the Government.*

3.3.11 Suggestions regarding restrictions on cropping patterns were also made but (as already mentioned) these seem to be inconsistent with the general approach, now widely prevalent, that the Government need not prescribe cropping patterns and that it should confine itself to supplying water in bulk and at the proper price on a volumetric basis to associations of farmers, leaving them to make the best use of the water. Moreover, any recommendation that sugar factories should not be licensed may lose much of its significance in the future with the progress of the current process of economic liberalisation and the possibility that the licensing requirement itself may disappear even for the sugar industry in the not too distant future. *What is necessary and important is that the proposed measures of low delta, controlled supply, etc., should be effectively operated and the discipline of the system maintained. If there is any weakening in this regard, and if unsound practices and illegitimate demands come to be accepted for whatever reasons, then the whole approach to irrigation in this project will get vitiated.*

3.3.12 The fears which have been expressed of possibilities of the waterlogging and salinisation of land in the command area are based on past experience of irrigation development in the country and are not groundless. The answer given is that the possibility has been foreseen and guarded against in the project. The following are the observations of the Ministry of Water Resources on this point:

> The reconnaissance level soil survey of the entire command area of the SSP indicates that the drainage intensity is good in most of the regions except in three regions (agro-climatic regions 4, 7 and 11) which cover hardly 20 per cent of the total culturable command area. A well planned intensive drainage network is being planned for these regions.
>
> Along with the irrigation network, the surface draining network in the command area has been planned. Natural drains are to be resectioned to cater to needs. Additional drains, where necessary, will be provided to take care of excess surface flow during monsoons. The system is planned to ensure drainage in a reasonable time not exceeding three days without damage to crops. Sub surface water table control includes limited water availability for irrigation by appropriate irrigation

planning to encourage ground water development through the private sector. If it does not come up to the desired level, ground water development through the public sector is proposed to be undertaken. Where the ground water is saline and cannot be used otherwise, ground water development through the public sector is planned to control the water table to fall below a specified level from the ground surface. Vertical drainage through pumping from wells and horizontal drainage through deep open drains are also planned. With this planning, the system will take care of excess water, if any, which could cause waterlogging and salinity.

Several measures like rotational water supply, lining of the canals, extensive irrigation instead of intensive irrigation, conjunctive use etc. have already been introduced in some of the major command areas in Gujarat State itself and have a proven record in checking waterlogging successfully.

It will thus be seen that all precautions have been taken in the planning, construction and operation of the project to ensure that the hazards of waterlogging and salinity do not occur even in seemingly vulnerable areas of the command. The project will be helpful in stalling the steady decline of the ground water in the North Gujarat area, thus correcting the environmental imbalance created over several decades in the past. Under the SSP, possible recharge is accomplished along with controlled surface irrigation and the water that is naturally recharged is planned to be used by farmers in a conjunctive use framework resulting in sustainable ground water regimes and affordable cost.

It will be seen that the planning of drainage, the resectioning of natural drains and the provision of additional drains where necessary are among the steps envisaged. *It is necessary to ensure that these plans are translated into realities. The Environmental Sub-group of the Narmada Development Authority will no doubt be monitoring this.* (As regards controlled distribution, low water allocation, etc., we have already recommended that this should be enforced).

NBA has made the point that various zones in the command area are known to be prone to water-logging and salinisation on the basis of preliminary soil surveys, and that further detailed command area investigations are yet to be completed. We regard these as cautions to be kept in mind and not as fundamental criticisms.

3.3.13 One important element in the precautions against the danger of waterlogging and salinisation has to be specially noted, namely the proposed ‘conjunctive use’ of surface water and ground water. ‘Conjunctive use’ is a much talked about but little practised idea. It is an accepted part of Government policy and is enshrined in the National Water Policy; but it has not been operationalised. It has not been an integral part of project planning so far; but an attempt in this direction is being made in the SSP. The conjunctive use of surface water and groundwater is a component of the approach to irrigation in this project. This is expected to guard against the over-use of canal water, and at the same time ensure vertical drainage through groundwater pumping. This means that a large number of tube wells will have to be sunk in the command area. It is envisaged that this should be primarily a private sector effort, supplemented by public sector investment in tube wells as may be found necessary. It is necessary to ensure that this actually happens. The expectation is that low canal water allocation, volumetric supply, and so on, will compel farmers to supplement this by going in for tube wells. This may indeed happen; on the other hand it is also possible that there may be pressures to increase allocations of canal water, or even unauthorised appropriation of water. Suitable measures are needed to ensure that the intended number of private tube wells do come up in the command area. Further, if it is proposed to invest in a large number of public sector tube wells, this too needs detailed planning and the allocation of the necessary funds. It is also necessary to work out in detail the integrated supplies from the two sources (canal and public sector tube wells), and their relative pricing.

3.3.14 Areas proposed for Irrigation (in lakh ha)

North Gujarat		South Gujarat		Saurashtra & Kachchh	
Ahmedabad	3.30	Vadodara	3.40	Surendranagar	3.04
Banaskantha	3.13	Bharuch	0.98	Bhavnagar	0.48
Gandhinagar	0.10	Kheda	1.16	Rajkot	0.34
Mehsana	1.50	Panchmahal	0.10	Kachchh	0.37
Sub-total	8.03		5.64		4.23
Total					17.90

The criticism is that adequate allocations have not been made to Saurashtra and Kachchh and that the lion's share of the water would go to areas which are already prosperous, namely Ahmedabad, Vadodara, Bharuch and Kheda. The reply given by the project authorities is to the effect that only a limited allocation of water (much less than had been demanded) has been given to Gujarat by the Tribunal; that this has to be used optimally with due regard to techno-economic considerations, and as far as possible by gravity flow; that it was never claimed that the whole of Saurashtra and Kachchh would be covered with irrigation from Narmada waters; and that within the available quantum of water, and subject to techno-economic considerations, whatever was possible has been allocated to Saurashtra and Kachchh. A further point made is that even the areas which are considered prosperous (Ahmedabad, Vadodara, Bharuch and so on) are water-short, and that about 75 per cent of the command area as a whole is drought-prone.

3.3.15 As against this, judging from Memoranda submitted to us by the Kachchh Development Forum and the oral presentation made to us by representatives of that Forum, there seems to be a sense of grievance in Kachchh that the allocation of Narmada waters to that area is very meagre. We have also received communications from Saurashtra to the effect that the claim that SSP will confer large benefits on Saurashtra and Kachchh is not justified and that these areas will receive only a very nominal benefit from the project.

(…)

3.3.23 *Having brought these facts to notice, we must leave it to the Gujarat Government to consider whether the provision for irrigation in Saurashtra and Kachchh can he enlarged, either by increased allocations of Narmada waters, or by adequate plans and provisions for harnessing the local water resource potential, or by a combination of both measures.**

(…)

3.4 Drinking Water

3.4.11 *In the light of the foregoing, we would recommend as follows*:

i) *A full list of all the 8,214 villages and 135 urban centres to be provided with drinking water should be announced as quickly as possible. The quantity of water required as per the norms indicated should also be clearly specified.*

ii) *The project authorities should state clearly the water-saving measures and the use-wise and area-wise re-allocations through which they propose to find the supplies needed. (The quantity needed for drinking water will no doubt be small in comparison with irrigation needs, and the adjustment required may not present great difficulty, but a clear announcement of the reallocations seems desirable with a view to answering criticisms and allaying anxieties).*

iii) *The agency or agencies which would be responsible for managing the actual water supply system for each village and each urban centre under the overall control or guidance of the GWSSB should also be identified and announced as early as possible.*

iv) *The detailed physical planning of the water supply project (off takes from the canal, filtration and purification plants, pumping, pipelines, etc.) should be completed (if not already done), and should be published as early as possible. The financing arrangements for the entire scheme should be fully worked out and made public. The works themselves should be completed within a pre-announced time-frame.*

v) *The detailed plans, the necessary physical arrangements and the funding needed for meeting the drinking water needs of the villages and towns in question during the next ten to twelve years (i.e., before the canal waters reach them), as also during the period of annual maintenance shut-down of the canals thereafter, should be firmed up and published as early as possible.*

---

* For more recent developments on the allocation of water to the Kachchh district, see *Kachchh Jal Sankat Nivaran Samiti v. State of Gujarat*, reproduced above at page 105.

vi) *Water-shed development and local storages and conservation, the recharging of groundwater aquifers, filling up of surface storages, etc., are very important with or without canal water from the SSP, and the necessary detailed plans, cost estimates and financial allocations should be made clear as early as possible.*

(...)

IV. Environmental Aspects

4.7 Recommendations

While it is clear from that note that many of the studies, plans and activities are behind schedule, there has also been a slippage in project construction, and it appears that it is still possible to retrieve lost ground on the environmental front through accelerated action. We would strongly recommend that this should be done. In particular we would draw attention to the following:

i) In so far as catchment area treatment is concerned, the condition imposed by the MoEF requires the treatment of erodable and highly erodable sub-watersheds adding up to an area of 7 lakh hectares. The project authorities have sought to make a distinction between 'directly draining' and 'freely draining parts of this, and this seems to have been accepted by the NCA. The 'directly draining' area is said to be of the order of 1.45 lakh hectares. The project authorities want to take responsibility for the treatment of only this part as a legitimate charge on the project and feel that the treatment of the rest of the area should be the responsibility of the Soil Conservation and Forest Departments. The MoEF seems to be willing to accept a phasing of the treatment, with 'directly draining' areas being treated in Phase I and the rest in the Phase II, but want the entire area to be treated, regardless of who bears the cost. Their point is that it is in the interest of the project that this should be done, as it is only through such treatment that the erosion of the catchment area and the consequent increase in the rate of sedimentation of the reservoir can be arrested. They further point out that the arresting of erosion has acquired some urgency as the waters have started rising, and eroded material will get trapped and can no longer be flushed away. *We do not wish to enter into the controversy regarding the debiting of costs, but would argue that both 'directly' draining and 'freely' draining areas should be treated in the interest of the project in a phased manner as acceptable to the MoEF: indeed, there seems to be no disagreement on this between the MoWR and the MoEF*. The *pari passu* condition in this regard has already been missed as the waters have started rising, and all that can now be expected is a time-bound Action Plan for avoiding further delays. However, there appears to be an inadequate appreciation of what catchment area treatment really involves. At least some parts of the catchment area would be private land, and even in government forest or revenue land there may be tribal communities or people practicing agriculture. *Action by governmental agencies alone would not be adequate. Measures of catchment area treatment, such as soil conservation or afforestation or other forms of greening would have to be carried out with the cooperation and through the agency of the people of the area. What is needed is the watershed development approach of the kind which has been successful in some places such as Sukhomajri or Ralegon Sindhi. A wide range of location-specific measures such as agro-forestry, horticulture, grass growing, fodder cultivation, tree-farming, and so on, may have to be adopted in the light of their suitability and acceptability. Innovative methods, a practical approach and social motivation are called for. The time-bound Action Plan referred to above will have to encompass all this.*

ii) *While compensatory afforestation seems to be making progress, the pace of plans and actions needs to be stepped up. On the whole we would leave it to the MoEF to monitor the progress of compensatory afforestation and have no special recommendations to make, except to commend here again the idea of people's participation in the afforestation programme, as well as the promotion of irrigated forestry in the command area to the extent possible. However, there is a possibility that serious difficulties may arise in regard to compensatory afforestation in the future, if forest lands keep getting released for rehabilitation purposes necessitating more and more compensatory afforestation, until a point is reached when it ceases to be implementable. That possibility needs to be obviated.*

iii) On flora and fauna we seem to face considerable uncertainty. There is little experience in this country in implementing remedial measures to minimise the impact of the creation of a reservoir on wildlife, plant genes, biodiversity, etc. With the best will in the world, it is going to be difficult enough to ensure that wildlife use the escape routes or migratory corridors provided and to guide them into the planned sanctuaries. *The least that we can do is to ensure that studies and action plans are ready in time and are put into operation in such a manner that valuable plant species are not lost and that wildlife does not drown or get cut off by water in large numbers. As the waters have started rising, there is urgent need to attend to these matters before it is too late.*

iv) In so far as fish (as a form of life) are concerned, fish ladders and elevators do not seem feasible in a project of this magnitude; given the kind of interference in the natural regime of the river that this project involves, some adverse impact on fish life seems inescapable. From the *commercial* point of view, i.e., from the point of view of fishing as an economic activity, it is argued that the temporary decline in the activity because of the creation of the dam will be more than made up in due course by the development of fisheries in the reservoir. While this may be true in total numbers, breeding for commercial purposes is likely to concentrate on certain varieties and it seems that there will be changes in the composition of the fish population. Moreover, the trapping of micro-nutrients by the reservoir will result in a decline in catches down-stream of the dam and will affect the lives of fisher folk who have been dependent on this livelihood. Efforts may be made to rehabilitate them through reservoir-fisheries, but that will be a different kind of fishing from what they have been accustomed to. It appears that both from the point of view of flora and fauna and from that of fisheries, the impact of the impoundment of water behind the dam cannot be fully determined and that remedial measures are difficult to design. *We hope that the authorities concerned at the state and central levels are fully alive to the complexities of the matter.*

v) Remedial action in regard to the public health consequences of water impoundment will have to be of two kinds, namely measures to prevent an increase in water-borne diseases and measures for the treatment of diseases when they occur. The general tendency is to attend to the provision of treatment facilities. However, it is understood that preventive measures are now receiving some attention. Some measures such as fluctuations in the water-level to hinder the growth of larvae are said to be under study. *We have no special recommendations to make in regard to this matter except to express the hope that it is receiving adequate attention on the part of the MoEF and the Ministry of Health.*

vi) It must also be noted that there may be a deterioration in water quality in the command area because of the use of fertilisers and pesticides in irrigated agriculture. So far as we know, this is not receiving much attention at the moment. *It may be appropriate to deal with the question of the treatment of agricultural effluents and the maintenance of water quality as a part of the command area development programme, but the conventional command area development approach does not include this. This is something that needs to be attended to.*

vii) Studies of the water-table and the possibility of waterlogging and salinity conditions are said to have been done up to Mahi and to be in progress in the rest of the command. There is a general claim that waterlogging and salinity problems will not arise in the command of the SSP because of the proposed automatic and semi-automatic control of canal supplies, the kinds of cropping patterns which are envisaged, and the promotion of the conjunctive use of surface water. *It appears to us that there is some danger of complacency here. It cannot be taken for granted that all these measures will work as planned and that waterlogging and salinity problems of the kind experienced in other projects will not arise at all in the SSP command.* In an earlier chapter, we have emphasised the need to ensure the strict enforcement of the control systems, conjunctive use, etc., which are planned. *We would recommend that these aspects must be constantly kept under study and proper remedial measures devised. The project authorities need to be on the alert to these problems emerging in the SSP command and to be able to respond promptly when they do.*

viii) The downstream effects of the creation of the dam would include the possible impact of a reduction of flows (as well as of micro-nutrients in the flows) on economic activities such as those of fishermen, boatmen, agriculturists, etc., downstream of the dam, as also possible shortages of water for domestic and industrial uses. A further impact would be the possible incursion of salinity in the coastal areas because of reduced flows in the river. *Some references have been made to remedial measures, but in general it appears to us the full impact of the creation of the dam on the down-stream side has not been adequately studied. This needs to be done as early as possible and detailed plans for countering those effects formulated on a top priority basis.*

V. Displacement Resettlement and Rehabilitation

5.8 Recommendations

In the light of the above discussion we would make the following recommendations:

5.8.1 *Given the reality of the project, every effort must be made to count the human and social costs of the project as fully as possible, and to provide remedial measures. For this purpose, there should be a complete census of all categories, groups, communities and individuals, affected in any manner whatsoever, including canal affected persons, communities downstream of the dam, groups and individuals providing supplies and services to others and so on, so that the data-base needed for planning is comprehensive. A number of category-specific rehabilitation packages should be worked out.*

5.8.2 *There should be a careful re-examination of the approach to and concepts of resettlement and rehabilitation in light of the actual difficulties which have been experienced with special reference to limitations on land-availability; difficulties resulting from the dispersal of close-knit groups or from resettlements far away from original homes; problems arising from the release of forest land for rehabilitation; secondary displacements which can arise from some of the environmental protection measures; and so on. Land-based rehabilitation should be extensively supplemented by other forms of rehabilitation such as the imparting of a variety of skills, assistance both financial and other for trade and self-employment, encouragement of entrepreneurship, and so on.*

5.8.3 *The responsibility for the resettlement and rehabilitation of the PAPs should be not merely that of the Government but also partly that of the beneficiaries of the project. The principle that a share in the benefits should be provided as a prior right to those who bear the social costs of the project, should be introduced through appropriate legislation. (If necessary, this should include restrictions on land sales in the command area). This principle should be adopted not only in the SSP but in all future projects.*

5.8.4 *With reference to the foregoing a Master Plan, with detailed time-schedules, should be got ready within the next six months.*

5.8.5 *There should be strict adherence to the NWDT and Supreme Court injunctions regarding rehabilitation well in advance of displacement. This will no doubt be monitored by the Rehabilitation Sub-Group of the NCA. The effective functioning of the Sub-Group should be ensured, and the NCA and the State and Central Governments should give the greatest importance to the observations and recommendations of the Sub-Group.*

5.8.6 *The administration of the rehabilitation package should be done in a humane manner. There should be special orientation programmes for every level of the bureaucracy and the police system engaged in this process.*

5.8.7 *Each State should establish an Ombudsman to whom PAPs and NGOs can take their grievances for redressal. The persons selected for this function should be such as command respect and are widely acceptable.*

5.8.8 *At all times the people should be kept informed of the steps contemplated. The flow of information should be full and timely. The government machinery should cooperate to the fullest with good and experienced voluntary agencies and ensure that these are not handicapped either by want of information or by a negative attitude on the part of the bureaucracy. Such voluntary agencies should be treated not as external to the Government but as extensions or arms of the Government for the satisfactory implementation of rehabilitation.*

Indian People's Tribunal on Environment and Human Rights, 2000

Source: The Indian People's Tribunal on Environment and Human Rights, Dammed Future? – Report of the Independent Inquiry into the Status of Rehabilitation of Project Affected Families (PAFs) of the Sardar Sarovar Project (SSP) in Maharashtra, Tribunal Headed by: Justice Rambhushan Mehrotra (Retired Judge, Allahabad High Court) (March 2000), Introduction, Chapter 6 and Chapter 7 reproduced.*

Introduction

The Indian People's Tribunal (IPT) was approached by the Punarvasan Sangharsh Samiti (PSS) to conduct an investigation into the status of the tribals who have voluntarily accepted government rehabilitation in the Nandurbar District of Maharashtra. The tribal people had been relocated here eight years ago from their ancestral lands in the Narmada Valley, as the dam had submerged their villages. On shifting to the new site they were promised, land for land, separate plots for adult sons, cash compensation for shifting to the new site, housing material as well as water supplies, schools, medical facilities and other facilities like parks and playgrounds.

However, even after almost a decade of residing in the new area they find they have been cheated and most of what they had been promised remains a dream. In

* Other reports of the Indian People's Tribunal on Environment and Human Rights concerning SSP include the Justice B.G. Kolse Patil (Retd) report *The Status of the Manibeli Oustees* (1993) and the Justice S.M. Daud (Retd) report *The Fate of the Gujarat Oustees, Narmada Valley Dispossessed, Hunted, Humiliated and Cast into Oblivion!* (1994).

1999 the Supreme Court that had earlier stayed construction on the dam pending the final hearing of the case changed its decision and allowed for the dam height to be raised by 5 metres. Fearing further displacement of people and having experienced the complete failure of the government's rehabilitation scheme, the Punarvasan Sangharsh Samiti approached the IPT to conduct an investigation.

The Tribunal comprised of Justice Rambhushan Mehrotra (Retd Judge, Allahabad High Court) and Professor Vijay Paranjpye, an economist and expert on the impact of large dams. Advocate Apoorva Kaiwar of the IPT accompanied them. The team visited the rehabilitation sites in Maharashtra from 19 to 21 March 1999. A series of public hearings were held where Project Affected Families (PAFs), Government officials in charge of the Rehabilitation and Resettlement (R&R) process in Maharashtra and representatives of the Punarvasan Sangharsh Samiti, Taloda, deposed before the Tribunal. Other citizens and prominent people's representatives were also invited to present their views before the Tribunal. The process of public hearing was carried out in a remarkably cordial and transparent manner, and the Tribunal members noted the forthright manner in which the Government officials as well as the oustees and their representatives made representations. On previous occasions when the IPT has conducted investigations, government officials have chosen to maintain a bureaucratic silence, hence their presence at this hearing was appreciated.

Terms of reference

1. To enquire into the process and status of the PAFs of SSP being settled in Maharashtra since 1992, in the light of the provisions of the Narmada Water Disputes Tribunal Award (NWDTA).

2. To hear the depositions made by the adivasi people amongst the PAFs who are being resettled in Taloda Tehsil of Nandurbar in Maharashtra, at Rozwa, Somaval, Amlibari, Akkalkuwa and Akrani Tehsils, and to record their statements.

3. To assess the preparations and the readiness of the Government of Maharashtra in terms of providing adequate land and associated infrastructure for rehabilitating all the oustees likely to be displaced by SSP in Maharashtra.

4. To make appropriate recommendations with regard to the above-mentioned terms.

Background

Sardar Sarovar Project on Narmada has remained controversial due to the various issues raised by people's organisations during the last 14 years. A major reason for controversy and the people's struggle has been the large-scale displacement and status of rehabilitation of the PAFs. However, the Central Government and the Governments of the States of Maharashtra, Madhya Pradesh and Gujarat have made claims of comprehensive plans and the best possible, legally perfect rehabilitation policy. The affected people and their organisation on the other hand have protested against human rights violations of the project-affected people as well as violation of their constitutional rights. They allege that the Narmada Water Dispute Tribunal Award (NWDTA) has been very poorly implemented. The parties have presented their views to the Supreme Court as well as at various forums in the civil society.

For almost four years, i.e. between 5 May 1995 and 18 February 1999, the Supreme Court had stayed all work related to the height of the dam which had been stopped at a reservoir level (height) of 80.3 metres. However on 18 February 1999, in the case *Narmada Bachao Andolan v. Union of India and others*,[1] the Supreme Court passed an interim order, to the effect that the height of the dam could be raised by 5 metres from 80.3 + humps to 85 + humps.[2] The operative clause of this order states, '[a]t this stage, however, we permit the State of Gujarat to raise the level of the dam to RL 85 metres, excluding the humps necessary for the maintenance of the safety of the dam'.

As a result of the order issued on 18 February 1999, approximately 220 additional Project Affected Families (PAF) would be displaced in Maharashtra and would need to be resettled before the reservoir level rose in June 1999. This order will have a serious impact on a few thousand families especially in the tribal villages in the Satpuras and Vindhya falling in the three riparian States affecting their fields and houses. The Court's order stipulates full and fair rehabilitation of all those to be affected at 88 metres as per NWDTA provisions.

Many people's organisations have raised allegations that the Governments of Maharashtra, Gujarat and India have made false claims concerning land availability in their affidavits, which led to the said interim order. These organisations claim that the rehabilitation of SSP oustees is not as per the policy and stipulations in the Narmada Water Disputes Tribunal Award (NWDTA) and is rather incomplete even in the case of the families already affected up to 80 metres, and that there is no master plan for rehabilitation of all categories of the oustees.

As Smitu Kothari puts it so aptly

> Beyond a point, it makes sense to reiterate what has been said dozens of times before. For over ten years (and longer if one takes into account the evolution of the project in the aftermath of the Narmada Water Disputes Tribunal Award that adjudicated on the sharing of the waters in the region and also gave consent to the SSP), the State and Central Governments have had time and enough resources to define a detailed, workable plan to comprehensively rehabilitate those who will be displaced or adversely affected by the project. And though extremely detailed plans exist for the physical structure of the dam and its infrastructure, even today

---

1 Writ Petition No. 345/94, SLP No. 3608/85, CA 6014/94, Writ Petition No. 104/97 and transferred case No. 35/95.

2 Humps means a speed breaker construction strip which could break/limit the speed of the waters gushing in to save the stilling basin of the dam from any damage which is now completed up to (85+3) 88 metres.

this rehabilitation plan does not exist. If the location of every inch of the dam and its infrastructure can be defined in great detail prior to digging the first shovel of earth, why isn't such a plan equally integral to the planning process?

Today, even project officials unofficially concede that the land to adequately rehabilitate those to be displaced by the project has not been found – and even if some of it is brought from landlords, the price will be exorbitant and the communities will be broken up and dispersed over a wide area. Besides, how justified is it that communities are not only broken up, not only dispersed but also relocated in geographical and linguistic areas significantly different from their own? And even when lands have been found, what about those who were dependent on it earlier?

And this is true of large projects all over the country. Particularly when the social and environmental costs are so enormous and, in many ways, so irreversible. I am not even going to into the continuing violations of basic civic rights by the State – the number of people falsely implicated, the continuing harassment, the use of repressive measures to break-up peaceful protest.[3]

Chapter 6: Conclusion

1) The Government of Maharashtra has not yet been able to satisfactorily resettle and rehabilitate even the oustees displaced by the partial completion of the dam at a height of 80.3 metres.

2) The land available for R&R is totally inadequate. The raising of the dam height to RL 85 metres will lead to displacement which completely contravenes the provisions of the NWDTA (1979), because the GoM simply does not have enough land to adequately resettle the oustees.

3) 'Zamin Dikhao Abhiyan' – The Tribunal also took cognizance of the statement made by Srimati Pratibha Shinde, claiming that when the PAFs took a delegation to the GoM officials in March 1999, the officials were unable to show them the land.

4) As per the Maharashtra Government R&R Policy (1991–92), one hectare of land has to be provided to the landless PAFs. Further, major sons (completing 18 years in 1987) are to be considered as 'khatedar' (separate family). However, all the submissions received showed that these conditions have been violated and that most major sons have not been considered as PAFs.

5) The NWDT Award clearly mentions that within two years of the declaration of the Award, the lands necessary for rehabilitation should be acquired. It is a matter of grave concern that the Government of Maharashtra has been unable to acquire adequate land during the 20 years since 1979. In the light of the government's failure, the IPT states that there has been a gross violation of the provisions of the NWDTA.

---

3 Lokayan Bulletin, May – August 1991, Editorial, p. 5, Published by Macro Graphics, New Delhi.

6) Another major issue is that oustees have been allotted forestland on which other tribal populations, referred to as 'encroachers' by government officials are living. Displacing the existing population to provide land to the oustees of the SSP, will in our opinion, only give rise to more problems rather than solutions.

7) Government officials spoke of their efforts and intentions to rehabilitate the oustees but were unable to answer specific questions; rather they confirmed the submissions made by the people and those of the supporters, politicians, reporters, etc.

8) The committee concludes that the claims made by the oustees are genuine, since none of the government officials present at the time of the hearing could give facts or information to the contrary. The committee makes a serious note of the fact that the existence of such large numbers of oustees without any land grossly violates the provisions of the NWDTA (1979).

9) Not only has the government not been able to provide land as per the NWDTA stipulations but it has violated every other clause such as providing cash compensation, housing material, medical facilities, schools and other benefits.

10) In no case were the people shifted a year before submergence as the NWDTA recommends. It is clear from the depositions, site visits and the study of relevant documents that there is a large gap between the model rehabilitation plan of the government and the ground reality. Sadly it is with this reality that the project affected families of the SSP are compelled to live.

Chapter 7: Recommendations

1) A major stipulation in NWDTA was to formulate the Master Plan for resettlement, with land details within two years, i.e. by 1981. That kind of 'Master Plan' is not yet ready. The Tribunal recommends that this Master Plan should be prepared and made available to the people before any further construction activity is undertaken.

2) An independent committee consisting of not just government officials, but also representatives of project-affected persons and of organisations working in the area be constituted to identify the project-affected communities and families. The committee should also ascertain the availability, identity and adequacy of land available for resettlement and rehabilitation.

3) Dereservation and clearing up of forestland is an environmentally damaging option for providing land for rehabilitation. Besides the identified forestlands are not free from claims from earlier occupants. Hence, further dereservation of forestland must be stopped.

4) All the stipulations regarding community resettlement must be adhered to strictly.

5) All the provisions of the NWDTA with respect to those already ousted due to the present height of the dam must be met with, for example, with regard to declaration of major sons as project-affected persons and allotment of land that is cultivable

6) The rehabilitation sites have not been provided with basic amenities. All efforts and resources must be deployed to ensure that the tribals already displaced are provided with fertile land, housing material, water supply and medical, education and public transport facilities. Further displacement will only compound the present dismal situation.

7) Apart from proper rehabilitation the project affected people of Nandurbar district must be compensated in view of the tardy rehabilitation facilities they have been forced to live with.

8) Government officials in charge of rehabilitation must be held personally liable for implementation of the rehabilitation plan and should be punished in case of lapses.

9) No further construction on the dam or an increase in height should be permitted until all the aforesaid recommendations are satisfactorily complied with.

Justice S.M. Daud Report, 2001

Source: Justice S.M. Daud, Report of the Chairman and other Non-official Members/Invitee, Committee to Assist the Resettlement and Rehabilitation of the Sardar Sarovar Project-Affected Persons, Government of Maharashtra, 29 June 2001 (extracts, pp. 1–2, 6–12, 39–55).

The Government has appointed this Committee under a Revenue & Forests Department Notification No. SSP 312000/PK4/R-5 dated 23 February 2001, which was later modified by further Notifications No. SSP 312001/PK4/R-5 dated 23 March 2001 and by a Corrigendum No. SSP 312001/PK4/R-5 dated 12 March 2001.

The Notification assigns to us various tasks which have been reduced to issues reading as follows:

1. Is it correct to say that there is land available for the resettlement of the PAPs in conformity with the Tribunal Award, the policy of the government and the verdict of the Supreme Court?

2a. Does there exist an irrigation facility that could be made available for the benefit of the PAPs?

b. If not, what recommendations should be made in this behalf?

3a. What is the impression of the Committee/Members/Invitees in respect of the villages earmarked for resettlement colonies?

b. If the resettlement colonies are wanting in certain respects, what recommendations should the Committee make to see that government's policy for rehabilitation is implemented?

4. Has the process of rehabilitation of affected tribals been in accordance with the NWDT Award and the government policies for rehabilitation as also the decisions given by the NCA?

To enable us to make appropriate recommendations, we have made extensive visits to the villages in the submergence areas as also the lands/sites/villages proposed or actually functioning as resettlement villages. The meetings held there for the tribal villagers were in the nature of 'gram sabhas' which came to be attended in large numbers by tribals from the affected villages situated far and near.

Documentation for Committee's Scrutiny

We regret that many documents, which we had sought from the government, have not been made available to us. While it may not be possible to say that this was done designedly, it certainly has hampered us to some extent in knowing the government's views on crucial matters. Some of the papers made available to us in a show of compliance with the requisitions were found to be irrelevant and not throwing any light on the subjects on which the Committee required illumination. As against this, the NBA (Narmada Bachao Andolan) has made detailed submissions and presented various documents covering almost the entire controversy. The government will find this material useful to carry out the different tasks mentioned in our report.

Issues

With a view to pinpoint the Terms of Reference with great accuracy we have considered it desirable to formulate issues. These issues give a fairly accurate picture of the range of inquiry to be conducted by us. While the witnesses/government officials/members of the NBA and PSS have been allowed to freely state their views without our asking them leading questions, we have tried to keep the oral submissions and depositions within the confines of the subjects covered by these issues. The issues are set out below with a short statement of our findings there. We must here state that the Annexures given to the notification constituting the Committee of which we are a part gives the erroneous impression of government having done everything possible to accommodate the rehabilitation needs of the tribals. It is with regret that we have to comment on the erroneousness of the impressions sought to be created by as crucial a document as the notification whereunder our Committee has been constituted. From the picture prevailing as on 7 February 2001 (from the document submitted to the Committee in April 2001 placed as Exhibit 'D') it would appear that practically all the oustee families including those affected at 90 m have been rehabilitated and if there be any remainder, their number is trifle. Our inquiry indicates the reverse to be the correct position. In other words the greater part of the oustees are yet to be resettled who are still without culturable land or a house plot. Even those brought into the rehabilitation sites, a large number of them face severe privations such as no land being allotted to them or those with allotments have to put up with less land or land unsuitable for cultivation, lack of water, etc. This deprivation goes to the extent of water required for drinking. If there be a shortage of drinking water it goes without saying that nothing like water for irrigation of the lands allotted to the tribals exist. And yet the figures supplied to us by the government seek to give the impression as if every facility prescribed by the Tribunal Award and the government policy has been provided to those who have been rehabilitated. Exhibit 'D', is a good example of the chasm between promises and reality. In the remarks column the stock comment is that some work has been completed, that further works are under progress, that they are proposed, etc. Thus on the government's own admission, fortified by the admissions which the officers were compelled to make before us, it is clearly established that the rehabilitation process is far from complete and lacking much the spirit that animates the Tribunal Award and the policies proposed by the government itself.

The Issues

Number 1. In response to Issue No. 1, whether there is land available for resettlement of PAPs, from what has been ascertained by us, it does not appear that land is available for the resettlement of the PAPs in conformity with the Award of the Tribunal, the policy of the Government and the verdict of the Supreme Court, specially in view of the fact that all those who have been shifted in the past or those presently affected at 90 m height of the dam have not been provided with suitable land.

Number 2(a). In response to Issue No. 2, requiring us to say if there exists irrigation facility that could be made available for the benefit of the PAPs, we say 'No'.

Number 2(b). Our recommendation in respect of the irrigation facilities would be that efforts should be made by the government within a limited time frame to provide separate irrigation facility to each agricultural land that is being allotted for cultivation.

Number 3(a). Our observation of the resettlement colonies we visited has been that all of them are full of shortcomings, specially in respect of quality and availability of suitable land to enable the resettled people to start life anew. The provision of civic amenities, including potable water, is not as it could be expected under the prescribed norms. Indeed, the government has made an effort, but that is not enough and will require more application of mind, resources and effort.

Number 3(b). Wherever the resettlement colonies are wanting in respect of facilities directed to be provided, we opine that the government rectifies the defects as a time-bound programme immediately, in the light of the recommendations that have been made herein.

Number 4. In response to this issue whether the process of rehabilitation has been in accordance with the NWDTA and the government policy for rehabilitation and the decisions of the NCA, we feel that it has not been wholly so. In fact, the attempted rehabilitation is substantially less and at times in breach of what has been required by the Tribunal Award, the government resettlement policy and the verdict of the Supreme Court. Those affected at 90 m were to be resettled by 31 December 2000; even that has not been accomplished. Moreover, NWDTA had insisted on rehabilitation preceding submergence. Obviously this has also not been complied with.

There are 33 villages which are in the submergence zone and had the sequence laid down by the Tribunal been adopted, i.e. rehabilitation first and eviction from areas marked for submergence later, many of the problems which the tribals and officialdom face today would not have arisen. What we find is that while the government is keen on shifting the tribals out of their villages, the same care and anxiety is not reflected in their programme for rehabilitation. Moreover, not everyone likely to be affected and so requiring rehabilitation has been accounted for.

Observations on resettlement and rehabilitation

The 33 villages to be submerged are from the two talukas of Nandurbar district, the talukas being Akrani and Akkalkua, 24 villages from Akrani and nine from Akkalkua. The tribals of these villages have been in occupation and settlement of that region from times immemorial. Except for one family of brahmins, who were functioning as priests at Shoolpaneshwar temple in Manibeli, the entire affected population is comprised of tribals.

The tribals, basically farmers, were not merely dependent on supplemental income from forest and river produce. A fair part of their income came from animal breeding and poultry keeping. They had unlimited access to pasturage available in plenty in all seasons and poultry had its full in the sense of multiple feed available in and close to the homes of tribals in the old villages.

Though living in seemingly secluded areas, it is not as if the tribals were completely cut off from the market. The collected forest produce and where necessary fish and melons obtained from or near the river, could be bartered or sold in the township of Taloda, Akkalkua, Molgi, Dhadgaon, etc., in case of need. The animals and poultry gave them the usual output of manure, eggs and reproduction of these species which enabled practically every tribal to have large flocks of cattle, goats and sheep and poultry, etc. These provided a valuable food supplement in lean times to the tribal and his family. In fair times a tribal could sell and buy his other requirements such as clothes, agricultural implements and necessities like salt and tea, etc.

The tribals have been in possession of these lands for many generations now. However, in Akrani the tribals are recognised as encroachers on these lands. These villages are still officially termed as forest villages and despite many resolutions and orders of the government to convert them into revenue villages; the task was initiated but aborted mid-way and now it looks like this process in these villages is being stalled. Akkalkua was formerly part of an estate whose owner has been described by the tribals as a Chieftain. The said Chieftain let his staff collect revenue or rent or whatever they could from the unfortunate tribals and they did so with impunity, there being no one to check their depredations. The tribals complain that the same state of affairs has continued even after the abolition of the rights of the Chieftain.

Attempts have been made for the rehabilitation of affected tribals on 4,200 hectares of forest lands specially converted for this purpose. From amongst those tribals who have opted to be resettled in their home State of Maharashtra, they have been brought over to the five resettlement colonies developed on this forest land, all in taluka Taloda of Nandurbar district. These resettlement colonies are Rozwa resettlement colony, Sardar Nagar, Rewa Nagar, Narmada Nagar and Deomogra Nagar. Indeed the Government of Maharashtra has made a good attempt in the matter of rehabilitation, but much still remains to be done.

Time has not stood still and the number of people yet to be rehabilitated has increased from year to year. The pressured officialdom has worked out schemes for rehabilitation in a haphazard manner. This has not been deliberate but on account of the pressure brought to bear upon them to come up with solutions without enabling a proper scrutiny of the numbers displaced and the assets required and available for resettling the oustees. Subsequently the tribals who have been shifted into the resettlement colonies set up by the government are facing numerous problems regarding land and civic amenities.

What this has led to is that patches of land meant for one PAF have been allotted to more than one such PAF with all the rivals virtually coming to blows. Another feature commonly found is that the patch of agricultural land said to be available for a PAF does not measure up to the promised 2 or 1 hectare. A part of it, sometimes a sizeable part, is what is known in the vernacular as 'potkharab'. This is on account of the soil being rocky, sandy, covered with streamlets and drains, covered with irremovable weeds and shrubs. Then we have come across cases of persons who on paper have been given lots of cultivable land, but find themselves resisted when they tried to take possession. Some of those offering claimed to be in possession as original dwellers since long and who questioned the right of the government to allot what they claimed to be their land to any of the Narmada oustees. The same story is repeated in the case of home plots. Some such plots have been allotted to more than one person leading to the usual trouble bordering on violence.

There are also instances of tribals who figure in the list of PAFs and have been shifted to the rehabilitation sites but have not been given their entitlements of agricultural land. It is said that land to be given to these tribals is not physically available. Non-availability of agricultural land to the PAFs is not something for which the tribals can be held responsible. Government has to make good this recompense if not in the rehabilitation site, somewhere in the close proximity thereof. This is in contradiction to the government's reasons for making ex parte allotments. It is said that such allotments are compelled because of the obstructive attitude of the tribals. If land is not available to accommodate those declared as PAFs, it is a contradiction in terms to say that the tribals refuse to move out from the villages earmarked for submergence for which reason the government is forced to make ex parte allotment of lands.

Apart from the defaults in the nature of omission, the authorities vested with the powers of rehabilitation have delayed acting with the promptitude expected of them. This has increased the problems of the tribals as they were either forced to stay in the villages to be submerged or be without a roof over their head and/or agricultural land which was to be their means of livelihood to begin with.

Some of the tribals still residing in the original villages have been deprived of the lands they were cultivating in the submergence areas without being placed in possession of the lands allotted to or allocable in the rehabilitation sites. Those who have received agricultural land, cultivable or otherwise, have not received legal titles for the same. And absence of legal titles concurrently with the allotment of land has often allowed errors and conflicts to creep in. This has reduced them to penury unknown to the unfortunates who had a supplemental source of livelihood in the form of forest produce, game, fish and fruits, etc. As of today, none of the resettled villages has a nearby forest or river which can enable them to get something of the past which appears to have been lost for ever.

As an example of the rehabilitation sites in Maharashtra we cite the situation we found in Rozwa. It is inhabited by about 300 to 400 families. Their homes are little better than hovels. They lack potable water, condemning the women folk to walk miles to fetch a pail of drinking water. The agricultural lands allotted to some of the residents were parched, it being one of the severest summers that the region had faced in recent times. To add to the woes of the residents the Maharashtra State Electricity Board, without making even the electricity available, had sent bills showing consumption of thousands of units of power and calling on the tribals to pay the charges. This mindlessness of officialdom is not an isolated incidence. No wonder the people at the rehabilitated village appear frustrated.

One of the contributory factors is the absence of a master plan which ought to have been formulated by the Maharashtra Government to ascertain the precise number of PAFs and the land requirements for their resettlement, its failure to ascertain from the PAFs their option as to the resettlement in Maharashtra and/or Gujarat, and the piecemeal measures taken by the government whenever a clamour arose either for expediting the SSP or for doing justice to the displaced tribals. A failure to prepare a master plan which was even found desirable as a prerequisite by NWDTA and the Supreme Court, is clearly a serious lapse giving rise to numerous problems including omission and underestimation of PAFs, vitiating the very process of resettlement and rehabilitation.

Recommendations

First and foremost, there should be a change in the definition of a PAF/oustee, to include all categories of people affected by the dam related works.

From the definition of an oustee offered in the NWDTA, it is clear that those ordinarily residing or cultivating land fall within it. Hence the application of an age-limit vis-à-vis cut-off date does not seem to have an appropriate basis in the NWDTA (i.e. only those who were 18 years as of 1 January 1987 are counted as PAFs.) What is also seen is that even those supposedly 'undeclared' by the government, have their own agricultural land and houses as well, which will be submerged due to the project. In fact most of them own 'Ghar Pattas' and have been paying the house tax regularly. Going by the above definition, all tribals owning property or otherwise who will be affected by submergence should be counted as PAFs and be offered the same entitlements in rehabilitation.

The cut-off date for major sons/unmarried daughters should be that on which the tribal inhabitant gets his resettlement entitlements in actuality as per the Award on his removal from his home and land both included in the villages to be submerged. The date of 1 January 1987 as assumed by the government to be the cut-off date has no bearing at all with the present realities of the life of the tribals.

Major sons and unmarried major daughters should be treated as individual units by themselves, entitling them to equal rehabilitation facilities identical to those meant for the recognised head of the family. This change should be effected in the resettlement colonies as well as in the original villages with immediate effect.

For ascertaining the acreage, which a displaced agriculturist amongst the tribals has in possession and hence is entitled to, unpublished official surveys of 1985–86 (which resulted in the Government Resolution of 1992), which do exist, should be treated as the base for completing the process of conferring land rights in the original villages. This should be completed presently before any further displacement takes place.

Further, on the land rights of the tribals, it ought to be said that it is the responsibility of the government to confer on them their due rights in respect of their land holdings in accordance with their own policy resolutions to that effect, and the Supreme Court ruling in the case of *Pradip Prabhu v. Government of*

Maharashtra, 1995, requiring the government so to do.* It does not appear that the Supreme Court's directions have been complied with. The government owes the tribals the rights that they have over these lands that they have possessed for generations. This would have to be done immediately and as a pre-requisite to any further displacement. The granting of rights would have to be carried out with the understanding that even those already shifted to the resettlement colonies would be accorded their rights in the original villages, and any suitable corrections in the rehabilitation entitlements would be made subsequently.

Much is made of the surveys of PAPs carried out by the government agencies for ascertaining the true number of tribal families to be affected. From the government side an attempt has been made to give finality to the last survey carried out. We are not satisfied with this approach. True, those not eligible for relief or rehabilitation should not be given that facility, but we find it very difficult to believe that tribals known for their honesty and integrity are falsely setting up claims so as to attain benefits to which they are not entitled. Such avarice and greed is fairly common amongst the plains' people, which may account for the government attitude. In the case of tribals, however, it is a misplaced suspicion, except possibly in a few cases.

The formation of 'tapu' (island) and isolated villages/hamlets becoming socially unviable units is common knowledge. The State policy is quite clear on this point, requiring affected persons from such villages/hamlets to be declared as 'affected' so as to become eligible for all rehabilitation entitlements. Since this is not being done, the government should carry out a survey with the help of the NGOs to ascertain the correct position in this respect in the region. The official list of affected tribal families should be subsequently prepared and verified in gram sabhas called in the affected villages, and the offers of similar resettlement entitlements should be made to them.

A fresh door-to-door census will prove the disparity between reality and the results of the government survey carried out in the past and hence this new survey ought to be carried out as a matter of urgency. For this purpose it is desirable that the government takes assistance from NGOs like NBA who enjoy the confidence of the people and from the elders of the villages to be submerged. This survey should be carried out prior to any further displacement or submergence or further increase in the height of the dam.

In the context of the resettlement of those tribals who are affected at the present height of the dam, we see many problems plaguing the process. On one hand there is no proper realistic survey of the number of tribals to be affected, while on the other there is no agricultural land available for their resettlement presently.

The process of ex parte allotment of lands is believed to be in keeping with the approval of the NCA. One seriously doubts if this is permissible having regard to

---

* This document is reproduced at page 267.

the Tribunal Award prescribing options to the oustees. It is of course true that the oustees cannot keep resisting and refusing eviction and rehabilitation for all times to come. But the reasons for rejection and resistance have to be considered and if found justifiable upheld, with the government being under an obligation to remedy the situation. Moreover, since the ex parte allotment is a 'deemed settlement' in the eyes of the NCA, this device ought not to be used by the government to inflate the number of settled PAFs in the context of a pre-requisite for the permission for further progress in dam construction where in fact they have no suitable lands to offer to all the recipients of such notices. The Committee had witnessed and intimated to the concerned authorities that such notices were being issued carelessly. There surely is the case where the offers are made but there is no availability of land in the resettlement colonies.

On the occasion of the first tour we had picked out at random three notices intimating the oustees of ex parte allotments of land/house plots to them. A physical verification showed that the notices were incorrect in material particulars and this forced the collector to concede that all the 145 notices that were then issued to those PAFs affected at 90 m could not be acted upon, and that those notices will be withdrawn until after proper scrutiny. The third tour was carried out after an interval of about a month, when instances of similar defects arising out of a wanton use of this method, whereby out of 145 notices issued earlier, the government had no land to offer to as many as 63 of them, came to our notice forcing the officers then present to concede the occurrence of mistakes. One hopes that this course will not be repeated in future. It is, however, learnt that without furnishing the compliance statement, as was expected in relation to the previous notices, fresh notices have further been issued to 71 PAFs from amongst the 145 PAFs stated earlier.

The PAFs having agricultural land in the submergence villages should be compensated acre for acre (within the ceiling limit) and at the least a 2 hectare plot of agricultural land and a house plot of 90 ft * 60 ft, as directed by the NWDTA. Building material should be given for the construction of a house to include tiles, bamboo, A/C sheets and such number of wooden posts as are considered necessary for setting apart the residential portion of the home from the non-residential. This has to be provided as per the State government policy immediately after shifting. The house plot should not be more than 1–2 km away from the agricultural plot allotted, except in exceptional cases with the written consent of the PAF prepared before the gaon sabha.

All agricultural plots allotted or to be allotted to the oustees to be made completely cultivable at the government expense. House plots, in a single new 'gaothan' of the oustees should be of such PAFs as are bound by ties of kinship and sub-tribal affinities as in the NWDTA. Every such 'gaothan' should have the community facilities mentioned in the Tribunal Award and the Government Resolutions on the subject.

The NWDTA has specified community resettlement by way of rehabilitation villages. The Government of Maharashtra, until now, has quite successfully pursued this principle.

Now that there is a need to purchase/acquire more land, it should be done in like manner as this principle is obligatory.

Each PAF allotted agricultural land should be immediately provided with the required quantity of water for irrigation purposes until some permanent arrangement in the shape of a water-bearing well is made separately available to the allottee of the agricultural land.

The right of the tribals in the new agricultural land and house plots will be in the nature of a permanent settlement. This is of particular importance since they should not be without legal titles to their lands in the new resettlement colonies as they were in their original villages. In the rehabilitation villages, grants of agricultural lands and house plots should not be merely physical. Those rehabilitated have to be supplied the relevant title deeds and revenue papers lest the dishonest revenue staff tries to tinker with the acquisitions the tribals have made after sacrificing so much. This should be done immediately lest they are again misdescribed as encroachers, trespassers, land grabbers or squatters and the lands then be re-acquired from them by the State itself. They will, however, not be allowed the right to sell or encumber this property though the right of inheritance will be kept intact. The tribals must be prohibited from selling, encumbering, leasing, etc. their agricultural lands except to recognised public institutions like the Revenue Department, nationalized banks, co-operative banks or credit societies and that for raising loans purely for agricultural purposes.

Excepting the resettlement village of Simamli in Gujarat, which offers a little satisfaction, the rest of the resettlement villages from Maharashtra in particular, visited by us, lack almost all the basic facilities required for habitation, specially quality and availability of suitable agricultural land. One cannot ignore the enormous number of complaints that the Committee came to hear from the aggrieved people about having been shifted to the new sites without being provided with compensatory agricultural land. One of the greatest shortcomings is that of non-availability of water even for domestic purposes like cooking and drinking. Even in Simamli, it is not as if everything is as it should be. People have complaints but comparatively speaking it appears to be the best of the resettlement habitats that we could visit.

In regard to the facilities for providing water at the resettlement colonies to irrigate crops and for potable purposes, the document dated 14 May 2001 and received by the Committee at its meeting at Nandurbar held on 20 May 2001 from one of the officers of the government, shows that near about 70–80 per cent work remains to be done, and this covers all the five resettlement villages. Needless to say this is an admission by the administration of the lack of a vital input in the rehabilitation package.

The present resettlement colonies are facing multiple problems in relation to those who have already been shifted there. Their basic problems, with regard to agricultural land and civic amenities, still remain unmet. While this situation prevails, these sites should not be burdened with further exacting demands in relation to new oustees to be shifted there. Unless those already shifted are fully cared for by resolving all their problems within a time bound framework, especially land related, there appears to be no room for further resettlement at these sites.

Moreover, it is considered that there are a certain number of people from the submergence villages who have been transported by the government to the rehabilitation sites. The government does not look upon them as PAFs and the shifting at government expense is claimed to be an act of compassion inspired by humanitarian ideals. We have seen very little of humanitarianism in the dealings of the government with the oustees. In fact, that certain people were shifted from the villages to be submerged is prima facie proof that they and their forefathers were natives of the same villages and were entitled to the prescribed acreage of land whether as recorded landholders or alleged encroachers. For this reason we recommend that the distinction between 'declared' and 'non-declared' oustees should be done away with. All those who are non-declared should be made entitled to the resettlement package of the government. This change should also be effected in the original villages.

It is surprising that towns in Gujarat as far away as Jamnagar are said to be getting Narmada waters, while those in the so-called resettlement village of Rozwa have to do without this precious liquid for days together. At times the rest of the resettled villages have to fetch potable water from distances ranging from 2–5 kilometres. Surely, this is a strange way of inflicting misery upon injury.

Every family of a displaced tribal has a right to a steady source of income from non-traditional sources. The project is going to need a number of employees to attend to various tasks. The tribals are not entirely illiterate or unskilled. The state of the villages to be submerged indicates a high level of cooperation between the residents for the mutual benefit of all. The tribals take little from the river and the forest and are aware of the need to renew the assets they partake of. That is why the 33 villages still have an unsullied forest covering and unpolluted source of water. Their care and concern shows that the tribals have an aptitude of which use can be made by the authorities to give unto the supposed beneficiaries what has been promised to them. Each tribal family should be having a member who can easily be absorbed in government service related to the project.

The residents of these 33 villages which are to be submerged are indigenous people who have been pushed into areas inaccessible or considered difficult by the invading plainsmen. Pushed to the extreme, the tribals have invested their labour and skills to eke out a fairly comfortable living. As they have been deprived of even this, it is but fair that civil society through its representatives, the legislature and executive, makes good the loss. How difficult it will be for the tribals to adjust to the new lives which they will be compelled to live cannot be really understood by those living on the labour of others. The rehabilitation villages do not have a perennial source of water or a forest nearby. As of today the tribals have very little acquaintance with the wisdom or craftiness necessary to survive in a market-dominated economy. At least two or more generations of the tribals will require sustenance in the shape of cash assistance to make up for the non-agricultural income which was at their disposal from the nearby river and forest. That is why we are going to suggest a fairly long period for the cash equivalent for the loss caused to the tribals by being deprived of fish, melons and forest produce, etc. This sustenance was available to every tribal without exception.

For the loss of the non-agricultural income the tribals should receive monetary compensation running over a period of at least 50 years. We are aware that what we prescribe is far less than what the tribals are entitled to. Their loss is almost irreparable. There should be an attempt to also focus on self-employment with natural resources by replacement of the lost resources. Forest lost cannot be replaced but food, fuel, fodder should be ensured at the resettlement colonies. Therefore, grazing land, fuel sources like fuel depots and adequate cultivable agricultural land wherein there be food for consumption and employment opportunities should be ensured. Land for land, minimum of 2 hectares, for each displaced family, declared or undeclared, would go a long way in achieving this.

But some amends must be made and we suggest that each project-affected nuclear family which we consider as including a man, wife and three children, all such children being below the age of 18 years, should get monetary compensation at the rate of Rs 2,000 per month. Every additional mouth should be entitled to an extra sum of Rs 400 per month. Children who have attained majority, whether with their parents or separated, shall be treated as a separate nucleus entitled to compensation on the above lines though, of course, if a daughter is married and is going to get compensation via her husband, it will not render her eligible to receive compensation separately. Marriages of the daughters to eligible PAPs (project-affected persons) will transfer their right to receive compensation from the father of the daughter to her husband into whose family she has moved.

For those who have been deprived of their agricultural land and houses because of submergence but have not yet been placed in possession of culturable land and house plots, we recommend compensation at the rate of Rs 3,000 per acre per annum right from the date of ouster until being placed in undisturbed and legal possession of the new agricultural land and house plots.

The monetary compensation suggested above, shall be index-linked and shall be payable with variations based upon the annual CPI (Consumer Price Index) – the base year for this purpose being 1987.

The tribals should be paid immediately their outstanding dues in respect of the loss of their standing crops and trees which have been submerged almost every year since 1993/1994, and for which the government has effected 'panchanamas' on every single occasion.

For the loss of the non-agricultural income, we deem it necessary to suggest that the tribals deserve compensation on this score. It may be said that they were encroachers/trespassers in the matter of collecting and selling forest produce, catching fish for either consumption or sale and raising different types of melons on the banks of the river, again for personal consumption or sale. It may also be said that the livestock and poultry that the tribals had is not something for which they deserve compensation as in any case, pasturage was from government lands. A closer look at the legal position would disabuse the common notion spoken of above. The tribals had acquired the right to collect forest produce, wood for construction and fuel, fish from the river, etc. That right is termed in law as a 'profit à prendre' called also a 'right of common'. In Black's Law Dictionary, 6th edition, profit à prendre is defined as:

> A right exercised by one person in the soil of another, accompanied with participation in the profits of the soil thereof. A right to take a part of the soil or produce of the land (...) the taking (profit) is the distinguishing characteristic from an easement. (…) It carries with it the right of entry and the right to remove and take from the land the designated products or profits and also include the right to use such of the surface as is necessary and convenient for the exercise of the profit.

These rights, the tribals, according to the depositions made before the Committee, have been exercising since times immemorial. Therefore, property has been taken away from them and for this they deserve compensation. Even in relation to agricultural lands it is not enough to say that the tribals were encroachers/trespassers vis-à-vis the Akrani lands which were forests governed by the Conservation of Forests Act of 1980. As a matter of fact, at one stage a survey was begun as early as in 1985, with the idea of converting the so-called forest land into 'bhoomidar' rights. Suddenly the process was later reversed by a non-speaking fiat of 1994. The survey begun at this stage was under the Maharashtra Land Revenue Code and when it was suddenly reversed, the reversal apart from being illegal, was by a non-speaking notification rescinding the work already done (Government Notification No. RB/DESK/II/LND/II/475/1994 dated 7 June 1994 issued by the Office of the Commissioner, Nashik Division). We do not see how the government is entitled to flout the laws in this fashion.

It appears that the matter has gone much further in the sense that the acreage, the boundaries and draft village papers had been virtually completed. Apparently, then it struck the government that there would be a contravention of the Conservation of Forests Act, 1980, which is a Central legislation. Assuming that this is so, it is not as if the Government of Maharashtra is helpless. We wish to make it clear that the SSP is an inter-State undertaking with the State of Gujarat being the biggest beneficiary. The Centre's assent to the project has come fairly late and therefore, the Union Government also cannot disclaim its responsibility by saying that it is a matter concerning States over which it has no control. Article 254(2) of the Constitution of India, lays it down that where a law made by the legislature of a State with respect to one of the matters enumerated in the concurrent list contains any provisions repugnant to the provisions of an earlier law made by Parliament or an existing law with respect to that matter, then, the law so made by the legislature of such State shall, if it has been reserved for the consideration of the President and has received his assent, prevail in that State. There is, of course, a proviso to this sub-article which enables the Parliament to enact at any time, any law, including a law adding to, amending, varying or repealing the law so made by the legislature of a State. Having regard to the equities, the least that can be done is to see that the displaced tribals get the necessary benefits to which they are entitled. To claim that they were near trespassers in forest land is a gross error. For generations these tribals have been cultivating the lands and prescriptive titles can be acquired even against the State. If the tribals and their forefathers have been in possession since the last at least 100 years if not more, it follows that they have become the owners of the said lands. The Forest Conservation Act of 1980 is not a parchment in granite which cannot be wiped out. It can be easily amended by State legislation, which legislation will have to undergo the process prescribed by the Sub-article (2) of Article 254.

The recommendations that we wish to make hereby are not only in relation to vital matters but also in relation to other matters, which may be considered as less important, but are nonetheless equally crucial for the everyday living of the tribals. For example, the tribals do require boats for crossing the river to meet their daily requirements. As the tribals in Manibeli have pointed out, as long as there are villages with tribals residing in the valley, government should provide free transport and boats, both ordinary and machine boats, depending on the distance for transport. Without this the tribals from almost all the villages in Maharashtra are facing a great difficulty in reaching out to places of market, health and other services, many of which are in Gujarat.

The schools presently in the submergence villages supposedly run by the government exist merely on paper. Complaints have been made that the teachers appointed to teach in these schools never visit the institutions, though are regular in the drawing of their salaries. For this reason, at some places, it is the Narmada Bachao Andolan (NBA) which is running schools (called 'jeevanshaalas') up to fourth standard, manned by local, not highly paid but dedicated teachers. These teachers are from the tribals themselves and the results are commendable. A good percentage of children pass fourth standard exams in good colours. Those who have passed out and gone to get education in the higher schools in places like Dhule, Malegaon, Dhadgaon, etc. have again shown high performance. The government should now support these schools with the existing staff also retained and all such facilities provided as are meant for educational institutions existing elsewhere in the State of Maharashtra.

Most of the tribals who appeared before us had scores of grievances against the officers with whom they had come in contact. The very fact that the government has not been successful in persuading all the residents of the villages to be submerged to move out to the resettlement regions, is proof that the oustees have not got their promised compensations whether in kind or cash. The government has not been able to convince the tribals of its capacity to carry out the task of rehabilitation as expected.

The great divide between officialdom and the NGOs like the NBA and PSS is most unfortunate. Whatever be the reasons, the misgivings have to be removed and this can best be done if the government in conjunction with the above-mentioned NGOs carries out the process of rehabilitation. Most important is that the two sides stop attributing improper motives to each other and cooperate to accomplish the difficult but necessary task of seeing that the displaced tribals get their dues as early as possible. In this context, where submergence and displacement have been made inevitable, it is now extremely necessary that at least those affected tribals must be fully rehabilitated through co-operation of all concerned.

The recommendations, which we make, should not be construed as a favour to the displaced tribals or a planned extravagance foisted upon a State facing financial stringency. The Sardar Sarovar Project is basically expected to yield benefits. It is but fair that those who are losing their lands and homes get a small share from the expected abundant yield. The Constitution of India in its very preamble speaks of several values, but places first that of justice in the social, economic and political context. If justice is required to be done to the water-starved regions of Gujarat, justice also requires that those who are making this possible get a small fraction of the goodies as their compensation, for what they are losing. The Directive Principles of State Policy may not be enforceable in any court, but are, to use the language of the Constitution, 'nevertheless fundamental in the governance of the country and it shall be the duty of the State to apply these principles in making laws' (see Article 37).

Again Article 46 of the Constitution enjoins the State to promote, with special care, the educational and economic interests. In particular of the Scheduled Tribes and to protect them from social injustice and all forms of exploitation.

Article 300A prohibits depriving a person of his property save by authority of law and law is not merely that is found in the different codes and GRs (Government Resolutions). One has to go by the spirit and this views the expression 'property' to include any proprietary interest including a precarious interest. Here we have seen that the tribals had the river waters, the forests and the rich-soiled Narmada valley agricultural lands for their exclusive enjoyment. Now that they are going to lose all this and more, it is incumbent upon the Indian people, through their representative governments, to compensate the losers to the extent permissible. As it is, all manner of people deserving or otherwise, get government handouts. Compensating tribals for the great loss they have or will sustain should not be viewed other than their just entitlement.

India is a signatory to an international convention to safeguard the interests of the tribals and recognise their age-old rights. The International Labour Organisation (ILO) Convention 107 – Convention Concerning the Protection and Integration of Indigenous and other Tribal and Semi-tribal Populations in Independent Countries – was adopted in 1957. India was one of the first countries to ratify ILO 107, doing so on 29 September 1958. The Convention, with a view to safeguard the interests of the tribals, provides in Article 11:

> The right of ownership, collective or individual, of the members of the populations concerned over the lands which these populations traditionally occupy shall be recognised.

Article 12:

> 1. The populations concerned shall not be removed without their free consent from their habitual territories except in accordance with national laws and regulations for reasons relating to national security, or in the interest of national economic development or of the health of the said populations.
>
> 2. When in such cases removal of these populations is necessary as an exceptional measure, they shall be provided with lands of quality at least equal to that of the lands previously occupied by them, suitable to provide for their present needs and future development. In cases where chances of alternative employment exist and where the populations concerned prefer to have compensation in money or in kind, they shall be so compensated under appropriate guarantees.
>
> 3. Persons thus removed shall be *fully compensated for any resulting loss or injury* (emphasis ours).

ILO 107, confirms the rights of tribals over traditional lands in their possession and the mandatory guidelines for the resettlement of those who are to be displaced. However, on both counts, it appears that the government has failed in regard to the tribals affected by the Sardar Sarovar Project. The land rights of the tribals are not only being denied but the tribals are being short-changed in resettlement too.

The tribals have lost or will be losing what was their hearth and home since umpteen generations. The loss they have sustained, quite apart from the disruption of family ties, and even if restricted to things material, is enormous. That loss has to be made good. In spite of government efforts, the rehabilitation process has fallen much behind again, as never has there been an attempt to make a detailed, perfect and realistic plan. A part of the loss which the tribals have sustained, apart from the loss of family ties to community resources, which is never counted in the official estimates, is also due to the non-compliance with the official policies. Now that the apex Court has permitted further construction of the dam, it should be remembered that the very apex Court has required the rehabilitation to be accomplished strictly in accordance with the norms governing resettlement as laid down by the NWDTA and State rehabilitation policy. Even the same has not been achieved with regard to those affected at 90 metres height of the dam. For this purpose we suggest that a Seven-Member Committee functions to oversee the process of rehabilitation. The Committee shall include a representative each from the NBA and PSS, and representatives of the tribal community each from the original villages and resettlement sites. The rest of the members shall be nominated by the government from their Revenue, Forest, Tribal Affairs and Rehabilitation departments. This Committee shall meet once a month for the next two years to come and thereafter once every three months until the physical rehabilitation is completed in its entirety. It is necessary that the tribals have an additional protection in the sense of their true representatives overseeing the work of rehabilitation. For this purpose we suggest the names of M/s Manikrao Gavit, a Member of the Parliament and also a Member of this Committee, and Shri K.C. Padvi, a Member of the State Legislature from Nandurbar. These two gentlemen shall participate in the Committee meetings irrespective of whether they continue to be returned or otherwise to the legislatures of which they are presently members. We have to make it clear that Shri Gavdit is not privy to the last recommendation.

To repeat, what we have recommended is not really an extravagance; it is but a small recompense for much that has been lost and that which will be lost in the years to come. Where lacking, the Government of Maharashtra should bring its R&R package up to the mark and it should consider improving its policy to make provisions advantageous to its own tribal citizens and superior to those pronounced elsewhere.

Government on its part should strictly comply with the conditions laid down by the Supreme Court for raising the height of the dam beyond 90 m. The present situation is such that there are still hundreds of families who are affected at 90 m and are yet to be offered any proper resettlement options. In this regard, the contradictions between the ground reality and the figures furnished by the government are eloquent and it will be necessary for the government to correct its figures. Since the earlier data has been placed before the Narmada Control Authority on the basis of which certain vital decisions were taken, the correct version should also be notified to the NCA. The government should not use incorrect information merely because a stamp of finality has been conferred upon it at the Mantralaya level. In any case, we would like to bring to the government's notice its responsibility to see that it does not accept any proposal to raise the height of the dam beyond its existing height of 90 m unless the obligations laid down by the NWDTA and Supreme Court are fulfilled as a whole in regard to those PAFs affected at 90 m and yet to be resettled. The government should also see to it that the process of shifting people conforms precisely to the permitted increase in the height of the dam.

The contradictions between official information coming from different sources indicates that all is not well with the collection of information by the government agencies. It is necessary that the government, for its own honour, garners accurate and incontrovertible evidence on basic facts like the number of project affected persons and families, exact acreage of agricultural lands required for the oustees, the number of house plots and building material to be made available, and the incidental details. If correct data be assembled, it will be a great aid in the rehabilitation of the displaced tribals. In collecting this information, the government should not pressurise the local officials by giving them impossible deadlines. The local officials are short of hands and the tasks they have to carry out are onerous. Increasing the pressure on them to furnish this or that detail by a deadline will only aggravate that problem. What we suggest is that hereafter any process of development proposed to be carried out by the government will be done in effective collaboration with the gram sabhas of the villages. It should be remembered that the tribals lack the skills to communicate with the officialdom, which also has a tendency to ignore their grievances. For this reason, assistance provided by the NGOs like the NBA should be solicited and welcomed. The gram sabhas of each tribal village, specially convened for the purpose, shall collect the relevant factual material within a reasonable period and verify its correctness or otherwise before the same is transmitted to the government. This information should prevail over the statistics prevailing at the Mantralaya level. The government should furnish every tribal with written information in the form of an official leaflet detailing entitlements to project-affected families.

The master plan, as described herein, should be presently prepared in the form of a draft and widely publicised amongst the people and their sympathisers and finalised after due deliberations.

The assistance to be rendered by or taken from the NGOs should be on a par with that provided in the policy documents for rehabilitating the population affected by the earthquake which took place on 30 September 1993 in the district of Latur in Maharashtra; where it was made obligatory on the part of the government agencies to seek meaningful assistance from and participation of the NGOs. Government should seek cooperation of the NGOs because they command the confidence of the people, unlike the government officials for whom, rightly or wrongly, there is distrust.

The recommendations in relation to compensation, whether in cash or kind, shall be in addition to and not treated as a substitute for compensation suggested by the Tribunal, the government and the NCA recommendations.

The recommendations that we may have made, if accepted, should be incorporated into either a legislation or a government resolution. This will prevent errors of omission and commission and give the administration, as also the affected people, an instrument to adhere to.

We hope the government will accept our recommendations in toto, and that in particular it shall publish this Report as early as possible.

We are grateful for the assistance rendered to us in the discharge of our duties by the officers of Maharashtra and Gujarat Governments, in particular those from the areas and sites we have visited. Further, we would like to thank the representatives of the NBA and PSS for the assistance rendered by them in enabling us to get a proper perspective of what has been done and remains to be done.

Jan Sunwai, 2002

Source: Along the Narmada – Jan Sunwai, 13–15 July 2002, Panel: Dr V. Mohini Giri, Admiral Vishnu Bhagwat, R. Venkataramani and Usha Ramanathan, Report written for the panel by Usha Ramanathan (extracts).*

Chapter (1) Introduction

The raising of the Sardar Sarovar Dam across the river Narmada, at Vadgaon in Gujarat, has been striven with conflict. Contending interests, and conflicting interpretations of what constitutes development, have provided hurdles that have, at times, appeared insuperable. In 1979, when the NWDT gave its Award, it had been predominantly occupied with matters of sharing of costs and benefits between the States that are directly affected by the proposed dam – Gujarat, Madhya Pradesh and Maharashtra – and a fourth which was intended to benefit from the project even if marginally – Rajasthan. In 1979 the Award represented the negotiation and settling of differences between states. The Tribunal was an interstate tribunal set up under the Inter-State Water Disputes Tribunals Act, 1956 which heard the States parties before it; and the Award was to bind all four States. The Award, however, also provided a formula for resettlement and rehabilitation in the limited context of deciding the costs that would have to be borne, and held the state of Gujarat responsible for meeting those costs.[1] In the years that followed, three significant changes occurred: two in the larger democratic domain, and the third in the Narmada Valley. These were:

- The increasing according of value to environmental concerns, along with a recognition of the potential for irreversibility where environmental matters are either unaddressed or unredressed;

---

* The full report is available online at http://www.ielrc.org/content/w0201.pdf.

1 Annexure 1, pages 20–21 of the Morse Committee Report.

- The emergence, and increasing visibility and articulateness, of civil society groups, non-governmental organisations and movements. This was also the time that the Supreme Court found an ally in judicial activism to give access to the courts, and to justice, through the device of public interest litigation.
- The third facet of change, which swirled in eddies in the Narmada valley, was the movements of resistance, and struggle, demanding a centrality for the displaced people in the planning for development in the valley, and in the execution of the plans generated.

The raising of the walls of the dam, and the resistance to its construction, has been through several phases. It was in 1987 that a conditional clearance was accorded to the Sardar Sarovar Project (SSP) by the Ministry of Environment and Forests in the Government of India. In 1992 an independent review of the SSP instituted by the World Bank, with Bradford Morse as its Chairperson, turned in a devastating report, categoric in its conclusion that: 'Environmental and social trade-offs have been made, and continue to be made, without a full understanding of the consequences. As a result, benefits tend to be overstated, while social and environmental costs are frequently understated. Assertions have been substituted for analysis'. This indictment, which followed a detailing of resettlement and rehabilitation in the three States, hydrology and water management, and environmental issues, was accompanied by a word of advice that 'the wisest course would be for the [World] Bank to step back from the Projects and consider them afresh'. The World Bank, by press release dated 22 June 1992, rejected the report to the extent that it '(did not) share the view that resettlement would be virtually impossible even if Maharashtra and Madhya Pradesh adopted the liberal resettlement package provided for displaced people by the State of Gujarat. Given the experiences so far, and the fact that most of the impact of submergence on people will not occur until 1997, there is still time to develop meaningful R&R packages and programmes in consultation with the affected peoples'. Despite this note of optimism, in 1993 the Indian government withdrew from the loan agreement.

In 1994 the issue was taken to court by the Narmada Bachao Andolan (NBA).[2] The court stayed work on the project in 1995. In 1999 it permitted construction of the main part of the dam from 80 m to 85 m. The government built 3-metre 'safety humps' beyond 85 metres. On 18 October 2000 the judges of the Supreme Court delivered their opinions. The judgment of the majority cleared the way for continued construction of the dam up to 90 metres height (…),* while asking for an action plan that would 'fix a time-frame so as to ensure relief and rehabilitation *pari passu* with the increase in the height of the dam'. There was, however, an inconsistency in the terms in which the Court spoke of the performance of the State governments, including in the matter of rehabilitation of the oustees (…).* As did other constituents of civil society, so too did we watch with a keenness and anxiety for willingness and capacity in the three States to rehabilitate those affected by the construction of the dam.

Three developments between the date of the judgment (18 October 2000) and early July 2002 raised our concern. The first was the *Report of the Daud Committee,* submitted to the Government of Maharashtra in June 2001. The Daud Committee was set up by the Government of Maharashtra as a result of agitation among the oustees in Maharashtra, to 'assist the resettlement and rehabilitation of the Sardar Project affected persons'. The Committee held open hearings, met the affected people, visited submergence villages as also rehabilitation sites – including two sites in Gujarat where Maharashtra oustees were to be rehabilitated – and consulted such documents to which the State government allowed them access. The findings of the Daud Committee are briefly set out in chapter 3; yet it may be here said that they are a severe indictment of the process, and efficacy, of rehabilitation in the State. Since then, the Government of Maharashtra has set up:

(1) a task force to survey the submergence villages in the State;

---

2 This was not the first case in the Supreme Court on the issues of dam construction and submergence. Three earlier interventions may be mentioned: In *Chhatra Yuva Sangarsh Vahini* (1985), this issue of loss of top soil was agitated in a writ petition by the oustees at the dam site in Vadgam. In *Pradeep Prabhu v. Government of Maharashtra* (1995), the Supreme Court required the government to confer land rights on the tribal oustees in respect of their land holding in accordance with the government's policy resolution. In *B.D. Sharma v. Union of India* (1991), the Supreme Court had directed: 'Rehabilitation should be so done that at least 6 months before the area is likely to be submerged, rehabilitation should be complete and should be in respect of homestead substitution of agricultural property and such other arrangements which are contemplated under the rehabilitation scheme'. The full order is reproduced at page 265.

* *See* paragraph 280 (Directions 1 and 2), *Narmada Bachao Andolan v. Union of India*, Writ Petition (Civil) No. 319 of 1994, Supreme Court of India, Judgment of 18 October 2000, reproduced at page 138.

* *Id.*

(2) a planning committee, to draw up a master plan, which is then to be implemented; and

(3) an overview committee, comprising government representatives, representatives from the movements in the valley, and experts to oversee this process.

The *changing numbers* constitute the second area of concern. Maharashtra had maintained that the PAFs that remained to be rehabilitated at 95 m was 17; after its task force had conducted its surveys, however, this was revised to around 1,000 PAFs. This revised figure is only illustrative of the statistical confusion that appears to dog the process (…). Set against the backdrop of the drastically altering figures of the displaced in the 245 villages expected to be affected by the submergence, [Table below] these escalating numbers reveal a whole population living on the edge, requiring governmental will and ability to restart their lives.

Estimates of the number of displaced	
1988–89	6,700 PAFs
1994	41,000 PAFs
March 2001	43,000 PAFs

The third concern surfaced with *the decision of the Narmada Control Authority to raise the height of the dam to 95 metres + 3-metre humps*. This decision was taken on 17 May 2002. Around 8,000 families in the three States are expected to face submergence, taking the tally to 19 villages in Gujarat, 33 villages in Maharashtra (almost all of these 57 villages are inhabited by adivasis) and 70 villages in Madhya Pradesh (the affected in 55 of these villages being predominantly adivasis). The Supreme Court had commented on the disinterest that the Madhya Pradesh Government had evinced in the matter of rehabilitation. The Daud Committee had spoken of the inadequacy of rehabilitation efforts in relation to the Maharashtra oustees. When, in May 2002, the NBA called upon the Supreme Court to consider the consequences of raising the dam height to 95 metres, the State of Maharashtra supported the stand of the NBA in its affidavit which asked that the present phase of construction be stayed. The resistance to submergence, and to being uprooted, has been reported from the valley, along with a growing resolve not to move from their habitat even if it means facing the flood that submergence will bring.

The people of the Narmada Valley have not had too many podia from which to speak and be heard. The NWDT decided a dispute that was between States, and the States and the Union Government were heard, although the decision was held to bind the affected people. The Narmada Control Authority has been constituted to act on the advice of its three constituent sub-groups: the Environment Sub-group, the Rehabilitation Sub-group and the Hydro Meteorological Sub-group; the NCA has no procedure to hear the people of the valley. It was only in 1999 that, acting on directions from the Supreme Court, Grievance Redressal Authorities (GRA) were set up in the three States, to be headed by retired High Court or Supreme Court judges. Yet, as the Madhya Pradesh GRA submitted to the Supreme Court in 2002, it has not been possible to conduct field verification due to infrastructural inadequacies. The affected people have, therefore, had to resort to extraordinary measures to lend decibels to their voices which will make them heard.

With the dam height having been raised to 95 metres (+3m), the imminence of submergence is upon the people of the valley. What stands between them and the deluge is the failure of the rains – which leaves them parched – and the non-closure of the sluice gates – which depends on engineering and political decisions.

It was in this configuration of circumstances that each of us saw an opportunity in NBA's invitation to us to visit the valley, and hear people facing submergence speak their version of the events that was so affecting their lives. Since the invitation had been extended to the resettlement authorities at the NVDA and Maharashtra and to each of the four of us, we hoped that it would be an exercise in exchanging points of view. This last was not to be, since the officials did not respond to the invitation. Listening to the women, men and children speak of their idea of development, law and justice led us to realise the importance of locating these ideas in mainstream discourse. They spoke to us with trust that we would help their words waft beyond the valley and be heard in wider spaces. This we attempt to do in recording (…) their questions, their challenges, their anguish, and their will to save their lives as they know them. They are in translation, and paraphrased where verbatim noting down was difficult, but they are close to the letter and spirit of their statements.

Each of us has been following the events in the valley over the years, and have attempted, in our own diverse ways, an unravelling of the issues around the construction of the dam, especially in the matter of submergence, and rehabilitation of oustees. Having perused the documents available to us, after hearing the people in the valley and the activists in the movements, we have essayed to set out some of the daunting issues that need to be addressed urgently even as further construction of the dam is under contemplation.

This is an effort to present the peoples' version, as emerging from narratives related in their own homelands. If we have not reached out to officials and agencies of State, it is not out of any disregard for their position. Instead, it is because many documents, including court proceedings, reveal the stand of the States and of the authorities and institutions that are involved in decision-making on the Sardar Sarovar Project. In contrast, we have often wondered how the individual, and community, position is articulated along the banks of the Narmada.

We have wondered too how the project has affected the lives of the women in the valley. It was in respectful acknowledgement of the importance of letting peoples speak for themselves, and with due regard to the stated position of the States involved, that this exercise got under way.

We spent three days in the valley, where public hearings were held. On 13 July 2002 a public hearing at Chikalda acquainted us with the anger and anxiety of the Nimad farmers, the residents of Chikalda, the traders and farmers of Nisarpur, a farmer from Sisagaon, a resident of Saurashtra, and a woman from Piprigaon. In the evening of the thirteenth, as we passed through Nisarpur, a town with a population of approximately 15,000, a public hearing happened, impromptu, standing testimony to the deep and urgent desire to be heard of a people threatened with imminent submergence. On the way to Nisarpur we visited the rehabilitation sites, set in undulating open spaces, with plot numbers painted on small stone posts planted in the earth. These were to be for the relocation of oustees from Kadmal and Nisarpur. Later in the night, a visit to Majra, an adivasi village, showed us the situs of a dispute, where a rehabilitation site, it was contended, was located on land that had belonged to an adivasi clan for generations; but since it had not been entered in the records it was being treated as government land and was now being used for resettlement of the adivasis.

On 14 July 2002 we reached Jalsindhi (via Hapeshwar). At Jalsindhi we heard residents of Jalsindhi, Kodamba, Chikalda, Bada Amba, Akadia, Jhandana, Anjanwara, Kakrana, Karya Bhadal and Sakarja villages.

On 15 July 2002, after a prayer meeting at Domkhedi, a public hearing was held across the waters at Nimgavan, Maharashtra with renditions by the residents of Nimgavan, Maal, Danel village in Akkalkuwa tehsil, Amlibari rehabilitation site in Maharashtra, Gamon village in Akkalkuwa tehsil and an original resident of Gadher, now settled in an R&R site in Gujarat.

The focus of our visit was on submergence and rehabilitation. We are aware that there are other issues of grave import, including the environment, the technical feasibility of the dam, and larger questions of the paradigms of development that remain unanswered in this domain. We have not ventured into these issues, and have therefore not explored or explained our understanding of the wisdom or wrongness of construction of the dam on these counts. The people spoke to us of their lives, their fears, their aspirations and their anger, and these constitute the context in which this report is set.

Chapter 4. A Web of Issues

The evidence that points to a complex of problems in the relationship between displaced peoples and the Sardar Sarovar Dam is increasingly incontrovertible. The narratives of peoples' experiences, official documents, court records and the voices of activists working in the difficult terrain that is the valley, all speak of the pain, exclusion and certain uncertainty that have become inextricably intertwined with the lives of many communities of peoples.

In respectful acknowledgment of the right of a people to life, going beyond (but not counting out) matters of shelter and livelihood to the way of life of communities, we venture to set out an array of issues that we encountered. Believing that

- people are not dispensable, and
- that they cannot be reduced to a mere cost in the development process,
- it is with an urgency that we ask that these issues are met, answered and resolved.

Master plan

The absence of a master plan, of submergence and of rehabilitation, met us at every turn. The NWDT had directed that a master plan of resettlement had to be in place within two years of the Award, i.e. by 1981. In 1987, while giving conditional clearance to the SSP, the Ministry of Environment and Forests was seeking details about '(i) rehabilitation master plan'. In 1988 it was estimated that 12,180 families would be affected in the three States, and the 'action plan' proposed then reckoned only with these numbers. By 1990 it had increased to 40,227 PAFs, and the numbers continue to rise. And this only accounts for those directly affected by the Sardar Sarovar Dam – not including the canal affected, pipeline affected. In 2001 the Government of Maharashtra was having to revise its estimates of affected people on the basis of the findings of the Task Force, set up after sustained pressure from the people. In 2002 the GoMP was revisiting its figures of PAFs, and even as it indicated that the number of oustees had escalated, it was suggesting a precipitous drop in the numbers who would need land to replace the land they were losing.

And, at different points in time, the Governments of Madhya Pradesh and Maharashtra were proclaiming that there was, indeed, no land to give to the oustees.

The significance of a master plan that says

- who will face submergence;
- when;
- what and where their rehabilitation will give to them (and what they will lose); and
- how their houses will be dismantled and reconstructed;

must be acknowledged and acted upon. *It may help to recognise that the development of a master plan is already late by at least 21 years.*

Survey

A house-to-house survey of the affected people is a prerequisite to assessing the extent of displacement, and computing the losses that the people may be expected to suffer. That the extent of submergence in MP was based on drawing contour lines on village maps is a stunning disclosure. As for explaining away the failure to conduct surveys on the resistance that the people put up, it must be said that this resistance was precipitated, and nurtured, by governmental policy and action which discounted people. Like person after person told us, 'why would we be fighting for 17 years if we were getting a fair and just deal?'.

A State cannot wish away its responsibility to take its people into account. Means of mediation, negotiation and revision of its own policies will have to replace repression or neglect in the arsenal of the State. Maybe it bears reiteration that discounting people, and what constitutes their lives, is not an option that is with the State.

Encroachments

The notion of encroachment ties in with ownership and private property. In predominantly tribal areas, where it is access to resources, and their use, which is a primary determinant of the relationship between a people and a resource, 'encroachment' is an alien phenomenon. While communities in tribal hamlets along the Narmada have been left to their own devices of property use and management all these years, land records have remained unattended, even as sweeping changes have been brought into the polity, such being the Forest Conservation Act, 1980. The neglect of tribal communities through the years cannot now be used as a reason for denying them a recognition of their life, needs and rights.

The idea of encroachment is plainly irrelevant to tribal communities, and policies dealing with recognition of their rights need to acknowledge this, even while

- *land records are updated*; and
- the nature of *common property resources* is understood.

In Maharashtra, while the administration has swung from promising conversion of forest villages into revenue land, to a silent squelching of this policy, the rights of tribal communities have been reduced by denial. In MP, the offer of cash compensation for land, denying the importance of providing land for land, echoes in the valley. Yet, travelling on the Narmada along the length of Jalsindhi, for instance, it is possible to appreciate the enviable stretch of land and resources in which a whole community shares and which, through years of drought and plenty, have helped them survive from generation to generation.

Protection of adivasi communities

The reduction of rights of adivasi communities, even to negation, is, apart from its evident injustice and inequity, also contrary to ILO Convention 107, adopted in 1957, to which India has been a party since 1958. The Convention Concerning the Protection and Integration of Indigenous and other Tribal and Independent Countries specifically provides:

> Article 11 – The right of ownership, collective or individual, of the members of the populations concerned over the lands which these populations traditionally occupy shall be recognised.
>
> Article 12.1 – The populations concerned shall not be removed without their free consent (…) except (…) in the interest of national economic development (…).
>
> 2. When in such cases removal of these populations is necessary as an exceptional measure, they shall be provided with lands of quality at least equal to that of the lands previously occupied by them, suitable to provide for their present needs and future development (…).
>
> 3. Persons thus removed shall be fully compensated for any resulting loss or injury.

Where it does suggest that 'where the populations concerned prefer to have compensation in money or in kind, they shall be so compensated', it adds, 'it shall be so compensated *under appropriate guarantees*' (emphasis added).

Ratifying the Convention was a statement of commitment; it cannot be allowed to degenerate into empty rhetoric.

Even as this year began, the Government of Madhya Pradesh set out its agenda for charting a new course for dalits in the twenty-first century. The Bhopal Declaration of 12/13 January 2002, which contains the government's statement of intent, spells it out in a 21-point agenda and reads, in part:

> 5. Ensure the restoration of the alienated lands to the tribals, restore their rights over forests and forest produce, provide them with compensation and rehabilitation measures, extend resources and capacity building measures for gainful utilisation of their lands and forests and those dalits displaced due to construction of dams/development projects and mining as shareholders of such enterprises.

This mandate must be respected This requires, too, a recognition of the unique relationship that the adivasi has with the land, the forest, the river and the seasons which, undisturbed by outside intervention for so long, has developed its own harmony. This is not a matter of idealising difference, but of recognising it.

There are other approaches to recognising the rights of the oustees which have the effect of *exclusion*. This calls to be acknowledged, and remedied.

Temporary and permanent submergence

Although the NWDT Award defines an oustee as including those 'likely to be submerged permanently or temporarily', and although it was decided in the 58th meeting of the NCA held on 18 December 1998 that 'there should not be any distinction between the temporary and permanent PAFs and resettlement of both will be prerequisite for the purpose of further raising the dam', the distinction has been maintained. In an affidavit dated 4 June 2002, the GoMP asserts that the number of villages where lands are *permanently affected or houses are temporarily affected* is only 30 with 1,883 PAFs. Apart from the swiftly changing numbers that we noticed in an earlier chapter, the loss and destruction that 'temporary submergence' brings may be witnessed in the submergence sites.

The categorising of oustees as 'declared' and 'undeclared', with the latter excluded from the kinder provisions of policy, is proving to be another prescription for exclusion.

Cut-off date

The policies endorse a distinct identity for grown up sons, when they reach the age of 18. The NWDT had directed that major sons as on 16 August 1978 ought to be extended R&R facilities. With conditional environmental clearance being given in 1987, the cut-off date was shifted to 1 January 1987.

But the project necessarily has to cross a range of barriers if it is to be constitutionally conceived and executed. In the meantime, the people grow older, even age. A cut-off date that is positioned in a distant past loses its meaning. The cut-off date will have to be consonant with the time that submergence, and displacement, occur. Applying the cut-off date of 1987 to submergence, displacement and resettlement that is happening in 2002, denies grown-up farmers the right to rehabilitation, including land – a vital component in the resurrection of their destroyed lives.

Tapus

Exclusion through non-recognition of partial submergence, or the isolation that submergence can bring, is stark in the matter of *tapus*. The undulating landscape ensures that islands of elevation stick out amidst swirling waters; yet state policy has willed it out of consideration. These raised pockets of isolation have to be drawn into the plans for rehabilitation.

Women

Women are, if anywhere, only on the margins of consideration for rehabilitation and compensation. While the interests of women have been clubbed with that of their men, we heard of the problems that widows face in having their claims to land, and to compensation, accepted.

It was also evident that, if women were given the right to decide, there would be many who would not settle for cash compensation. They spoke of the impermanence of money, as indeed did many men. But the women spoke further of the temptations, often in the nature of drink, which would entice men to accept cash compensation, and induce them to blow it away in drink.

The role that the woman needs to play in decision making in the matter of compensation and rehabilitation has been grossly underestimated, and must be re-written into policy.

The otherwise displaced

We met people from Kevadia, where six villages were displaced in 1961 to make way for a colony for the staff working on the dam. They were paid a small sum in compensation then, perhaps equal to compensation for the crops they lost in that year. They remain castaways on the fringes of their own lands. Their loss, and their sense of loss, continues, uncompensated, with restitution nowhere in sight. There are those who will be displaced by the canal that will lead away from the reservoir to irrigate, tribals of 108 villages who will have to make way for the Shoolpaneswar bear sanctuary, those whose lands have been lost in the cause of compensatory afforestation, even where no forest has been reared (…). All of these communities of peoples displaced by activity ancillary to the dam have remained beyond the pale of rehabilitation policy and practice. The restricted regime of the Land Acquisition Act, 1894 is their only remedy.

The inequality in treatment between oustees of the Sardar Sarovar Project who are from Maharashtra and Madhya Pradesh who decide not to move across the border into Gujarat, and those who are willing to so move, has been set out earlier (…). The complete wiping out from rehabilitation policy of communities displaced by ancillary projects is an even more unqualified form of exclusion. It was the NWDT which negotiated a place for oustees in Maharashtra and Madhya Pradesh, and this happened because the costs had to be borne by Gujarat, and this had to be factored into the Award. There was no law then, and none now, to compel the State to recognise the losses that such projects generate for the displaced. While this may appear, to the project authorities, to be pragmatic and cost-reducing, there is no running away from the truth that losses are indeed generated to large numbers of people affected by the projects, leading to an unaccounted escalation of actual (including social) costs. It is imperative to recognise that, while this may help suppress real costs, it will only serve to help the project free ride on the oustee, and on those who are within range of the ripple effect that will inevitably result.

Ex parte allotments

The practice of making ex parte allotments is so evidently an exercise in completing paper work, and not in rehabilitation, that there can be little question that it should be immediately halted.

For one, ex parte orders negate the right of an oustee to select the site for rehabilitation; this is a basic, non-derogable right of the oustee.

For another, where the same parcel of land is allotted to a large number of oustees, it is evident that this is intended only for the record. And, where the State avers that it has resorted to the practice because oustees are refusing to cooperate in the move away from the submergence area, it is a reason that fails to convince since, by the State's own admission, there is no land on which to rehabilitate.

It is also difficult to sustain a logic which denies the right to protest and resist when faced with what they see as proven failure, and abdication, by the State. The effort that is needed is to help the people move from distrust to faith, and to provide proof of the State's capacity to perform in keeping with its promises, and with equity in view.

Pari passu

The understanding of rehabilitation to happen *pari passu* with the construction of the dam seems to have lapsed in the recent past. Even in its whittled down version, rehabilitation *is to be completed* at least 6 months before the construction for raising the height of the dam can commence. How, then, was the decision to raise the dam to 95 m made in January 2002,

- even as the GoMP was revising (upward) the estimated number of affected persons at the 95m level as being a 'few thousands' more than the 35,716 families that was earlier in the reckoning, and the submergence area as being 2,703 ha more than the 20,822 ha earlier estimated;
- the Government of Maharashtra was admitting to not having enough land on which to rehabilitate, and the numbers of those not yet rehabilitated at the 90 m level had risen from 17 to 1,000 PAFs by the State's own finding, and
- the affected people were anticipating a greater spread of submergence if the rains were to come?

It raises legitimate doubts about whether an engineering, or political, or perhaps even bureaucratic decision had displaced the rights, interests and needs of the oustees.

The rehabilitation of oustees is not a nominal exercise, but a matter of relocating, reconstructing and restarting new lives in alien environs. It is dependent on the capacity of the State to find land, provide amenities, and facilitate the shift. This exercise, clearly, cannot be reduced to a matter of record from a matter of fact.

Grazing land

The GoMP, in its attempts to find land on which to resettle the oustees, has made a policy decision to reduce the grazing land (charnoi) in the State from 5 per cent to 2 per cent, and deploy the 3 per cent in rehabilitating the oustees. The oustees are worried. On the one hand, the host community may resent the taking away of grazing land. On the other, the cattle that the oustees would take with them – their bank, as they think of their cattle – would need grazing pastures In any case, grazing land is generally degraded and is unlikely to prove productive. Is this a policy made in desperation? May be it should be opened to question again.

- The watering down of the *pari passu* provision;
- the improperly selected rehabilitation sites which, in some cases are at a level lower than the submergence site; and
- the denial of relief to oustees who had complained to the GRA about the inhospitable and rocky land that was being offered to them for cultivation;

are each causes that could result in further dislocation of an already displaced population. It is difficult to overemphasise the importance of averting situations of multiple displacement – one coerced move is traumatic enough.

Where compensation is a part of the rehabilitation package, the procedure for its recovery evidently needs to be made less onerous. Especially where the procedure leads to the oustee losing a part of the amount because of delayed payment where time is of the essence of a contract of sale of land between a landowner and the oustee, this needs revamping.

Perceptions

There is a definite difference of perception in how adivasi life is viewed. The Supreme Court, for instance, in its 2000 judgment said: 'The tribals who are affected are in indigent circumstances and who have been deprived of the modern fruits of development such as tap water, education, road, electricity, convenient medical facilities, etc'.* The court evidently saw the move out of this region of deprivation as an upward thrust. The adivasis seemed to think differently. They were resentful that a government which had neglected them for over 50 years was now wanting to take away from them what they did have. In their land and their way of life, they seemed to see the wealth that promises survival at a minimum, and plenty when the weather gods were with them. There is a chasm between these perceptions which needs to be bridged by understanding. Such understanding may also clarify why the adivasi will not move unless there is productive land to move to; and will find majdoor (waged labour) entirely unacceptable, even demeaning.

Resistance and protest

Ever since the NWDT treated the SSP as a matter between three States (+ Rajasthan), the people likely to be affected by the project have had to battle hard to

---

* *See* paragraph 177, *Narmada Bachao Andolan v. Union of India*, Writ Petition (Civil) No. 319 of 1994, Supreme Court of India, Judgment of 18 October 2000, reproduced at page 138.

get information, to resist, to be heard and to make a decisive impact on issues affecting their lives. Throughout the jan sunwais,

- the percolation of information to the people at large,
- the awareness of the issues at stake,
- the strength of informed opinion, and
- the willingness and capacity to participate in decision making was striking.

The people we met, and who spoke to us had faced submergence, or it was a peril that was lurking around the next turn in their lives. It was not that they were enemies of the project; but the project threatened them, and the fabric of their very existence. It is their resistance and protest that compelled the State to recognise the spiralling numbers of the displaced, way beyond the estimate in 1988–89 of 6,148 PAFs. It has pushed the State into a more realistic awareness of losses, which has in some measure been translated into the policy of rehabilitation. It is doubtful that, without the resistance from the oustees, the inertia of the State could have been dislodged. The oustees, otherwise stated, did not just melt away.

Against this backdrop of struggle for recognition, the frequency and familiarity with which people – men and women – spoke of the lathi and jail was deeply disturbing. The wryness with which 'and now they call us Naxalwadi', was said, was testimony to the criminalising of protest – protest, which is an essential aspect of a living democracy, and which, in the context of the happenings in the valley, would find itself justified.

Perhaps a less repressive and more reasoned treatment of protest would see decision-making becoming more participative, with the State demonstrating willingness and capacity to provide alternatives for displaced people.

The numbers of the population to be directly displaced by the dam has been growing even on the official record: from 6,148 PAFs in 1988–89 to 41,000 in 1994, to 43,000 in March 2001; and the numbers continue to rise. The unavailability of land on which to rehabilitate those in the submergence zone has been acknowledged by both the Governments of Madhya Pradesh and Maharashtra. It is a problem that will not disappear, and has to be confronted and surmounted *before* the homes and lands of the people in the area are let to submerge. Rehabilitation of those in the submergence zone is a daunting task for any State, but it is a task that has to be taken on and completed before submergence; that is the minimum that the State would need to do for a people who are having to pay the price of development. In addition, it is constitutionally mandated. It is imperative that this exercise in rehabilitation be done with equity in policy, and with compassion in practice. The displaced are not adversaries of the State, nor are they obstacles to the continuance of the project. They are, instead, potential victims of a mega project. Recognising this, and acting to protect the displaced from being reduced to victims, is a responsibility that the State will have to undertake.

The scale of the problem is still revealing itself. In the context of the changing dimensions of the recognised extent of displacement, the costs and losses that are being generated by submergence ask to be computed again.

Chapter 8

Financial Aspects

Narmada River Development (Gujarat) Sardar Sarovar Dam and Power Project – International Development Association Loan Agreement, 1985

Source: Credit Number 1552 IN, Development Credit Agreement (Narmada River Development (Gujarat) Sardar Sarovar Dam and Power Project), Agreement of 10 May 1985 between India and the International Development Association.

Agreement of 10 May 1985, between India, acting by its President (hereinafter called the Borrower) and International Development Association (hereinafter called the Association).

Whereas (A) the Borrower has requested the Association to assist in the financing of the Project described in Schedule 2 to this Agreement by extending the Credit as hereinafter provided;

(B) the Borrower has also requested the International Bank for Reconstruction and Development (the Bank) to provide additional assistance towards the financing of the Project and by an agreement of even date herewith between the Borrower and the Bank (hereinafter called the Loan Agreement (Narmada River Development (Gujarat) Sardar Sarovar Dam and Power Project)), the Bank is agreeing to provide such assistance in an aggregate principal amount equivalent to two hundred million dollars ($200,000,000) (hereinafter called the Loan);

(C) the Narmada Water Disputes Tribunal (as hereinafter defined) has adjudicated upon the water dispute among the following States of India: the State of Gujarat, the State of Madhya Pradesh, the State of Maharashtra and the State of Rajasthan regarding the inter-State river Narmada and the river valley thereof, establishing procedures, *inter alia*, for the cost-sharing contributions for, and benefits to be derived from the Sardar Sarovar Project (as hereinafter defined) and the Narmada Sagar Reservoir Project (as hereinafter defined);

(D) the Borrower and the Association intend, to the extent practicable, that the proceeds of the Credit provided for in this Agreement be disbursed on account of expenditures of the Project before disbursements are made of the proceeds of the Loan provided for in the Loan Agreement (Narmada River Development (Gujarat) Sardar Sarovar Dam and Power Project);

(E) the Project, except for part P thereof, will be carried out by the State of Gujarat, the State of Madhya Pradesh and the State of Maharashtra, with the Borrower's assistance and, as part of such assistance, the Borrower will make available to the State of Gujarat, the State of Madhya Pradesh and the State of Maharashtra, part of the proceeds of the Credit as hereinafter provided, and part of the proceeds of the Loan as provided in the Loan Agreement (Narmada River Development (Gujarat) Sardar Sarovar Dam and Power Project);

(F) Part P of the Project will be carried out by the Borrower with the assistance of the Narmada Control Authority;

(G) by an agreement of even date herewith among the Association, the Bank and the State of Gujarat, the State of Gujarat has agreed to undertake certain obligations in respect of the carrying out of the Project;

(H) by an agreement of even date herewith among the Association, the Bank and the State of Madhya Pradesh, the State of Madhya Pradesh has agreed to undertake certain obligations in respect of the carrying out of the Project;

(I) by an agreement of even date herewith among the Association, the Bank and the State of Maharashtra, the State of Maharashtra has agreed to undertake certain obligations in respect of the carrying out of the Project;

(J) the Borrower has also requested the Association to assist in the financing of the Narmada River Development (Gujarat) Water Delivery and Drainage Project described in Schedule 2 to the Development Credit Agreement (Narmada River Development (Gujarat) Water Delivery and Drainage Project) of even date herewith between the Borrower and the Association and by such agreement the Association is agreeing to provide such assistance in an aggregate principal amount equivalent to one hundred forty nine million five hundred thousand Special Drawing Rights (SDR 149,500,000).

(K) the implementation of the Project and of the Narmada River Development (Gujarat) Water Delivery and Drainage Project referred to in (J) above is closely interlinked and each such Project constitutes an integral part of the Sardar Sarovar Project, which project forms part of the overall implementation of the Narmada River Development Programme; and

Whereas the Association has agreed, on the basis, *inter alia*, of the foregoing, to extend the Credit to the Borrower upon the terms and conditions hereinafter set forth, in the Gujarat Project Agreement of even date herewith among the Association, the Bank and the State of Gujarat, in the Madhya Pradesh Project Agreement of even date herewith among the Association, the Bank and the State of

Madhya Pradesh and in the Maharashtra Project Agreement of even date herewith among the Association the Bank and the State of Maharashtra;

Now therefore the parties hereto hereby agree as follows:

Article I: General Conditions; Definitions

Section 1.01. The parties to this Agreement accept all the provisions of the General Conditions Applicable to Development Credit Agreements of the Association, dated 30 June 1980, with the same force and effect as if they were fully set forth herein, subject, however, to the following modification thereof (said General Conditions Applicable to Development Credit Agreements of the Association, as so modified, being hereinafter called the General Conditions), namely, that paragraph 9 of Section 2.01 is deleted and the following substituted therefor:

> 9. The term Project means Narmada River Development (Gujarat) Sardar Sarovar Dam and Power Project as described in the Development Credit Agreement (Narmada River Development (Gujarat) Sardar Sarovar Dam and Power Project) and as the description thereof may be amended from time to time by agreement among the Borrower, the Association and the Bank.

Section 1.02. Wherever used in this Agreement, unless the context otherwise requires, the several terms defined in the Loan Agreement (Narmada River Development (Gujarat) Sardar Sarovar Dam and Power Project), in the General Conditions and in the Preamble to this Agreement have the respective meanings therein set forth and the following additional terms have the following meanings:

a) 'Gujarat' means the State of Gujarat, a State of India, acting by its Governor;

b) 'Madhya Pradesh means the State of Madhya Pradesh, a State of India, acting by its Governor;

c) 'Maharashtra' means the State of Maharashtra; a State of India, acting by its Governor;

d) 'Rajasthan' means the State of Rajasthan, a State of India, acting by its Governor;

e) 'Decision' means the Final Order and Decision dated 7 December 1979, rendered by the Narmada Water Disputes Tribunal and published in the Gazette of India on 12 December 1979;

f) 'Narmada Control Authority' and the acronym 'NCA' mean the Authority established by the Borrower's Notification dated 10 September 1980, as amended to the date of this Agreement, issued pursuant to the provisions of Section 6-A of the Borrower's Inter-State Water Disputes Act, 1956 (33 of 1956) and of Clause XIV of the Decision;

g) 'Narmada Review Committee' means the Committee established by the Borrower's Notification dated 10 September 1980, as amended to the date of this Agreement, issued pursuant to the provisions of Section 6-A of the

Borrower's Inter-State Water Disputes Act, 1956 (33 of 1956) and of Clause XIV of the Decision;

h) 'Sardar Sarovar Construction Advisory Committee' means the Advisory Committee established by the Borrower's Resolution No. 22/7/80-P.I. dated 4 September 1980, issued pursuant to the provisions of Clause XIV of the Decision;

i) 'Narmada Water Disputes Tribunal' means the Tribunal established by the Borrower on 6 October 1969, pursuant to the provisions of the Borrower's Inter-State Water Disputes Act, 1956, to resolve the matter of water disputes among Gujarat, Madhya Pradesh, Maharashtra and Rajasthan regarding the interstate Narmada river and the river valley thereof;

j) 'Gujarat Project Agreement' means the agreement, of even date herewith, entered into among the Association, the Bank and Gujarat for purposes of carrying out the Project, as the same may be amended from time to time, and such term includes the Schedules to the Gujarat Project Agreement and all agreements supplemental to the Gujarat Project Agreement;

k) 'Madhya Pradesh Project Agreement' means the agreement, of even date herewith, entered into among the Association, the Bank and Madhya Pradesh for purposes of, carrying out the Project, as the same may be amended from time to time, and such term includes the Schedule to the Madhya Pradesh Project Agreement and all agreements supplemental to the Madhya Pradesh Project Agreement;

l) 'Maharashtra Project Agreement' means the agreement, of even date herewith, entered into among the Association, the Bank and Maharashtra for purposes of carrying out the Project, as the same may be amended from time to time, and such term includes the Schedule to the Maharashtra Project Agreement and all agreements supplemental to the Maharashtra Project Agreement;

m) 'Narmada River Development Programme' means the multistate programme for the development of the resources of the Narmada river to be implemented, as the case may be, by Gujarat, Madhya Pradesh, Maharashtra and Rajasthan, and consisting of multi-purpose projects, *inter alia*, hydroelectric power projects, irrigation projects, and municipal, industrial and domestic water projects, including the Sardar Sarovar Project and the Narmada Sagar Reservoir Project;

n) 'Sardar Sarovar Project' means the project whose salient features are described in Schedule 4 to this Agreement, to be implemented, in accordance with the provisions of the Decision, by Gujarat, Madhya Pradesh, Maharashtra and Rajasthan, portions of which shall be implemented under the Project described in Schedule 2 to the Development Credit Agreement (Narmada River Development (Gujarat) Water Delivery and Drainage Project), and under the

Project described in Schedule 2 to the Development Credit Agreement (Narmada River Development (Gujarat) Sardar Sarovar Dam and Power Project);

o) 'Narmada Sagar Reservoir Project' means the project to be implemented by Madhya Pradesh pursuant to the provisions of the Decision;

p) 'Dam Safety Panel' means the panel of experts to be established by Gujarat pursuant to the provisions of Section 2.18 of the Gujarat Project Agreement;

q) 'Oustee' means any person, whether landed or landless, who since at least one year prior to the date of publication of the notification under Section 4 of the Borrower's Land Acquisition Act, 1894, as amended to the date of this Agreement, has been ordinarily residing, or cultivating land, or carrying on any trade, occupation, or calling or working for gain, in Gujarat, Madhya Pradesh and Maharashtra, and who would be displaced from his usual habitat due to the carrying out of the Project;

r) 'Narmada Development Department' means the Department established by Gujarat on 7 March 1984, responsible for the overall implementation of the Sardar Sarovar Project;

s) 'Narmada High Power Committee' means the Committee established by Gujarat on 29 September 1978, for the purposes set forth, *inter alia*, in Section 2.08 (a) of the Gujarat Project Agreement;

t) 'Narmada' Planning Group' means the Group established by Gujarat, within the Narmada Development Department, for the purposes set forth, *inter alia*, in Section 2.08 (b) of the Gujarat Project Agreement;

u) 'Central Procurement Unit' means the unit established by Gujarat, within the Narmada Development Department, on 31 May 1984;

v) 'Management Information Systems Cell' means the cell to be established by Gujarat pursuant to the provisions of Section 2.09 (b) of the Gujarat Project Agreement;

w) 'Land Acquisition and Rehabilitation Cell' means the Cell established by Madhya Pradesh, within its Narmada Planning Agency, on 9 February 1984, for purposes of planning, coordinating and implementing, within Madhya Pradesh's State boundaries, the resettlement and rehabilitation of the oustees;

x) 'Narmada Development Division' means the Division established by Maharashtra, within its Irrigation Department, on 5 January 1981, for purposes of planning, coordinating and implementing, within Maharashtra's State boundaries, the resettlement and rehabilitation of the oustees; and

y) 'Narmada Main Canal' means the canal under Part B (i) of the Sardar Sarovar Project which traverses Gujarat up to the Rajasthan border, about 438 kilometres long, with designed capacity of about 1,133 cumecs at its head and with a full supply level of about 91.46 metres at its head, including the related structures therein.

Article II: The Credit

Section 2.01. The Association agrees to lend to the Borrower, on the terms and conditions set forth or referred to in this Agreement, an amount in various currencies equivalent to ninety nine million seven hundred thousand Special Drawing Rights (SDR 99,700,000).

Section 2.02. The amount of the Credit may be withdrawn from the Credit Account in accordance with the provisions of Schedule I to this Agreement, as such Schedule may be amended from time to time by agreement between the Borrower and the Association, for expenditures made (or, if the Association shall so agree, to be made) in respect of the reasonable cost of goods and services required for the Project and to be financed out of the proceeds of the Credit.

Section 2.03. Except as the Association shall otherwise agree, procurement of the goods and civil works required for Parts A, B, H, K and N of the Project and to be financed out of the proceeds of the Credit shall be governed by the provisions of the Schedule I to the Gujarat Project Agreement.

Section 2.04. The Closing Date shall be 30 June 1995, or such later date as the Association shall establish. The Association shall promptly notify the Borrower of such later date.

Section 2.05. (a) The Borrower shall pay to the Association a commitment charge at the rate of one-half of one per cent (1/2 of 1 per cent) per annum on the principal amount of the Credit not withdrawn from time to time. The commitment charge shall accrue from a date sixty days after the date of this Agreement to the respective dates on which amounts shall be withdrawn by the Borrower from the Credit Account or shall be cancelled.

(b) The commitment charge shall be paid: (i) at such places as the Association shall reasonably request; (ii) without restrictions of any kind imposed by, or in the territory of, the Borrower; and (iii) in the currency specified in this Agreement for the purposes of Section 4.02 of the General Conditions or in such other eligible currency or currencies as may from time to time be designated or selected pursuant to the provisions of that Section.

Section 2.06. The Borrower shall pay to the Association a service charge at the rate of three-fourths of one per cent (3/4 of 1 per cent) per annum on the principal amount of the credit withdrawn and outstanding from time to time.

Section 2.07. Commitment charges and service charges shall be payable semiannually on January 1 and July 1 in each year.

Section 2.08. The Borrower shall repay the principal amount of the Credit in semiannual instalments payable on each January 1 and July 1 commencing 1 July 1995, and ending 1 January 2005, each instalment to and including the instalment payable on 1 January 2005, to be one-half of one per cent (1/2 of 1 per cent) of such principal amount, and each instalment thereafter to be one and one-half per cent (1-1/2 per cent) of such principal amount.

Section 2.09. The currency of the United States of America is hereby specified for the purposes of Section 4.02 of the General Conditions.

Article III: Execution of the Project

Section 3.01. (a) The Borrower declares its commitment to the objectives of the Project as set forth in Schedule 2 to this Agreement, and, to this end, without any limitation or restriction upon any of its other obligations under this Agreement, the Borrower shall: (i) cause Gujarat, Madhya Pradesh and Maharashtra, to perform in accordance with the provisions of the Gujarat Project Agreement, the Madhya Pradesh Project Agreement, and the Maharashtra Project Agreement, all their respective obligations therein set forth, and shall cause Rajasthan to perform its obligations towards the implementation of the Project as provided for in the Decision; (ii) shall take and cause to be taken all action, including the provision of funds, facilities, services and other resources, necessary or appropriate to enable Gujarat, Madhya Pradesh and Maharashtra, each to perform the aforesaid obligations, and shall not take or permit to be taken any action which would prevent or interfere with such performance; and (iii) with the assistance of NCA, carry out Part P of the Project with due diligence and efficiency and in conformity with monitoring and evaluation practices, and shall provide, or cause to be provided, promptly as needed, the funds, facilities, staffing, services and other resources required for the purpose.

(b) For purposes of assisting Gujarat, Madhya Pradesh and Maharashtra in carrying out the Project, the Borrower shall make available to Gujarat, Madhya Pradesh and Maharashtra out of the proceeds of the Credit and in accordance with the Borrower's standard arrangements for development assistance to the States of India and the cost-sharing provisions of the Decision, an aggregate amount equivalent to ninety nine million five hundred thousand Special Drawing Rights (SDR 99,500,00).

Section 3.02. Without limitation to the provisions of Section 3.01 of this Agreement, the Borrower shall take, or cause to be taken, all such action as shall be necessary, including, if required, the provision of forest land within the meaning of the Borrower's Forest (Conservation) Act, 1980, for enabling Gujarat, Madhya Pradesh and Maharashtra each to implement, within their respective State boundaries, the Project.

Section 3.03. The Borrower shall, within the powers available to it, take all such action as shall be necessary to maintain at all times, the Narmada Control

Authority, the Sardar Sarovar Construction Advisory Committee, and the Narmada Review Committee, each with such powers, functions, responsibilities, staffing, funds and membership as shall be required to enable each of them to accomplish its purposes as set forth in the Decision, and shall not take, or cause to be taken, any action which shall prevent or interfere with such purposes.

Section 3.04. In order to assist the Borrower in carrying out Part P of the Project, the Borrower shall, by 30 September 1985, employ, or cause to be employed, and, thereafter, assign to NCA, a social scientist whose qualifications, experience and terms and conditions of employment shall be satisfactory to the Association, such social scientist to be selected in accordance with principles and procedures satisfactory to the Association on the basis of the 'Guidelines for the Use of Consultants by World Bank Borrowers and by the World Bank as Executing Agency published by the Bank in August 1981.

Section 3.05. The Borrower shall, on June 1 of each year, commencing on 1 June 1986 and thereafter until 1 June 1995, furnish to the Association semiannual and annual reports, of such scope and in such detail as the Association shall reasonably request, regarding the rehabilitation and resettlement of the oustees under Part O of the Project.

Article IV: Remedies of the Association

Section 4.01. For the purposes of Section 6.02 of the General Conditions, the following additional events are specified pursuant to paragraph (h) thereof:

(a) Gujarat shall have failed to perform any of its obligations under the Gujarat Project Agreement.

(b) As a result of events which have occurred after the date of this Agreement, an extraordinary situation shall have arisen which shall make it improbable that Gujarat will be able to perform its obligations under the Gujarat Project Agreement.

(c) Madhya Pradesh shall have failed to perform any of its obligations under the Madhya Pradesh Project Agreement.

(d) As a result of events which have occurred after the date of this Agreement, an extraordinary situation shall have arisen which shall make it improbable that Madhya Pradesh will be able to perform its obligations under the Madhya Pradesh Project Agreement.

(e) Maharashtra shall have failed to perform any of its obligations under the Maharashtra Project Agreement.

(f) As a result of events which have occurred after the date of this Agreement, an extraordinary situation shall have arisen which shall make it improbable that Maharashtra will be able to perform its obligations under the Maharashtra Project Agreement.

(g) Rajasthan shall have failed to perform any of its obligations under the Decision which shall have affected materially and adversely the carrying out of the Project.

(h) The Borrower, Gujarat, Madhya Pradesh, Maharashtra and Rajasthan, individually or collectively, or any other authority having jurisdiction, shall have taken any action resulting in the dissolution or disestablishment of NCA or in the suspension of its activities.

(i) Gujarat, Madhya Pradesh, Maharashtra and Rajasthan, individually or collectively, or any authority having jurisdiction shall have taken any action to suspend, abrogate, repeal, waive any provision of the Decision in such a manner as to affect materially and adversely the ability of either the Borrower Gujarat, Madhya Pradesh or Maharashtra to carry out the Project or any Part thereof.

(j) (i) Subject to subparagraph (ii) of this paragraph:

(A) the right of the Borrower to withdraw the proceeds of any loan or grant made to the Borrower for the financing of the Project shall have been suspended, cancelled or terminated in whole or in part, pursuant to the terms of the agreement providing therein for, or

(B) any such loan shall have become due and payable prior to the agreed maturity thereof.

Subparagraph (i) of this paragraph shall not apply if the Borrower establishes to the satisfaction of the Bank that: (A) such suspension, cancellation, termination or prematuring is not caused by the failure of the Borrower to perform any of its obligations under such agreement; and (B) adequate funds for the Project are available to the Borrower from other sources on terms and conditions consistent with the obligations of the Borrower under this Agreement.

Section 4.02. For the purposes of Section 7.01 of the General Conditions, the following additional events are specified pursuant to paragraph (d) thereof:

(a) that any event specified in paragraphs (a), (c), (e) and (g) of Section 4.01 of this Agreement shall occur and shall continue for a period of 60 days after notice thereof shall have been given by the Association to the Borrower;

(b) that any event specified in paragraphs (h) and (i) of Section 4.01 of this Agreement shall occur;

(c) that any event specified in paragraph (j) (i) of Section 4.01 of this Agreement shall occur, subject to the provisions of paragraph (j) (ii) of such Section;

(e) that any event specified in paragraphs (a) and (c) of Section 4.01 of the Development Credit Agreement (Narmada River Development (Gujarat) Water Delivery and Drainage Project) shall occur and shall continue for a period of 60 days after notice thereof shall have been given by the Association to the Borrower; and

(f) that any event specified in paragraphs (d) and (e) of Section 4.01 of the Development Credit Agreement (Narmada River Development (Gujarat) Water Delivery and Drainage Project) shall occur.

Article V: Effective Date; Termination

Section 5.01. The following events are specified as additional conditions to the effectiveness of the Development Credit Agreement (Narmada River Development (Gujarat) Sardar Sarovar Dam and Power Project) within the meaning of Section 12.01 (b) of the General Conditions:

(a) that all conditions precedent to the effectiveness of the Loan Agreement (Narmada River Development (Gujarat) Sardar Sarovar Dam and Power Project) have been fulfilled, except for the effectiveness of this Agreement; and

(b) that all conditions precedent to the effectiveness of the Development Credit Agreement (Narmada River Development (Gujarat) Water Delivery and Drainage Project) have been fulfilled, except for the effectiveness of this Agreement.

Section 5.02. The following is specified as additional matters, within the meaning of Section 12.02 (b) of the General Conditions, to be included in the opinion or opinions to be furnished to the Association:

(a) that the Gujarat Project Agreement has been duly authorised or ratified by Gujarat, and is legally binding upon Gujarat in accordance with its terms;

(b) that the Madhya Pradesh Project Agreement has been duly authorised or ratified by Madhya Pradesh, and is legally binding upon Madhya Pradesh in accordance with its terms ;

(c) that the Maharashtra Project Agreement has been duly authorized or ratified by Maharashtra, and is legally binding upon Maharashtra in accordance with its terms;

(d) that Gujarat has taken all the necessary governmental action required for adopting and thereafter implementing the rehabilitation and resettlement plan pursuant to the provisions of Section 2.10 of the Gujarat Project Agreement;

(e) that Madhya Pradesh has taken all the necessary governmental action required for adopting and thereafter implementing the rehabilitation and resettlement plan pursuant to the provisions of Section 2.07 of the Madhya Pradesh Project Agreement; and

(f) that Maharashtra has taken all the necessary governmental action required for adopting and thereafter implementing the rehabilitation and resettlement plan pursuant to the provisions of Section 2.07 of the Maharashtra Project Agreement;

Section 5.03. The date 6 November 1985 is hereby specified for the purposes of Section 12.04 of the General Conditions.

Article VI: Representative of the Borrower; Addresses

Section 6.01. Any Secretary, Additional Secretary, Joint Secretary, Director, Deputy Secretary and Under Secretary of the Department of Economic Affairs in the Ministry of Finance of the Borrower is designated as representative of the Borrower for the purposes of Section 11.03 of the General Conditions.

Section 6.02. The following addresses are specified for the purpose of Section 11.01 of the General Conditions:

For the Borrower: The Secretary to the Government of India, Ministry of Finance Department of Economic Affairs, New Delhi, India. Cable address; ECOFAIRS, New Delhi, India.

For the Association: International Development Association, 1818 H Street, NW, Washington, DC 20433, United States of America. Cable address: INDEVAS Washington, DC. Telex: 953-313546 Telex: 440098 (ITT) 248423 (RCA) or 64145 (WUI).

In Witness Whereof, the parties hereto, acting through their representatives thereunto duly authorised, have caused this Agreement to be signed in their respective names in the District of Columbia, United States of America, as of the day and year first above written.

World Bank, Office Memorandum from Ibrahim Shihata to Joseph Wood, 1993

Source: The World Bank, Office Memorandum from Ibrahim F.I. Shihata (Vice President (Legal) and General Counsel of the World Bank) to D. Joseph Wood concerning 'Narmada', 30 March 1993.

In the brief Board discussion today of the cancellation of the Bank's Loan for the Narmada Project, the impression was left that the Government of India is no longer legally obligated towards the Bank to carry out its obligations under the Loan agreement.

As I explained verbally yesterday, this is not the case. Section 6.06 of the General Conditions applicable to all Bank loans (and IDA credits) which constitute an integral part of each loan (and credit) agreement provides that '[n]otwithstanding any cancellation or suspension, all the provisions of the Loan Agreement and the Guarantee Agreement shall continue in full force and effect except as specifically provided in this Article'.

The above conclusion also applies in case the loan (or credit) is fully disbursed (which, I understand, is the case of the IDA credit financing the same project).

I hope that the above will be borne in mind by staff in their further pursuance of this matter.

World Bank – Project Completion Report, 1995

Source: World Bank, Project Completion Report – India – Narmada River Development – Gujarat Sardar Sarovar Dam and Power Project (Credit 1552-IN/Loan 2497-IN), Report No. 14159, 29 March 1995 (extracts).

Evaluation Summary

Background and objectives

1. The Sardar Sarovar Dam and Power Project and the Gujarat Water Delivery and Drainage Project form part of the Narmada River Development Plan, whose overall objective is to harness the Narmada river through a series of dams, for irrigation, water supply and energy. The arrangements for sharing and utilising water by the four benefiting States (Madhya Pradesh, Gujarat, Maharashtra and Rajasthan) were defined by the Narmada Water Disputes Tribunal (NWDT) in December 1979. Under this Award, Gujarat was allocated 11,000 Mm^3 representing about one-third of the 75 per cent dependable usable annual water flow, and was authorised to implement a project comprising: (i) the Sardar Sarovar Dam, (ii) a 1,200 megawatt (MW) river bed power house (RBPH), (iii) a 250 MW canal head power house (CHPH), (iv) a canal system to irrigate 1.87 m. ha in Gujarat and 70,000 ha in Rajasthan, and (v) a water supply system for about 30 million people in the drought-prone areas of Saurashtra and Kachchh. On that basis, Gujarat prepared the Sardar Sarovar Project in 1980, which was to be implemented in various phases over a period of about 20 years. Phase I comprised: (i) the Sardar Sarovar Dam, (ii) the RBPH, (iii) the CHPH, (iv) the first 144 km of the main canal up to the Mahi River, and (v) a network of branch and distribution canals and drainage system to irrigate about 450,000 ha.

2. The original components of the Sardar Sarovar Dam and Power Project (Cr. 1552-IN/Ln. 2497-IN), to be financed within 110 months, included: (i) the Sardar Sarovar concrete gravity dam and its spillway gates, (ii) the underground river bed power house and selected electro-mechanical equipment, (iii) the irrigation bypass tunnel, (iv) the Narmada Basin hydro-meteorological network, (v) a management information system, (vi) training, and (vii) technical assistance. In the same period, the Borrower's own resources were to finance: (a) the canal head power house and its five 50 MW turbine generating sets, (b) the Garudeshwar Weir, (c) the Vadgam saddle dam, (d) the headworks reservoir regulation system, (e) the resettlement and rehabilitation (R&R) of about 10,758 projects affected families (PAFs), and (f) a monitoring and evaluation system for the R&R activities. The six 200 MW reversible turbine generating sets were to be financed through a separate credit

arrangement with the Government of Japan. The project did not finance any environmental components, nor did the Credit/Loan finance any of the R&R activities that became so critical during project implementation.

3. The Water Delivery and Drainage Project (Credit 1553-IN) was processed in parallel with this one and is the subject of a separate Project Completion Report. The two projects were presented to the Board concurrently on 7 March 1985. The Staff presentation highlighted the R&R programme for the projects. The Directors queried the institutional arrangements for that programme and the relatively undefined state of the environmental studies. Their other concerns related to the project size, future Narmada Basin development, and being kept informed (during implementation) about the projects' social and environmental effects. Taken together, these Board concerns revealed the early exposure of all parties in the Bank to the R&R and environmental issues, and the attention afforded to them by the EDs. Management agreed and complied to a specific Board request to keep it informed of the projects' progress on social and environmental issues through many written responses to questions and a number of briefings of EDs.

Project design and organization

4. Project preparation took four years from identification in 1979 to pre-appraisal in 1983. The project was prepared by the Narmada Planning Group (NPG) with assistance from the United Nations Development Program (UNDP) and the Bank. Project concept and design closely follows the NWDT Award. NPG conducted supporting studies and analysed the project's financial and economic benefits. The Narmada Development Department, in charge of project implementation, prepared detailed designs and cost estimates for the major project structures.

5. Appraisal took a lengthy 17 months (March 1983 to August 1984) and several Bank missions. Negotiations took place in November 1984 and Board Presentation in March 1985. R&R policies and related institutional, monitoring and evaluation arrangements were described in the SAR. The legal documents included three covenants on health, 16 dealing with R&R, and three with environment including the party States' obligation to implement an R&R plan satisfactory to the Bank, plus a work plan on the project's environmental effects (see Evaluation Summary, Attachment 1).

6. In May 1988, responsibility for project implementation was transferred from the Narmada Development Department, a government agency, to the SSNNL, a parastatal corporation organised along functional lines,

Hydrology

7. Bank and Borrower reviews deemed the basic hydrological data used for project design to be satisfactory. The NWDT has set the annual 75 per cent dependable water availability for the project at 28 MAF (34,580 Mm^3) with 65.2 per cent allocated to MP, 32.1 per cent to Gujarat. 1.8 per cent to Rajasthan, and 0.9 per cent to Maharashtra. With first priority given to water supply, second to irrigation

and third to power generation. Water availability for firm power generation at the river bed power house will decline sharply – and could drop to practically nil when MP will use its full water allocation. However, the river bed power house will still be able to produce peaking power generation due to its reversible turbine generators. Revised estimates show that MP is likely to use only 70 per cent of its allocation by 2024, when the NWDT Award will be reviewed. The canal head power house will continue to produce 50 MW of power generation throughout the life of the project

Project implementation

8. General. The Credit became effective in January 1986, nine months after Board approval, but construction was delayed until July 1987, pending environmental clearance from India's Ministry of Environment and Forests (MoEF). The project implementation schedule was revised in December 1989 – with irrigation and power generation expected to start in July 1995, after dam construction had reached 112.5 m. The dam, with its spillway gates, was then expected to be fully completed by January 1998. However, the Credit/Loan closed on 29 March 1993, two years and three months ahead of time, at the Borrower's request

Civil engineering

9. The Sardar Sarovar Dam was to be completed in 110 months (by June 1996), but was initially slowed by difficulties in importing and commissioning key construction equipment, especially a 330 m^3/hour concrete batching plant and two 1,600 m, 28 ton capacity cable ways. Procurement, installation and commissioning of this equipment took about 42 months, rather than 26 as stipulated in the contract. Thereafter, concrete placement increased considerably, reaching a peak of 1.0 Mm^3 in FY 1990/91. High standards of construction and quality were maintained throughout due to the establishment by SSNNL of an Expert Committee on Cement and Concrete, the creation of a Quality Control-Quality Assurance organization, and an effective Management Information System.

10. The River bed Power house (RBPH), the second largest civil works component financed under the project, was to be completed in 93 months, i.e. by April 1995. Though the construction of this exceptionally large underground power house had a good start, it suffered delays due to unexpected physical problems. The first was unstable slopes in the collection pod, followed by the development of cracks in the machine hall walls. The latter problem required the assistance of international consultants provided by the Bank. The completion of the RBPH now depends on procurement of six 200 MW reversible turbine generating sets that were to be provided by a Japanese consortium under a credit agreement with Overseas Economic Cooperation Fund of Japan (OECF). Though three turbine generating sets were being manufactured in Japan, they could not be delivered because the Government of Japan withdrew its support for the project in 1991 – citing deficiencies in R&R. The Borrower has, since then, been actively negotiating an alternative financing arrangement with the supplier. However, the financing issue

is still unresolved and the date of completion of the RBPH with its turbine generating sets remains uncertain.

11. Construction has not begun on the irrigation bypass tunnel, which is to supplement the discharge of the CHPH to provide water to the main canal. Its invert level was originally fixed at reservoir level 93.6 m, but Bank appraisal and supervision missions recommended lowering it to reservoir level 86.00 m to optimise Gujarat's diversions, particularly in drought years. The Government of Madhya Pradesh (GoMP) opposed lowering the invert level to preserve its own rights over power generation and to prevent Gujarat from using more than its share of water. Because the invert level has still not been decided, the designs for this key project component cannot be finalised. This structure will require a minimum of three years to complete, and any further delay will cause serious construction problems.

12. The Narmada Basin hydro-meteorological (hydromet) network for flood forecasting and reservoir operation is another important project component to be financed under the Credit/Loan, but which has not yet been implemented. In 1988 international experts recommended meteor-burst – a system not previously used in India – as a viable dedicated telecommunication system. This new technology was endorsed by the Telecom Commission, GoI, and a two-stage bidding procedure was initiated in 1991 to procure the required equipment. However, just as price bids were being sought, the National Informatics Centre, GoI, operator of a nationwide satellite telecommunication system, declared that its system would be cheaper and more reliable under Indian conditions. As a result, the bidding process was halted and the Bank agreed to change the specifications so that satellite telecommunications could compete with meteor-burst.

13. The SSNNL established an efficient Management Information System to monitor the progress of civil works, and prepared an operation and maintenance manual satisfactory to the Bank.

14. The canal head power house, though not financed under the Credit/Loan, made good progress and its first 50 MW conventional T/G set will be ready for commissioning by August 1995. However, the construction of the Garudeshwar Weir had not been initiated at the time of the PCR.

Environment and R&R – Background

15. Board approval was granted in 1985, in the absence of formal forest clearance from the MoEF. Forest clearance and an environmental conditional clearance were issued in 1987. As discussed in Annexes 1 and 2, the Bank's dialogue with GoI/States on environmental and R&R issues which started as early as 1985 and continuing through credit closure, including a major R&R mission in 1989, resulted in substantial strides being achieved on these two fronts. The project's R&R and environmental aspects had come under heavy criticism from national and international entities, including non-governmental agencies (NGOs). In response,

in September 1991 the Bank commissioned an Independent Review. The resulting report (June 1992) questioned both the project's economic viability and the Bank's ability to follow its guidelines. It highlighted potential environmental problems relating to sedimentation, back water effects and downstream impacts, and drew attention to specific R&R inadequacies.

16. Independent Review. In June 1992 the Independent Review report criticised the Bank, the three party States, and the Central Government for poor implementation of the R&R programme. The report offered three main criticisms of the Bank:

a. despite the existence of explicit operational guidelines with respect to involuntary resettlement and tribal people, the Bank failed to insist on proper preparation of R&R plans by the Borrower and, accordingly, failed to appraise adequately the resettlement components of SSP;

b. the appraisal team and the legal documents did not address adequately the special needs of tribal people affected by the project, because they were mostly encroachers on government land and therefore legally treated as landless; and

c. during the last few years of project implementation, Bank efforts to compensate for the lack of an adequate appraisal have helped to bring about some improvements, particularly in Gujarat; however this 'incremental approach' failed to achieve all the changes needed, particularly in Madhya Pradesh (MP).

17. On the Borrower's side, the principal criticisms related to:

a. the lack of sufficient baseline data gathered on the affected population, which made it impossible to prepare effective resettlement plans;

b. the lack of consultation with the affected population, and failure to inform them of their resettlement options and rehabilitation packages;

c. the lack of compliance with the provisions of the NWDT and the Bank's legal agreements; and

d. institutional weaknesses, poor implementation particularly in MP, and inadequate linkage between dam construction and R&R implementation.

18. The Independent Review report also gave a detailed description of R&R implementation in the three States and drew attention to specific issues. In its concluding remarks, the report indicated that R&R could not be implemented as planned and recommended that the Bank 'step back' from the project to re-evaluate the situation and institute improvements.

19. Management response to the Independent Review. The Bank's Management, in its response of 23 June 1992, generally agreed with the description of R&R experience in the States, and also agreed that: (i) adequate R&R plans had not been prepared by the Borrower for appraisal by the Bank, (ii) while the development of the resettlement policy in Gujarat was indeed an achievement, R&R policy in the

other two States – Maharashtra and Madhya Pradesh – needed urgent improvement, and (iii) implementation needed to be further strengthened in all three States. The Management Response explained that the Bank's supervision strategy had focused on obtaining incremental progress through: (a) improvements in resettlement policy, planning and implementation in Gujarat, (b) efforts to replicate Gujarat's R&R policy and standards in the other two States, and (c) concentrating its supervision efforts on MP during the remaining years up to 1994, when R&R will commence in that State (see Evaluation Summary, Attachment 3 for a selected list of some of the key steps in the incremental approach followed). Nevertheless, Management concluded that the incremental approach did not yield the results expected and was too slow in the light of the time before submergence. It supported a new approach embodied in a monitorable action plan which became the benchmarks subsequently approved by the Board. Finally, the Bank believed that the establishment of these measurable benchmarks for addressing the situation was preferable to stepping back, as recommended by the Independent Review. In retrospect, the incremental approach proved to be a necessary process toward the Borrower's adoption of the changes in R&R policies and implementation that were so strongly advocated by the Independent Review and the Board, but was not a sufficient step to achieve the desired result soon enough. Stronger actions to enforce the Bank guidelines would have been warranted in the earlier stages of project implementation.

20. At the time of setting of the benchmarks the Bank secured the Borrower's assurance that a comprehensive environmental action plan would be prepared, and that GoI would promptly correct the weaknesses in its R&R programme. In October 1992 the Board agreed to continue support for the project, subject to Borrower's compliance by 31 March 1993 with performance benchmarks related to environmental and R&R issues. On 29 March 1993, however, the Borrower requested cancellation of the outstanding loan amount of US $181.5 million – while reiterating at the same time its commitment to meeting the benchmarks and agreed standards for R&R and environmental implementation,

Environment

21. After the Independent Review, the Borrower made concerted efforts on a number of environmental fronts. Compensatory afforestation has progressed well and is expected to be completed prior to submergence. Catchment area treatment is underway and, for selected priority areas, should be completed on time. Plans for preserving cultural sites and protecting wildlife are in various stages of implementation,

22. The consultants recruited to assist the Borrower in meeting the benchmarks believe the remaining environmental matters cited by the Independent Review do not pose serious threats to either the environment or local residents, except in the estuary downstream of the dam where some fishing communities may need R&R and water may need to be released for environmental control at a later stage.

Resettlement and rehabilitation

23. The Borrower now estimates that the Sardar Sarovar Dam will affect about 41,014 families/131,245 persons in 245 villages (compared with SAR estimates of 10,758 families/67,340 persons). However, about 20,000 families (of which 45 per cent major sons) will lose only houses and/or no land and would therefore be able to remain within or close to the village area beside the reservoir. The balance would be resettled mostly in Gujarat in the command area of SSP. The new figures count adult sons as separate families, and also reflect population growth and better surveys. About 82 per cent of all project-affected families (PAFs) are in MP, 11 per cent in Gujarat, and about 7 per cent in Maharashtra. Of these PAFs, 97 per cent are tribal people in Gujarat, 100 per cent in Maharashtra, and 29 per cent in MP mostly living at subsistence levels. The project's impact on tribals in all three States as well as the Nimad Plains of Madhya Pradesh are discussed in detail in Annex 2 Section V. About 23 per cent PAFs in Gujarat and up to 47 per cent in MP are estimated to be landless.

24. Because this was the first time that such high standards of R&R had been applied to a project in India, all concerned underestimated the complexity and time required to resettle and rehabilitate over 100,000 people. Ultimately, however, due to the efforts by the Bank and also the role played by State governments, Government of India, project authorities, independent M&E agencies, NGOs and the Independent Review, this project has been the source of many improvements in R&R. With respect to R&R policies, the principle of land-for-land was applied on a large scale, adult sons (and daughters in Maharashtra) were recognised as separate families, tribal people on government forest lands were treated as landed families, and additional assistance was provided for subsistence and housing construction. On the implementation side, Gujarat developed a unique mechanism for acquiring replacement agricultural land at market price through Land Purchase Committees. State R&R units were established, as well as a central project coordinating unit and monitoring mechanism. In MP, progress was made with the help of technical assistance and a computerised management system for R&R.

25. The R&R component had to face a number of difficulties in part due to the Bank's failure to follow its own guidelines at appraisal such as the lack of early consultation with oustees and the preparation of detailed R&R plans by the States or to other reasons which became known only during project implementation. However, the lack of consultation is probably one of the major reasons for problems encountered in carrying out socio-economic surveys, preparing satisfactory R&R policies and plans and implementing them in a timely manner. Some other difficulties encountered were: (i) estimating the precise number of PAFs through house-to-house socio-economic surveys; (ii) applying the principle of land-for-land where land markets are limited, and without creating secondary displacement; (iii) linking dam construction with R&R without incurring huge costs through construction delays, and defining when R&R has been completed;

(iv) predicting permanent and temporary submergence levels years in advance, in areas with highly variable rainfall; (v) deciding the size and location of resettlement centres; and (vi) integrating resettlement centres into host communities. Most of these issues were resolved satisfactorily except for the survey of oustees of some villages of the Akrani Taluka in Maharashtra whose resolution continues to be hampered by dam opponents. In particular, Land Purchase Committees were established in Gujarat, a Submergence Schedule for the 1 in a 100-year flood was prepared by the Narmada Control Authority (NCA) on the basis of the revised construction schedule, of December 1989, approved by the Sardar Sarovar Construction Advisory Committee a certification process was established to link dam construction with R&R implementation, resettlement site selection and development was improved, and steps were taken to integrate resettlement centres into host communities with the assistance of NGOs.

26. A long learning process led to what can now be considered a reasonably well-structured programme. Although implementation still exhibits weakness and further improvements will be required in the selection and construction of resettlement sites, the acquisition of replacement land, and overall programme management, the SSP R&R programme has set new policy and implementation standards which can be applied in future resettlement programmes. There remain concerns about the extent of land options being offered for Madhya Pradesh oustees who wish to remain in Madhya Pradesh rather than take the Gujarat package.

27. The Board's performance benchmarks on R&R required the Borrower to: (i) assess the number of PAFs and analysis of tribal people affected, (ii) prepare comprehensive R&R plans in each State, (iii) strengthen institutional arrangements and the role of NGOs, and (iv) acquire land in Maharashtra and MP. The Board agreed to continue its support for the project subject to Borrower compliance with the benchmarks by 31 March 1993. Based on documentation submitted by GoI in August 1993, the Borrower appears to have met a substantial part of the benchmarks. The benchmarks still to be fully met are: (i) the completion of socio-economic surveys in Maharashtra, (ii) the acquisition of 2,000 ha of land in MP (about 1,892 ha have been identified but not acquired), and (iii) an inter-State agreement on the sharing of R&R cost. In addition, the R&R plan for Maharashtra is still based on a provisional number of oustees, and that of MP requires improvements with regard to the total number of oustees and those willing to resettle in Gujarat

Project achievements

28. Physical achievements. The Credit/Loan closed 27 months ahead of schedule, on 29 March 1993. By then, the Sardar Sarovar Dam was about 50 per cent complete and the river bed power house about 55 per cent complete, but construction had still not begun on the irrigation bypass tunnel. The Narmada Basin hydro-meteorological network was to be retendered so the national satellite

communication system could compete with the meteor-burst system. Satisfactory operation and maintenance (O&M) manuals had been prepared and an efficient computerised management information system was in place. As for project components not eligible for Bank financing, good progress had been made on the canal head power house and the main canal headworks. However, financing had not been secured for the turbine generating sets for the RBPH, and construction of the Garudeshwar Weir had not yet begun. According to the revised implementation schedule of December 1989, irrigation and power generation could start in the monsoon of 1995. However, this date is likely to slip for the RBPH due to slow R&R implementation in Maharashtra and MP, and delays in financing the turbine generating sets.

29. Notwithstanding some shortfalls in SAR targets, the project set excellent engineering, construction, and management standards by introducing many innovative features. Highlights include the use of a modern concrete plant and two high capacity cable ways, the development and implementation of a comprehensive computerised management information system (MIS), and SSNNL's establishment of a Dam Safety Panel and an Expert Committee on Cement and Concrete.

30. Financial results. At appraisal, the cost of the entire Sardar Sarovar Project, including physical and price contingencies, was estimated at Rs 136,407.0 m. (US $11,367.2 m.) at 1985 price levels, while the estimated cost approved by the Planning Commission, GoI in October 1988, was Rs 64,060.4 m., at FY 1986–87 price levels. During project implementation, the total cost of SSP was re-estimated at Rs 145,985.1 million in 1992 prices, i.e. a 128 per cent increase over the cost approved by the Planning Commission, but only 7 per cent above the appraisal estimate. In US Dollar terms, the revised cost of SSP is estimated at US $5,576.5 million in 1992 prices, compared to the appraisal cost of US $11,367.2 million. The reduction in cost is due to the large devaluation of the Rupee against the US Dollar, which fell from Rs 12 at appraisal to about Rs 26 in 1992.

31. The IDA Credit of SDR 99.7 m. (US $100 m.) and the IBRD Loan (2497-IN) of US $200 m. (equivalent to a total Credit/Loan amount of US $300 m.) were signed on 10 May 1985. Bank Group financing amounted to about 18 per cent of total project costs, net of taxes and duties. No co-financing arrangements were made but it was expected that the six turbine generating sets of the RBPH would be financed through a separate credit arrangement with the Government of Japan.

32. Disbursements started two years late in 1987 after the project had obtained environmental clearance from GoI. Thereafter, disbursements continued to be slow for another two years but picked up momentum in 1990 when the concreting plant and cable ways became operational. The Credit was fully disbursed in September 1992. Only US $18.49 m. (9 per cent) of the Loan had been disbursed when the Borrower decided to terminate the Loan. Thus, only US $151.79 m. (51 per cent)

had been disbursed against the total Credit/Loan amount of US $300 m., by the date the Loan was cancelled.

Revised economic analysis

33. The economic analysis was revised in 1990 and updated again in 1993 for this PCR. Like the SAR, it covers the project's irrigation, power and D,M&I components. It reflects changes in quantities, prices and phasing since 1984 and pays particular attention to environmental costs and benefits (such as estuary changes, fisheries costs and benefits, fuelwood, and wildlife) which might have been underestimated or omitted in the original analysis. The revised analysis has an economic rate of return (ERR) of 12 per cent – slightly lower than the 13 per cent in the original SAR, but within the acceptable range for a project of this type. The scale of benefits is large, relative to any feasible alternatives, with substantial multiplier effects as well. Had the SSP project been conceived as a power project alone, the ERR would have been 14 per cent. The command area, taken alone, exhibits a somewhat lower ERR. If expenditure to date is treated as a sunk cost, the ERR is 17 per cent. The ERR estimate of 12 per cent incorporates a five-year delay in the Narmada Sagar Dam's coming on stream.

Sustainability

34. The Dam and Power Project has adhered to high construction standards. If the Nigam can maintain present levels of quality control/quality assurance, the project should attain the 100-year-plus life expectancy for such projects. However, ultimate sustainability depends upon adequacy of O&M. Though India's track record in this area is not good, the exceptional nature of this project and the standard of excellence developed by SSNNL may be able to chart new directions for O&M.

35. The preparation and implementation of the Narmada Basin Environmental Management Plan are crucial for the long-term sustainability of the project. As part of this plan, the impact of the dam on the downstream communities, fisheries and Bay of Khambat are to be further investigated. Though the Borrower has so far rejected the option of releasing a minimum flow of water in the lower Narmada, the PCR feels strongly that this option should be thoroughly studied and possibly considered prior to the revision of the NWDT Award in 2024. Unless this is done, the environmental sustainability in the Gulf of Khambat, immediately downstream of the dam, can not be guaranteed.

The Bank's and the Borrower's performance

36. Overall, both Bank and Borrower performance were excellent in the field of engineering, but mixed or poor in the R&R, environmental and health fields at the preparation and in the early stages of implementation. The Bank responded to emerging problems by appointing an Independent Review. Though its findings and recommendations were not fully accepted by either the Bank or the Borrower, the Independent Review Report had beneficial effects on their respective efforts, and

they each acknowledged and addressed a number of shortcomings cited in the Report. The Bank improved its policy on disclosure of information while the Borrower improved to a large extent its policies and performance on R&R and environment. However, there remains some inadequacies in R&R planning and implementation particularly in MP. The Bank could have taken a stronger stand in the earlier stages of project implementation to enforce conditionalities in these two fields.

Lessons Learned

A. General lessons

37. The first lesson is the need to ensure that all government clearances have been obtained before negotiations are finalised, particularly in a sensitive area such as environment. The project had not been cleared by the Planning Commission nor by the Ministry of Environment and Forests at the time of Board approval (see Evaluation Summary, Attachment 2). As a result, the lack of clearance led to at least a two-year delay in project implementation. .

38. The second lesson is the need for the Bank to apply more rigorously its guidelines on environment and R&R during the preparation, appraisal and supervision stages. One of the major criticisms that emerged from the Independent Review was that the project had been approved without a proper environmental impact assessment of the dam and power complex, and also without adequate policies and plans for the R&R of project-affected persons. Bank policies on the environment and R&R were still relatively new at the time of project formulation and appraisal and much less stringent than current requirements with respect to operational requirements. Clearly, the Bank's performance was wanting in terms of quality at appraisal of the R&R and environmental plans. The Bank should have taken a stronger stand in the application of its guidelines during the early stages of project implementation. Legal covenants on R&R and environment should be strong and specific and the Bank must also be willing to enforce rigorously its conditionalities, which it failed to do early enough under the project.

39. The third lesson concerns the need to be cautious when financing a large multi-purpose project with a development period exceeding 10 to 15 years. In view of the difficulties in predicting the performance of such projects, it would be wise for the appraisal mission to select smaller, self-contained portions of the overall programme, such as the dam and power complex, or at least to recommend, as is the Bank's current practice, a thorough mid-term review at an appropriate point to enable updating of data on the basis of actual performance and to enable required adjustments to be made. It is difficult to predict the final outcome of a project so far in advance. In the case of SSP, the Bank was over-confident throughout project design and implementation that this project would be implemented as planned. However, this outcome was – and remains – uncertain: (i) there are always unforeseen delays due to technical and financial constraints, (ii) R&R and

environmental issues have and will continue to hamper project implementation, (iii) there is still uncertainty about adequate funding for the Narmada Sagar Dam in MP and its pace of implementation, the final size of the command and the important role that the domestic, municipal and industrial component is likely to play in the future.

B. Lessons on resettlement and rehabilitation

40. The first lesson concerns the need to ensure the quality of the R&R process at project outset. The R&R component could have been improved early on through prior consultations with PAFs to gain a better understanding of land acquisition issues, transaction costs and the special needs of tribal people. The R&R training programme should have been developed with more input from NGOs which had local knowledge of the PAFs. Additionally, a detailed review of the R&R process itself in the project's early stages would have improved performance.

41. The second lesson relates the need to empower the Borrower and executing agencies to take greater ownership for implementation of the R&R process. The Bank should have confirmed that the Borrower and executing States would assume greater ownership of the R&R implementation process. In the case of the SSP, this would have required GoI to supervise the execution of R&R policies while the Bank played an 'arms-length' role, simply holding the Borrower to the terms of legal agreements. The SSP operated in a particularly difficult context, however, because GoI entered into financial agreements with the Bank, while the States actually executed the R&R policies. The alternative of establishing a common authority which in turn would have taken charge of all operational issues as regards R&R was not considered under the NWDT Award and was not possible under the Indian Constitution.

42. The third lesson is with regard to the need for the Bank to develop better communication and decision-making policies for R&R. There were major political difficulties to be faced in achieving R&R objectives across three States. Facilitating communications among the States should have been set as an early priority; decision-making should have been made more explicit and timely. In addition, alternative views should have been both recognised and encouraged. A qualified sociologist/anthropologist should have been hired to handle R&R issues for the Bank. The SSP lacked this needed expertise and authority at the early stages of appraisal and supervision.

43. The fourth lesson concerns the importance of maintaining strong ties with NGOs. Through the development of the R&R policy for the SSP, the Bank recognised the important contribution that NGOs could make towards devising a more acceptable policy. During the early years of project implementation in Gujarat, NGOs played an increasingly useful role. NGO complaints should have been responded to by the Borrower rather than the Bank.

44. The fifth lesson concerns the need to go through a lengthy and sometimes complex learning period. In the case of SSP, neither the Bank nor the Borrower had anticipated the complexity and magnitude of issues that emerged during implementation. Even if the R&R component had been prepared to the standards prevailing at the time of appraisal, it is likely that further adjustments would have been required during implementation. NGOs played an important role in drawing the attention of the Borrower and the Bank to the inadequacies of implementation, and particularly to the importance of assuring equitable treatment for all categories of oustees, whether they be landed, landless or tribal. The legal covenants must therefore be flexible enough to accommodate any improvements to the policies and plans.

45. Finally, the sixth and last lesson relates to the means and procedures required for satisfactory R&R implementation. This project has been the source of many improvements in R&R policies and implementation, which are now being put into practice in other projects. With respect to R&R policies, the principle of land-for-land was implemented on a large scale, adult sons were recognised as separate families, tribal people living in government forest lands were considered as landed families, and the need for additional assistance for subsistence and house construction was recognised. On the implementation side, Gujarat developed a unique mechanism for acquiring replacement agricultural land, at market prices through Land Purchase Committees. Well developed R&R units with central monitoring cells were established. Specific lessons on means and procedures required for satisfactory R&R implementation are given in Part I, paragraphs 16.11 to 16.16.

46. In summary, all future projects with major R&R aspects should conform to the upgraded standards that ensure: (i) baseline studies are adequately completed, (ii) specific project policies are agreed upon, (iii) institutional arrangements have been made for project implementation, and (iv) proper monitoring and evaluation arrangements through independent agencies have been made. Such projects should have a separate R&R component financed under the Credit/Loan. Precise legal covenants should be agreed and enforced. The Bank should use highly qualified R&R specialists to supervise implementation and provide recommendation to implementing agencies. Bank guidelines require such rigour.

C. Lessons on environment

47. While there is increasing evidence that many of the possible negative environmental impacts have been exaggerated by critics and a number of secondary environmental benefits ignored, there are still several important lessons.

48. The first lesson is that the Bank should not approve a project, particularly one that involves a large darn, without a proper environmental impact assessment and management plan. Strict adherence to the Bank's Operational Directive No. 4.01

on environmental assessment should now ensure that environmental issues are addressed more thoroughly and at a much earlier stage.

49. The second lesson is that no project should be approved until a formal environmental clearance is obtained from the Borrower.

50. The third lesson is that, in the case of a project with a large environmental impact, a mitigatory environmental component should be made an integral part of the project supported by strong legal covenants.

51. The fourth lesson relates to the need for the Bank to use highly qualified environmental specialists to supervise the environmental component and, as needed, to provide guidance to the Borrower in the resolution of special environmental issues.

52. The fifth lesson is that ample use should be made of remote sensing and computer modeling technologies for satisfying environmental impacts of large projects such as SSP.

World Bank – Operations Evaluation Department – Memorandum to Executive Directors on the Sardar Sarovar Dam and Power Project, 1995

Source: World Bank, Operations Evaluation Department, Memorandum to Executive Directors on the Sardar Sarovar Dam and Power Project, 1995.

Subject: Project Completion Report on India – Narmada River Development (Gujarat) Sardar Sarovar Dam and Power Project (Credit 1552-IN/Loan 2497-IN)

1. Attached are Parts I and II of the Project Completion Report (PCR) on the Sardar Sarovar Dam and Power Project (SSDPP Credit 1552-In/Loan 2497-IN). Part I was prepared by the South Asia Regional Office. Part II was contributed by the Borrower who, supporting most of the analysis in Part I, disagreed with some of its interpretations and conclusions. Supporting annexes and Part III (228 additional pages) are not included but are available on request. The PCR should be read in conjunction with the PCR on the parallel Water Delivery and Drainage Project (Credit 1553-IN).

2. The SSDPP and the Water Delivery and Drainage Project were part (the first phase) of the Narmada River Development Plan, which aimed to harness the Narmada river through a series of dams and associated investments for irrigation, energy and domestic/municipality/industry (DMI) benefits in three States – Gujarat, Maharashtra and Madhya Pradesh. SSDPP was approved in March 1985 and was due to close in June 1995. From 1989, many national and international non-governmental organisations expressed serious concern regarding

unsatisfactory arrangements for the resettlement and rehabilitation (R&R) of the project affected families (PAFs) displaced by the project as well as for the project's broad environmental implications. The President of the Bank commissioned an Independent Review (IR) of the project in September 1991. The ensuing report in June 1992 was critical of the Bank's and the Borrower's performance, especially with respect to R&R and environmental planning. The IR pointed out that the Bank had not followed its own guidelines, and recommended that the Bank 'step back' from the project rather than continue with financing of its planned implementation. Following a review mission by the Bank in July 1992, Management endorsed most of the IR findings, but considered that the most promising course of action would be to take advantage of Government of India (GoI) agreement to link progress of dam construction to associated R&R requirements. In October 1992 Management presented for the Board's approval a GoI action plan which would address the issues raised in the IR. The Board endorsed the plan which included a commitment by Management to suspend disbursements if the GoI failed to meet specific performance benchmarks by 31 March 1993. However, progress was uneven and on 29 March 1993, GoI requested the Bank to cancel the remaining Loan amount (US $181.5 million of US $200 million). The IDA Credit (US $133.3 million) has been fully utilised by September 1992. At the same time, the Executive Director representing India reaffirmed GoI's continued commitment to complete the project and to meet agreed R&R and environmental standards. Performance under the action plan, as reported by GoI in August 1993 and in subsequent communications, indicates that significant progress has been made in meeting benchmarks but that important R&R requirements have not yet been met (mostly due to difficulties in Maharashtra and Madhya Pradesh). The project implementation schedule, revised in 1989, is unlikely to be met.

3. The basic rationale for the project is sound. The arrangements for basic development of the largest unexploited water resource in India, involving three riparian States, had evolved over a 10-year period through submissions to the Narmada Water Disputes Tribunal (NWDT) of GoI. The comprehensive basin development and management plan that resulted, with its binding inter-State agreements for water allocation and institutional and financial arrangements for programme execution, was a substantial achievement. The scheme promises to make an important hydroelectric energy contribution to India's western grid and to provide water for irrigation and DMI consumption to about 30 million people in the severely drought-prone State of Gujarat, where a wide array of endemic diseases is traceable to malnutrition and unsafe domestic water supply.

4. Although delayed and incomplete, the project has made substantial progress in terms of civil works, not only in quantity of construction but also in the quality of work; achievements are described in detail in the PCR. A slow start-up associated with delayed GoI environmental and forestry clearances and the need to import and mobilise major construction equipment led to a revised construction schedule in 1989 with a completion date of 1998.

5. Positive aspects of physical implementation have been overshadowed by major deficiencies in planning and management of R&R and environmental impact analysis. In brief, the Bank did not follow its own R&R guidelines (OMS 2.33 of 1980 and OMS 2.34 of 1982) designed to ensure appropriate treatment of landed and landless PAFs. The NWDT agreement had obliged Gujarat to pay to Madhya Pradesh and Maharashtra all costs of R&R for oustees (persons cultivating, residing or working in the submerged area), with replacement agricultural land being provided for 'landed' oustee families. An R&R plan was prepared and approved at negotiations but it did not meet the requirements of existing Bank guidelines. Insufficient attention was afforded to R&R in 1989, and progress picked up arrangements. A Bank mission provided increased attention to R&R in 1989, and progress picked up after 1990, with further advances after the IR was issued in 1992. Approximately 41,000 oustee families were identified against 10,700 in the SAR which did not have the benefit of adjusted surveys and had excluded certain categories of oustees (adult sons, landless and encroachers). However, substantial obstacles still remain. While Maharashtra has recently adopted satisfactory policies, implementation problems persist in that about half of the oustee villages refuse to cooperate to allow final definition of numbers and R&R arrangements. In Madhya Pradesh, the rehabilitation grant to landless families and adult sons is less attractive than in the other States and implementation capacity is weak. An earlier assumption that most PAFs from this State would move to Gujarat is now in question, possibly obliging considerable additional land procurement, a process which has been cumbersome in the past. NGOs have been usefully involved in the Gujarat R&R programme, but in the other States confrontation continues between authorities and some local NGOs. The persistence of difficulties is likely to affect the pace at which the infrastructure programme (and subsequent reservoir filling) can proceed.

6. The Bank did not follow its own environmental guidelines which came into existence before the final appraisal mission in August 1984 (OMS 2.36 of May 1984). This OMS required environmental impact studies to have been done and an evaluation made at appraisal of measures to be taken to avoid or mitigate serious environmental risks. The SAR included a brief description of the anticipated environmental consequences, and indicated that a work programme for assessing and managing fisheries, forestry, wildlife and public health effects would be undertaken and submitted to the Bank for its approval. Project costs did not appear to specifically include any environmental programme except as a training component and as a watershed management study, and environmental measures were not financed under the loan/credit. The Board approved the project in March 1985 with legal documents requiring the Environmental Work Plan to be submitted for Bank approval by December 1985. The formal GoI forestry clearance had not been obtained at that time. The forestry clearance was obtained only in September 1987 and an environmental clearance (which became necessary by a GoI 1986 Conservation Act) was obtained in June 1987; both clearances were given subject

to a number of important conditions in view of the unavailability of appropriate plans. Deadlines in 1985 and 1989 for completion of the Environmental Work Plan were allowed to lapse and ultimately the Plan was submitted in August 1993 following severe criticism in the IR of 1992 and subsequent studies by ODA-funded consultants.

7. Following the IR, the Borrower made significant efforts on a number of environmental fronts. The compensatory forestry programme has progressed, catchment area treatment is underway, and plans for preserving cultural sites and protecting wildlife are in various stages of implementation. After a long delay, each of the States has carried out in-depth baseline health surveys in the project area and set up health care facilities for resettled and permanent populations; State programmes for malaria control have been strengthened, although staffing and funding of these operations is not yet satisfactory. Prompted by the IR, consultants re-examined the projected sedimentation rate in the Sardar Sarovar Reservoir. This study confirmed that sedimentation would be greater than originally calculated, but that it would not be a major problem for the combined Sardar Sarovar and Narmada Sagar complex. Concerns about the impact on the estuary downstream of the dam were investigated by the ODA-funded consultants in 1993. This indicated that changed flow patterns are likely to affect the spawning of hilsa, but further work is needed to determine the significance of hilsa in marine catches by fishermen; a local institute continues to work on this matter. Continued implementation of the Environmental Work Plan, vigilance in monitoring and responding to fisheries and socioeconomic developments in estuarine areas, and increased support for the health programme could yet help avoid significant detrimental effects.

8. The PCR re-estimates the economic rate of return (ERR) on the projects' irrigation, power and DMI water investments to be 12 per cent, compared with 13 per cent in the SAR. The analysis includes the costs and benefits of Gujarat Water Delivery and Drainage Project as an integral part of the first phase in the basin development as in the SAR. It also includes environmental costs and benefits and R&R costs which were not adequately covered in the SAR, adjustments for actual construction costs and updated future costs, a share of the costs of upstream Narmada Sagar Dam which is required to enable regulation of water for irrigation and power generation up to the quantities projected (but with an assumed 5-year delay in construction and consequent reduced power generation), a higher (more realistic) value for DMI water delivery, and an assumption that some irrigation and (very minor) energy benefits would commence in 1995. However, despite the approximate adherence to the 1989 construction schedule so far, further delays of at least two years in both the irrigation and energy generation schedules are likely. The pace of construction must be adjusted to progress in the R&R programme which has encountered persistent difficulties (para. 5). Floods in 1994 damaged the stilling basin which will need major reconstruction. There are uncertainties in financing arrangements for the overall development and for the foreign exchange

purchase of turbine generators for the river bed power house (after the cancellation of an Overseas Economic Cooperation Fund (OECF) loan by the Government of Japan). Gujarat and Madhya Pradesh have not yet agreed on the invert level for the irrigation bypass tunnel (lowering it would benefit Gujarat's diversions, particularly in drought years, but at the expense of Madhya Pradesh's rights), which has delayed initiation of tunnel construction. This has meant that the base case ERR scenario prepared for the PCR is already out of date. Sensitivity analysis shows that a delay of four years in the commissioning of the river bed power house would reduce the ERR by a further one per cent, and some other delay scenarios continue to provide a return in excess of 10 per cent (...). The stability of the ERR projections in the PCR scenarios is due to the multi-purpose nature of the investment and to benefit delays being associated with investment cost delays. Nevertheless, the robustness of the ERR calculations is clouded by many uncertainties.

9. The PCR is based on a hydrological model which a team of experts recently appointed by the GoI has not endorsed in full (see Memorandum on the Water Delivery and Drainage Project). PCR scenarios do not examine the economic implications of a failure to construct the complementary Narmada Sagar Dam. This dam is an integral part of the basic development package agreed by the three States, and the construction contract has already been let (May 1992), but the fiscal implications of its development are substantial and its resettlement dimensions even larger than those of Sardar Sarovar. Delays in the completion of Narmada Sagar due to financing and/or R&R difficulties cannot be ruled out. To examine the implications of the omission of Narmada Sagar Dam, a new water modelling exercise would have to be undertaken in which the alternative of a series of upstream smaller dams to complement the Sardar Sarovar investment would have to be considered; major changes could be expected in the amount of water available for energy and for the command area in Gujarat which would, of course, influence the scale of the water distribution investment which could be made in that area. In the absence of such a model, the PCR uses a proxy to reflect the implications of extended delays in the completion of Narmada Sagar – 25 per cent reduction in power and 20 per cent reduction in irrigation benefits throughout the project life. Consequently, while the alternative scenarios appear to confirm the economic viability of the overall investment, uncertainty remains concerning the hydrological assumptions and the scale and sequencing of project benefits.

10. Institutional improvements were substantial in some areas of the project but weak in others. Overall, the institutional development impact is rated as modest. Certainly, advances in engineering design and construction quality control were significant, and an effective management information system to monitor the civil work component was established and is being expanded to financial and administrative aspects. Good progress has been made in operation and maintenance (O&M) plans, although these will need to be updated when the remote control and monitoring system for the operation of the main canal is available (the consultancy

contract had not been signed at Loan cancellation, and bilateral technical assistance is being sought) and there is no evidence of institutional development in terms of formation of local irrigation groups or of providing such groups with financial autonomy. While the planning and implementation capacity of R&R has been improved, implementation still needs fundamental improvement in Maharashtra and Madhya Pradesh, and health services continue to be affected by inadequate funding and staffing.

11. Bank appraisal and supervision performance has been unsatisfactory. Adherence to the Bank's guidelines could have avoided many of the negative consequences of the project – social unrest, diversion of resources (by GoI and the Bank) to address the consequences of inadequate early attention to environmental and R&R requirements (especially in PAF consultation), delayed benefits, and funding difficulties including the cancellation of the Bank and OECF loans. As a development investment, however, the project outcome is rated as marginally satisfactory based on PCR assumptions. This rating is affected by more than the usual uncertainties. It assumes that the hydrological parameters of the project will be fully funded and the complementary Narmada Sagar Dam constructed. On this basis, the ERR is likely to be in excess of 10 per cent. The substantial benefits to be obtained from this major investment of over US $1.5 billion through energy generation and DMI distribution, as well likely irrigation benefits in the command area (see Memorandum on the Water Delivery and Drainage project), should be sufficient to outweigh the significant negative aspects of the projects and to compensate for all resettlement costs based on Bank guidelines. It is emphasised, however, that the likely outcome is less favourable than anticipated at appraisal, and has been affected by many avoidable social costs and deficiencies, especially in the areas of resettlement and of environmental planning, on both of which decisive action is still pending.

12. The sustainability of the project achievements is rated as uncertain. The project has adhered to high construction standards and the institutional capability in this regard is now well established. But much will depend on effective O&M for which India does not have a good track record. Action is needed to staff, equip and train an appropriate O&M unit in advance of the dam and power complex operation. The hydrometeorological monitoring network which is needed for dam safety, flood control and efficient reservoir operation had not been tendered at the time of the PCR. Realisation of the Environmental Management Plan will be crucial for sustainability, especially with respect to monitoring, and responding to, the effects of the scheme on downstream fisheries and communities. These effects, *inter alia*, will need to be addressed in the revision of the NWDT Award in 2024. The Borrower is adamant that it will continue with the Narmada River Basin Development Plan, but, as previously noted, the rate of progress will depend on the Borrower's ability to deal with the social dimensions of the plan and to allocate to it the necessary managerial, financial and technical resources.

13. The PCR provides a detailed assessment of project implementation and shortcomings. With few exceptions (noted below), it is comprehensive, and it acknowledges major performance shortfalls. Concerns of the Executive Directors (EDs) at the 1985 Board presentation are noted, as is the commitment by Management to keep the Board informed on the social and environmental aspects of the project through written reports and briefings. The PCR deals with the major issues raised by the IR. The extent of compliance by GoI with the action plan agreed in late 1992 (as a basis for continued Bank support) is elaborated. As indicated earlier, significant uncertainty about the rate at which investments will be completed, and possible variations in hydrological parameters and water use scenarios, limit the robustness of PCR estimates of the ERR. The PCR fails to assess the scope or usefulness of M&E data and socioeconomic surveys as baseline information for impact analysis. In the discussion on 'Bank Performance', the PCR does not address the missed opportunities in potential use of selected, reputable NGOs to assist it in meeting the Bank's guidelines in this aspect of the project. While the PCR indicates that the Bank should have taken 'stronger action' to correct the R&R and environmental deficiencies during implementation, it does not make a forthright judgment that the Bank's enforcement of covenants was delinquent.

14. In its part of the PCR the Borrower is more confident that the Bank in predicting some aspects of future performance and scheduling, and highlights the high socioeconomic benefits which will be obtained by the intended provision of both irrigation and DMI water in the drought-prone areas of Gujarat.

15. Considering the extraordinary amount of investigation and lesson learning which has taken place with respect to environmental and resettlement dimensions of the project, there would be little value in carrying out further evaluation of these aspects at this stage. A performance audit is planned, however, in order to assess the realism and the sensitivity to risks of the re-estimation of project costs and benefits. In due course, OED will undertake impact evaluations of socioeconomic effects (including health aspects) on PAFs and the impact of basin development on command area populations.

Chapter 9

International Documents

Official Documents

International Labour Organisation Convention No. 107, 1957

Source: Convention No. 107 Concerning the Protection and Integration of Indigenous and Other Tribal and Semi-Tribal Populations in Independent Countries, Geneva, 26 June 1957 (extracts).

Part II. Land

Article 11

The right of ownership, collective or individual, of the members of the populations concerned over the lands which these populations traditionally occupy shall be recognised.

Article 12

1. The populations concerned shall not be removed without their free consent from their habitual territories except in accordance with national laws and regulations for reasons relating to national security, or in the interest of national economic development or of the health of the said populations.

2. When in such cases removal of these populations is necessary as an exceptional measure, they shall be provided with lands of quality at least equal to that of the lands previously occupied by them, suitable to provide for their present needs and future development. In cases where chances of alternative employment exist and where the populations concerned prefer to have compensation in money or in kind, they shall be so compensated under appropriate guarantees.

3. Persons thus removed shall be fully compensated for any resulting loss or injury.

ILO, Individual Observation Concerning Convention No. 107, 1998

Source: Committee of Experts on the Application of Conventions and Recommendations: Individual Observation Concerning Convention No. 107, Indigenous and Tribal Populations, 1957 India (ratification: 1958), 1998 (extracts).

4. On the question of the Sardar Sarovar Dam and Power Project, the Committee recalls that thousands of tribal people are being displaced from their homes by this very large project. Over a number of years, the Conference Committee and the present Committee have requested the government to take urgent measures to bring its resettlement and rehabilitation policies for tribal people into line with the Convention. The Committee had noted that the information provided by the government on the progress of the resettlement and rehabilitation of the tribal populations affected by this project up to April 1996 indicated that substantial differences continue to exist among the States of Gujarat, Maharashtra and Madhya Pradesh.

5. The Committee requested information on the progress of rehabilitation and resettlement policies of the three above-mentioned States and on the manner in which the allocation of resettlement land takes into account the amount of land previously occupied by the displaced tribal population (the legal concept of 'traditional occupation'), including any measures taken or envisaged to compensate for different kinds of land use. The Committee remains concerned by the difficulty encountered in acquiring land for resettlement and providing compensation, in particular in Maharashtra and Madhya Pradesh. It requests the government to keep it fully informed of the progress achieved in this case.

6. The Committee takes note of the explanations provided by the government representative in the Conference Committee, in particular that the differences in the progress of resettlement and rehabilitation between the various States were explained by the fact that measures were only taken as the project advanced every year, and the construction of the dam was linked to the implementation of resettlement and rehabilitation measures in order to ensure successful rehabilitation before the lands in question were submerged. The Committee also notes that the progress of rehabilitation is evaluated by a rehabilitation committee, under the chairmanship of the Ministry of Welfare and reporting to the Supreme Court of India. Families affected by the project are given priority which accounts for the substantial differences between States. The Committee requests the government to keep it fully informed of the development of this project and the resettlement and rehabilitation of the people concerned, as well as of the progress achieved by the States in acquiring land for these purposes. It also requests the government to provide a detailed report on the compensation measures taken in each State and on the number of people who have already been settled and rehabilitated. Please also

provide information on the number of people which the government expects are still to be displaced.

ILO, Individual Direct Request Concerning Convention No. 107, 2005

Source: ILO, Committee of Experts on the Application of Conventions and Recommendations: Individual Direct Request concerning Convention No. 107, Indigenous and Tribal Populations, 1957 India (ratification: 1958), 2005 (extracts).

12. For some years the Committee has been raising the question of the Narmada Valley hydroelectric project, that has displaced many thousands of people already, and that will displace many more in the future. It notes in this regard a decision by the Supreme Court of India in October 2000 that construction of the dam could proceed, in spite of widespread concern over its effects. The Court's order stated, *inter alia*:

> 2. As the Relief and Rehabilitation Sub-group has cleared the construction up to 90 metres, the same can be undertaken immediately. Further raising of the height will be only *pari passu* with the implementation of the relief and rehabilitation and on the clearance by the Relief and Rehabilitation Sub-group. The Relief and Rehabilitation Sub-group will give clearance of further construction after consulting the three Grievances Redressal Authorities.
>
> 5. The reports of the Grievances Redressal Authorities, and of Madhya Pradesh in particular, show that there is a considerable slackness in the work of identification of land, acquisition of suitable land and the consequent steps necessary to be taken to rehabilitate the project oustees. We direct the States of Madhya Pradesh, Maharashtra and Gujarat to implement the Award and give relief and rehabilitation to the oustees in terms of the packages offered by them, and these States shall comply with any direction in this regard which is given either by the NCA or the Review Committee or the Grievances Redressal Authorities.

13. The government has provided details in its report indicating that a majority of the more than 40,000 'project-affected families' has now been resettled, or that provision has been made for their resettlement, as the number of people displaced by rising water levels continues to rise. Concern continues to be expressed, however, that resettlement has resulted in breaking up communities, and in compensation in amounts of land lower than the amounts previously available for use by many of these communities, and that the availability of suitable land is more limited than the need. The committee requests the Government to provide further information on the progress of resettlement in its next report.

14. In addition, as earlier noted, this hydroelectric project is only one of many development projects in India resulting in the displacement of tribal communities from their lands, and the information available to the Committee indicates that not all such projects have been subject to the same kind of rehabilitation plans as the

Narmada Valley project. It notes the information provided in the report with respect to two specific projects, and requests the government to provide in its next report a comprehensive overview of the situation in this regard around the country.

United Nations Commission on Human Rights, Report of the Special Rapporteur on the Situation of Human Rights and Fundamental Freedoms of Indigenous People, 2003

Source: United Nations, Commission on Human Rights, Fifty-ninth session, Human Rights and Indigenous Issues – Report of the Special Rapporteur on the Situation of Human Rights and Fundamental Freedoms of Indigenous People, Rodolfo Stavenhagen, submitted in accordance with Commission Resolution 2001/65, UN Doc. E/CN.4/2003/90 (extracts).

44. The Sardar Sarovar Dam in India is the largest of 30 large, 135 medium and 3,000 small dams to be built to harness the waters of the Narmada river and its tributaries, in order to provide large amounts of water and electricity for the people of Gujarat, Maharashtra and Madhya Pradesh. With a proposed height of 136.5 m, the government claims that the multipurpose Sardar Sarovar Project (SSP) will irrigate more than 1.8 million hectares and quench the thirst of the drought-prone areas of Kachchh and Saurashtra in Gujarat. Others counter that these benefits are exaggerated and would never accrue to the extent suggested by the government. Instead, the project would displace more than 320,000 people and affect the livelihood of thousands of others. Overall, due to related displacements by the canal system and other allied projects, at least 1 million people are expected to become uprooted or otherwise affected upon completion of the project. Indeed, the development surrounding the Narmada river has been labelled 'India's greatest planned human and environmental disaster', a far cry from former Prime Minister Nehru's idealisation of dams as the 'secular temples of modern India'.[1]

45. Two thirds of the over 40,000 families expected to be displaced by the reservoir's creation will be tribal people or adivasis, belonging to different groups collectively referred to as Bhils. Displacement of adivasis from their traditional lands and resources due to the creation of reservoirs, canals and reforestation projects significantly impacts on the ability of adivasis to fully enjoy their human rights. They live mainly in 14 villages in Gujarat, 33 in Maharashtra and around 53 in Madhya Pradesh. The adivasis are largely self-sufficient, growing their own food and collecting fuel, building materials, fodder, fruits and other resources from the forests and common lands around their villages, as well as relying on water and fish from the river. Resettlement away from their territory means the destruction of

---

1 John R. Wood, 'India's Narmada River Dams – Sardar Sarovar under Siege', 33/10 *Asian Survey* 968 (1993).

their lifestyles and village organisation. One farmer whose village will be submerged commented: 'the forest is our moneylender and banker. From its teak and bamboo we built our homes. From its riches we are able to make our baskets and cots. (...) From its trees we get our medicines'.[2]

46. In the early 1990s opponents of the dam staged a series of non-violent protests (*dharnas* and *satyagraha*), prompting the World Bank, after commissioning an Independent Review which underlined the flaws in the project, to withdraw its remaining funding for it (the Bank cancelled $170 million remaining on its loan of $450 million). Work on the dam continued nevertheless, despite attempted judicial restraint, and by the summer of 2002 the water level in the reservoir rose much higher than initially expected, threatening many more people and villages with flooding. The government's rehabilitation and resettlement measures for 'oustees' (displaced persons) appeared to be insufficient, generating a number of protest activities by the affected villagers within the rising waters themselves. Protest against the project has remained strong and the Narmada Bachao Andolan NBA) (Save the Narmada Movement) has been particularly instrumental in fostering awareness and dissent. Many activists and tribal people continue to maintain that they will never abandon their land to the dam, even if it means *doobenge par hatenge nahi*: death by drowning.[3]

47. Multipurpose dams surely stimulate economic activity and have the potential for bringing benefits to large sectors of the population. The problem is whether these benefits are designed to reach the indigenous peoples who provide the land on which such projects are established, and how. It is estimated that the SSP will enable the irrigation of 1.8 million hectares of land in Gujarat alone. Irrigation facilitates the production of food and other crops, which could significantly improve food production in drought-prone areas. However, it appears that much of this area is unsuitable for irrigation because of waterlogging and salinisation. Moreover, some of the designated water is likely to be consumed by sugar plantations before reaching more needy farms further away from the dam. Other potential benefits of the irrigation scheme and electric power generation from the dam are unlikely to benefit the adivasi population.[4]

48. Adivasis were not involved nor consulted in the dam construction process, on the premise that the project and the displacement of people was to serve a 'public purpose' which would provide a 'development opportunity' to the affected population. While some local governments did involve non-governmental organisations, an observer notes that 'while NGOs can play an important

---

2 Amrita Patwardhan et al., 'Dams and Tribal People in India', paper contributed to the World Commission on Dams, prepared for thematic review 1 February 2000, p. 13.

3 Wood, *op. cit.*, p. 978.

4 Patrick McCully, 'Sardar Sarovar Project: An Overview', May 1994, reproduced at www.narmada.org/sardar-sarovar/irnoverview940525.html.

supportive role they cannot substitute [for] the voice of the affected people, nor can they replace what is the basic responsibility of the State'.[5]

49. Only the adivasi population who live in the area that will be submerged in the reservoir (considered as project affected peoples, or PAPs) are eligible for compensation and resettlement. However, many more will be affected indirectly, yet they are often not considered as PAPs and are therefore ineligible for rehabilitation. This would include people on islands that would be marooned and in areas affected by canals, dykes, the creation of a new wildlife sanctuary and a reforestation scheme to compensate for tree loss, and resettlement schemes on traditional adivasi lands. Adivasi territory has also been affected by the construction of a colony to house the workers and officials engaged in the construction work and administration of the dam. All of these secondary consequences have displaced adivasi villages and affected their lives and livelihoods. Patwardhan comments that 'displacement needs to be viewed as a process rather than an event which starts much before the actual physical displacement and continues for a long time after uprooting has taken place', and concludes that the current situation is symptomatic of the 'gross underestimation of the human costs of large dams'.[6]

50. Whereas State governments have offered comprehensive resettlement and compensation packages to 'landless' adivasis displaced from their homes, observers point out that in practice adivasis have not fully benefited from them. The promised lands in Gujarat did not materialize or were of poor quality, whereas in Madhya Pradesh the government had no resources to resettle displaced adivasis. Moreover, resettlement has been delayed for many years and it is reported that 75 per cent of the displaced people have not been rehabilitated.[7] To the extent that the law does not recognise customary rights to land and that therefore adivasis may be considered 'encroachers' on government land, they have not received adequate compensation for their losses. In common with other indigenous peoples, adivasis have a unique and close relationship with the land and its resources. Compensation packages treat land as property, whereas for adivasis, their land is intrinsically linked to their culture and livelihood. It appears that the government has failed to deal with the numerous non-quantifiable losses experienced due to the dam such as loss of access to religious sites and social disintegration.[8] Displacement due to the SSP has led to the fragmentation of adivasi communities as well as loss of cultural identity. Resettlement areas are often unsuited to the communal lifestyle of adivasis, particularly if they have been resettled in communities of non-tribal people who reject the tribal way of life or have had to move to the cities.[9]

---

5 Patwardhan, *op. cit.*, p. 11.
6 *Ibid.*, p. 12.
7 *Ibid.*, and McCully, *op. cit.*
8 Patwardhan, *op. cit.*, p. 9; see in particular the section by Chaube in Patwardhan.
9 *Ibid.*, p. 19.

51. Involuntary displacement readily leads to a violation of several economic, social and cultural rights. Despite claims to the contrary, resettled adivasis have generally had to suffer a reduction in their standard of living, the loss of livelihood resources, and a reduction of health standards, a situation that stands counter to articles 11 and 12 of the International Covenant on Economic, Social and Cultural Rights. While in displaced communities government has established schools for the population, there are reports that due to economic hardship many children cannot afford to stay in school, whereas the curriculum appears to be ill-adapted to the cultural and language needs of adivasi children (Article 13).[10] There have also been reports of violence and the use of force by the police upon protesters and resisters to displacement, in violation of the International Covenant on Civil and Political Rights.[11] The NBA recently called for protest of the decision by the Narmada Control Authority[12] in May 2002 to allow the dam height to rise to 95 m even though over 35,000 families displaced when the dam height reached 90 m have still not been resettled.[13] In a recent urgent appeal to the Prime Minister of India the Habitat International Coalition reported that 'submergence due to the monsoons and raising the dam's height have destroyed the crops and homes of SSP-affected villages in Nandurbar District (Maharashtra) and Jhabua District (Madhya Pradesh), rendering the villagers homeless. These people now face a severe food and drinking-water shortage'. It also reports that the Maharashtra Government indicates an increase in the number of project-affected persons at the 95 m level, and admits that the government does not have enough land for rehabilitation of the affected persons.[14]

52. The Sardar Sarovar dam and other similar projects on the Narmada river raise a number of complex issues. Originally, the interests and aspirations of the affected adivasi population were not considered in the project design and implementation. As a result of continued lobbying by tribal and human rights organisations, the Government of India now recognises that the issues raised by the affected communities must be taken into account. Yet the implementation of measures intended to mitigate the negative effects and increase the benefits of the project for the adivasi population has lagged behind and is considered insufficient by the people involved. The Special Rapporteur recommends that the human rights of the adivasis be included as a foremost priority in the implementation of this development project and others of its kind. Only with the full and informed consent of the tribal people concerned will truly human rights-centred development, as

---

10 *Ibid.*, p. 22.

11 *Ibid.*, pp. 7–8.

12 NCA was established by the Narmada Water Disputes Tribunal to oversee the implementation of the dam. NCA is comprised of government officials from Gujarat, Maharashtra, Madhya Pradesh and Rajasthan.

13 http://www.narmada.org/sardarsarovar.html.

14 Letter dated 22 November 2002 from Joseph Schechla, Coordinator, Housing and Land Rights Network, Habitat International Coalition, to the Primer Minister of India.

recommended by the General Assembly, become possible. An immediate step would be to halt any further rise in the reservoir's water level until the outstanding issues of rehabilitation and resettlement are fully solved to the satisfaction of the affected population, through constructive dialogue and negotiation between the parties. India could also signal its commitment to the human rights of its adivasi population by ratifying ILO Convention No. 169 and approving the draft United Nations Declaration on the Rights of Indigenous Peoples. Alternative ways of involving the adivasis in the project should also be considered. It has been suggested that they should be considered as partners in the project, with their investment being their natural resources. Adivasis qua investors would be entitled to share in the project's benefits.[15]

Independent Assessments

Lawyers Committee for Human Rights Report, 1993

Source: Lawyers Committee for Human Rights for the Narmada International Human Rights Panel, Unacceptable Means – India's Sardar Sarovar Project and Violations of Human Rights, 1993.

I. Introduction

Sardar Sarovar, a massive dam under construction on the Narmada river, is designed to provide hydro-electric power, irrigation water and drinking water to western India. More than 100,000 people, in 245 villages, will be affected by the submergence caused by Sardar Sarovar; another 140,000 families are expected to be affected by the resulting canal and irrigation system.[1] Most of those in the submergence area are tribal Indians or adivasi, whose culture and way of life are threatened by resettlement. NGOs in India, led by the NBA have challenged the project because of its adverse social, economic and environmental impact.

Since 1989 there have been numerous serious violations of civil and political rights occurring in the context of the Sardar Sarovar Project.[2] Those engaged in peaceful protests have been unlawfully arrested and detained. Police have used excessive

---

15 The information on the Sardar Sarovar project is taken from numerous documents on file at OHCHR and, among other sources, at www.narmada.org.

1 'Sardar Sarovar', Report of the Independent Review (Resource Futures International, Ottawa: 1992), p. xii. [This part of the Report of the Independent Review is reproduced above at page 331.]

2 See generally Asia Watch, 'Before the Deluge: Human Rights Abuses at India's Narmada Dam', 17 June 1992.

force in breaking up demonstrations and acts of civil disobedience. There are many reports of the torture or mistreatment of those arrested. As grassroots opposition to the project has become better organised, repression by the Indian authorities has intensified. NGOs that have organised opposition to the dam have been subjected to politically motivated harassment and intimidation and their members subjected to unlawful arrest.

State authorities have acted in violation of India's obligations under international law, including the International Covenant on Civil and Political Rights. They have also violated the UN Code of Conduct for Law Enforcement Officials and the UN Basic Principles on the Use of Force and Firearms by Law Enforcement Officials. While the Lawyers Committee does not adopt a position either for or against the Sardar Sarovar Project, it holds the State Governments in Gujarat, Maharashtra and Madhya Pradesh, as well as the Government in New Delhi, responsible for human rights violations committed in bringing about its completion.

International attention has focused on the project since 1985, when the World Bank approved a loan of $450 million for the construction of Sardar Sarovar, of which approximately $280 million has been disbursed. Because of widespread international concerns raised by the project, the World Bank commissioned an Independent Review, known as the Morse Commission, to undertake a comprehensive assessment of Sardar Sarovar as regards the resettlement and rehabilitation of persons affected or displaced by the dam and the project's environmental impact. In a report issued in June 1992, the Morse Commission found obvious flaws in both the social and environmental effects of the project and called on the World Bank to 'step back from the Projects and consider them afresh'.[3]

A World Bank memorandum of 11 September 1992 rejected either a suspension or reappraisal of the Sardar Sarovar Project and urged ongoing support, subject to 'continued improvement over the next six months, to be monitored against actions already agreed and benchmarks for assessing progress in implementation'.[4] On 23 October 1992 the majority of the Bank's Board of Directors voted to continue funding the project with these stipulations. Six of the Bank's directors, representing 42 per cent of the vote, called unsuccessfully for a suspension of funding.

The performance benchmarks announced by the World Bank may have had the effect of exacerbating the human rights situation in the Narmada Valley. The benchmarks, described in a World Bank memorandum of 19 November 1992, call on the Government of India to show by 31 March 1993, 'satisfactory improvement

3 'Sardar Sarovar', supra note 1, p. xxv.

4 The World Bank, 'Sardar Sarovar (Narmada) Projects: Review of Current Status and Next Steps', 11 September 1992, p. 15.

of the data describing the number of project affected people (PAPs) in the reservoir area, including an analysis of affected tribal people' and 'satisfactory demonstration of improved consultation practices'. According to the memorandum, should the Bank management conclude, on the basis of a review mission to India in April 1993, that 'project performance has failed substantially to meet the benchmark tests, steps would be taken to suspend disbursements'.

State authorities in India appear to have accelerated the pace of surveying in villages scheduled for submergence to meet the benchmark for improved data on the number of project affected people. Likewise, they have promoted 'information camps' in an apparent effort to improve 'consultation'. Groups opposed to the Sardar Sarovar Project have expressed concern regarding the surveying efforts and the information camps. They contend, for instance, that the surveying is typically done in a heavy-handed manner to convince the villagers that their resettlement is imminent and inevitable. They also assert that the information camps have been *pro forma*, with publicity shots and attendance by important politicians taking the place of genuine discussions and dissemination of information.

In its efforts to fulfil the performance benchmarks set out by the World Bank, the Indian authorities have resorted to 'unacceptable means' and violations of human rights have multiplied. Peaceful demonstrators in Kakrana in October 1992 and in Bombay in November 1992 were beaten by police. Officials in Baroda unjustifiably denied a rally permit to an anti-dam organisation in late December and then arrested demonstrators. Persons protesting government 'information camps', designed to promote the consultation process required by the World Bank, have been unlawfully arrested and detained. Tribal villagers in Anjanwara, Madhya Pradesh who are resisting resettlement were beaten and fired upon. Members of an NGO in Madhya Pradesh have been the target of official harassment, politically motivated arrest, and torture in detention. The Lawyers Committee urges the Federal and State Governments in India to conduct independent and impartial investigations of human rights violations and prosecute those responsible.

On 30 March 1993 the Government of India asked the World Bank to cancel the undisbursed portion of the World Bank loan for the Sardar Sarovar Project. According to a World Bank press release, the Indian government stated that although 'significant progress has been made' in meeting performance benchmarks for resettlement and rehabilitation and the environment 'despite difficult internal complications unrelated to the project, (...) under the circumstances, it wished to proceed with the project on its own, without the financial assistance under the Bank loan'.[5]

In recent years the World Bank has recognised 'good governance' as essential to sustained development. According to the World Bank's 1992 report 'Governance

---

5 World Bank, 'Indian Government Cancels World Bank Loan For Sardar Sarovar (Narmada) Project', Bank Release No. 93/651, 30 March 1993.

and Development', poor governance includes the '[f]ailure to establish a predictable framework of law and government behaviour conducive to development, or arbitrariness in the application of rules and law'.[6] Ibrahim F.I. Shihata, World Bank Vice-President and General Counsel has written:

> The close relationship between environment, human rights and development has been well recognised at long last in the Bank and elsewhere. (...) The enjoyment of civil and political rights and the accomplishment of economic development both constitute necessary elements for human progress.[7]

These concerns are especially important with respect to the Sardar Sarovar Project because the alleged violations of human rights are not merely a general country concern, but have been intimately connected to the project itself.

Unless the Federal and State Governments in India take appropriate action, human rights violations in the context of the Sardar Sarovar Project are likely to continue. It is important that the World Bank, after years of funding Sardar Sarovar, not abdicate responsibility for the project's impact on human rights. The Lawyers Committee urges the World Bank to ensure Indian government compliance with international human rights norms in the Narmada Valley and elsewhere. Sardar Sarovar is only one of many World Bank funded projects in India. Continuing human rights violations in the context of Sardar Sarovar should compel the Bank, in the context of its requirements for 'good governance', to reconsider its other efforts in India.

II. Cases [not reproduced]

III. Conclusions

India is party to a number of important human rights treaties, including the International Covenant on Civil and Political Rights[36] and the Convention against Torture and Other Cruel, Inhuman or Degrading Treatment or Punishment.[37] As a member of the United Nations General Assembly it has endorsed the UN Code of Conduct for Law Enforcement Officials.[38] In the context of the Sardar Sarovar Project, Indian authorities have repeatedly violated these international standards.

---

6 The World Bank, 'Governance and Development' (Washington: 1992).

7 Ibrahim F.I. Shihata, 'The World Bank and Human Rights: An Analysis of the Legal Issues and the Record of Achievements', paper presented to the International Third World Legal Studies Association Panel, Miami, 8 January 1988.

36 General Assembly Resolution 2200 A (XXI), 16 December 1966. India acceded to the Covenant in 1979.

37 General Assembly Resolution 39/46, 10 December 1984.

38 General Assembly Resolution 43/173, 9 December 1988.

a) Restrictions on the Right to Peaceful Assembly

Article 21 of the Covenant on Civil and Political Rights recognises the right of peaceful assembly: 'No restriction may be placed on the exercise of this right other than those imposed in conformity with the law and which are necessary in a democratic society in the interests of national security or public safety, [or] public order ("ordre public")'.

The authorities have unjustifiably denied or prevented peaceful anti-dam demonstrations outside government buildings or official meetings. The denial of a permit to the NBA in Baroda in late December 1992 and the arrests of demonstrators in Omkareshwar in November 1992 appear to be politically-motivated efforts to prevent the NBA from peacefully assembling.

b) Excessive Use of Force

The UN Code of Conduct for Law Enforcement Officials states that '[l]aw enforcement officials may use force only when strictly necessary and to the extent required for the performance of their duty'.[39] The Commentary to the Code of Conduct emphasizes that the use of force should be 'in accordance with a principle of proportionality'.[40] The UN Basic Principles on the Use of Force and Firearms by Law Enforcement Officials state that:

> Law enforcement officials, in carrying out their duty, shall, as far as possible, apply non-violent means before resorting to the use of force and firearms. They may use force and firearms only if other means remain ineffective or without any promise of achieving the intended result.[41]

Furthermore:

> [i]n the dispersal of assemblies that are unlawful but non-violent, law enforcement officials shall avoid the use of force or where that is not practicable, shall restrict such force to the minimum extent necessary.[42]

Even when involved in 'the dispersal of violent assemblies, law enforcement officials may use firearms only when less dangerous means are not practicable and only to the minimum extent necessary'.[43]

The Indian police forces have acted in violation of the Code of Conduct for Law Enforcement Officials and the Basic Principles on the Use of Force and Firearms by Law Enforcement Officials. In Bombay on November 14; Manibeli, Maharashtra, on January 3; and Anjanwara, Madhya Pradesh in late January 1993, the police used excessive force, including beatings with lathis and rifle butts, and

---

39 General Assembly Resolution 34/169, 17 December 1979, Article 3.

40 *Id.*

41 Adopted at the Eighth UN Crime Congress on the Prevention of Crime and the Treatment of Offenders, A/Conf. 144/28/rev.l, Principle 13.

42 *Ibid.*, Principle 4.

43 *Ibid.*, Principle 14.

(in Anjanwara) the firing of rifles, against persons who were peacefully, whether lawfully or unlawfully, expressing their opposition to the Sardar Sarovar Project.

c) Arbitrary Arrest and Prolonged Detention

Article 9 of the Covenant on Civil and Political Rights states that '[n]o one shall be subjected to arbitrary arrest or detention'. In addition, anyone who is arrested 'shall be informed, at the time of arrest, of the reasons for his arrest and shall be promptly informed of any charges against him'. Anyone arrested or detained on a criminal charge 'shall be brought promptly before a judge'.

Persons arrested for participating in anti-dam demonstrations have been commonly arrested without being informed of the basis for their arrest. In Kakrana, Madhya Pradesh, those arrested during a confrontation on October 30 were charged only for their alleged role in an entirely different incident. Those arrested for participating in a demonstration in Omkareshwar on November 22–23 were never notified of the charges against them in a court hearing, but were merely informed by the jailer upon their release from detention. Following the events in Anjanwara in late January 1993, more than two dozen activists and supporters of the Khedut Mazdoor Chetna Sangath (KMCS), an organisation that works on behalf of tribal people in Madhya Pradesh, were arrested on the basis of a blanket arrest warrant. Available evidence strongly suggests that the charges were without basis and were politically motivated.

d) Infliction of Torture and other Mistreatment

Article 7 of the Covenant on Civil and Political Rights states that '[n]o one shall be subjected to torture or to cruel, inhuman or degrading treatment or punishment'.

The problem of torture appears particularly prevalent in Madhya Pradesh. In several cases investigated by the Lawyers Committee, police have inflicted torture on persons detained in connection with their opposition to the Sardar Sarovar Project. The most common form of torture includes beatings with lathis (canes) and fists and by kicking. There has also been psychological humiliation and threats. Detainees have also been unlawfully handcuffed while in detention. Those arrested in all three States appear to be routinely denied access to food, and sometimes water, for long periods of time. For instance, those arrested in Chimalkhedi, Maharashtra on January 19 reported that they were denied food and given little water for nearly two days. Persons who suffer torture appear to have difficulty getting access to proper medical treatment, even on these instances when they are presented before a doctor.

e) Curtailment of the Right to Free Expression and Association

Articles 19 and 22, respectively, of the Covenant on Civil and Political Rights provide for the rights to freedom of expression and association.

Authorities in Madhya Pradesh have repeatedly asserted that the KMCS, a registered trade union, is an organisation that endorses violence or is 'Naxalite' (an armed group operating in rural Andhra Pradesh). The government has also taken unlawful steps to deprive the KMCS of its office and access to local goods and services, as well as conducting unjustified arrests of its members. All together, these official actions seek to deprive the organisation of its ability to communicate its opposition to the Sardar Sarovar Project and exercise its rights to free association.

IV. Recommendations

The Lawyers Committee for Human Rights urges the Government of India and the State Governments of Gujarat, Madhya Pradesh and Maharashtra to take those steps necessary to uphold the fundamental human rights of persons in the Sardar Sarovar affected areas, in particular those expressing opposition to the project. These include the rights not to be subjected to arbitrary arrest, unlawful detention or torture and other cruel, inhuman or degrading treatment. Also to be protected are the rights to freedom of expression, association and assembly. Official action should be taken to end the excessive use of force by the police against peaceful protesters. The authorities should conduct prompt, impartial and independent investigations of alleged human rights violations and prosecute those officials responsible.

On 30 March 1993 the Indian government announced that it would cancel the remaining undisbursed portion of the World Bank loan for Sardar Sarovar. Considering the important human rights and environmental objections to the project raised by persons in India and elsewhere, the Indian government's explanation that Bank funding was cancelled because of 'difficult internal complications unrelated to the project' (an apparent reference to communal violence that followed the 6 December 1992 destruction of the Ayodhya mosque) is insufficient. The Lawyers Committee calls on the World Bank to undertake a full review of the Sardar Sarovar Project in light of its 'good governance' policy.

Furthermore, the Lawyers Committee urges the World Bank not to abdicate its long-term responsibility with respect to the impact of the project on the human rights situation in western India. The Bank should ensure that the completion of Sardar Sarovar is consistent with its standards of good governance, in particular, respect for human rights and the rule of law. Sardar Sarovar is only one of many Bank-funded projects in India. The Bank should consider withholding funding for these or future projects should human rights violations in the Narmada Valley continue.

Chapter 10

Additional Resources

List of All Documents Reproduced in Chronological Order

Letter from Mr Chitale, Secretary, Ministry of Water Resources to Mr Vergin, World Bank Concerning the Morse Report, 1992, partly reproduced at page 353.

Government of Gujarat, Comment on the Report of the Independent Review Mission on Sardar Sarovar Project, 1992, partly reproduced at page 355.

Parliamentary Questions on Environmental Impact Assessment, 1993, reproduced at page 80.

Letter from the Secretary, National Commission for Scheduled Castes and Scheduled Tribes, 1993, reproduced at page 267.

World Bank, Office Memorandum from Ibrahim Shihata to Joseph Wood, 1993, reproduced at page 421.

Lawyers Committee for Human Rights Report, 1993, partly reproduced at page 450.

Report of the Five Member Group, 1994, partly reproduced at page 357.

World Bank – Project Completion Report, 1995, partly reproduced at page 422.

World Bank – Operations Evaluation Department – Memorandum to Executive Directors on the Sardar Sarovar Dam and Power Project, 1995, reproduced at page 435.

Pradip Prabhu v. State of Maharashtra, 1995, reproduced at page 267.

Supreme Court Order Restraining Further Construction of the Dam, 1995, reproduced at page 127.

ILO, Individual Observation Concerning Convention No. 107, 1998, partly reproduced at page 444.

Supreme Court Order Allowing Construction up to 85 Metres, 1999, reproduced at page 128.

Notification Setting up the Grievances Redressal Authority of Madhya Pradesh, 2000, reproduced at page 94.

Affidavit of the State of Madhya Pradesh, 2000, reproduced at page 130.

Narmada Bachao Andolan v. Union of India, Majority Judgment, 2000, reproduced at page 138.

Narmada Bachao Andolan v. Union of India, Minority Judgment of Justice S.P. Bharucha, 2000, reproduced at page 228.

Narmada Bachao Andolan v. Union of India, Review Petition, November 2000, reproduced at page 244.

Indian People's Tribunal on Environment and Human Rights, 2000, reproduced at page 357.

Justice S.M. Daud Report, 2001, partly reproduced at page 377.

Narmada Bachao Andolan v. Union of India, Review Orders, March 2001, reproduced at page 262.

Narmada Valley Development Authority, Proposal for Amending the Terms of Resettlement and Rehabilitation under NWDT, 2001, reproduced at page 299.

Narmada Control Authority, Consideration of Proposal for Special Rehabilitation Package, 2001, reproduced at page 301.

Narmada Valley Development Department, Terms of Special Rehabilitation Package, 2001, reproduced at page 302.

Jan Sunwai – Report, 2002, partly reproduced at page 395.

Narmada Bachao Andolan v. Union of India (Petition 328 of 2002) Supreme Court Order, September 2002, reproduced at page 269.

United Nations Commission on Human Rights, Report of the Special Rapporteur on the Situation of Human Rights and Fundamental Freedoms of Indigenous People, 2003, partly reproduced at page 446.

Narmada Bachao Andolan v. Union of India (Petition 328 of 2002) Supreme Court Order, April 2004, reproduced at page 269.

Order of the Grievance Redressal Authority, Madhya Pradesh, 2004, reproduced at page 270.

Narmada Bachao Andolan v. Union of India (Petition 328 of 2002) Supreme Court Judgment, March 2005, reproduced at page 277.

Kachchh Jal Sankat Nivaran Samiti v. Gujarat, 2005, reproduced at page 105.

ILO, Individual Direct Request Concerning Convention No. 107, 2005, partly reproduced at page 445.

Resettlement and Rehabilitation Sub-group, Clearance for Construction up to 121.92 Metres, March 2006, reproduced at page 304.

Narmada Control Authority, NCA Decision, March 2006, reproduced at page 311.

Group of Ministers, Note on the Assessment of Resettlement and Rehabilitation Sites and Submergence of Villages of the Sardar Sarovar Project, April 2006, reproduced at page 314.

Union of India, Status of Resettlement and Rehabilitation, April 2006, reproduced at page 320.

Status of Resettlement and Rehabilitation, Submission of Shri Shanti Bhushan, April 2006, reproduced at page 321.

Supreme Court Order of 17 April 2006, reproduced at page 322.

Supreme Court Order of 8 May 2006, reproduced at page 323.

Shunglu Committee Report, July 2006, partly reproduced at page 325.

Supreme Court Order of 10 July 2006, reproduced at page 328.

Selected References on the Sardar Sarovar Project

Books and booklets

Alagh, Y.K., Pathak, M. & D.T. Buch, *Narmada and Environment – An Assessment* (New Delhi: Har Anand, 1995).

Alagh, Y.K. et al., *Economic Dimensions of the Sardar Sarovar Project* (New Delhi: Har Anand, 1995).

Alvares, C. & Billorey, R., *Damming the Narmada – India's Greatest Planned Environmental Disaster* (Penang: Third World Network, 1988).

Baviskar, A., *In the Belly of the River – Tribal Conflicts Over Development in the Narmada Valley* (New Delhi: Oxford University Press, 2nd ed. 2004).

Chenoy, K.M. et al., *Narmada Rehabilitation – Fact and Fiction* (New Delhi: Indian Social Institute, 2006).

Dhawan, B.D. ed., *Big Dams: Claims and Counter Claims* (New Delhi: Commonwealth Publishers, 1990).

Drèze, J. et al. eds, *The Dam and the Nation – Displacement and Resettlement in the Narmada Valley* (New Delhi: Oxford University Press, 1997).

D'Souza, D., *The Narmada Dammed – An Inquiry into the Politics of Development* (New Delhi: Penguin, 2002).

Dwivedi, R., *Conflict and Collective Action – The Sardar Sarovar Project in India* (New Delhi: Routledge, 2006).

Environmental Services Group, *Dams on the Narmada: A People's View* (New Delhi: World Wildlife Fund India, 1986).

Environmental Services Group, *Dams on the Narmada: The Official View* (New Delhi: World Wildlife Fund India, 1986).

Fisher, W.F. ed., *Toward Sustainable Development? Struggling over India's Narmada River* (Armonk, NY: M.E. Sharpe, 1995).

Jain, L.C., *Dam v. Drinking Water: Exploring the Narmada Judgment* (Pune: Parisar, 2001).

Joshi, V., *Rehabilitation is Possible: The Sardar Sarovar Narmada Project* (Ahmedabad: Tax Publications, 2000).

Khagram, S., *Dams and Development – Transnational Struggles for Water and Power* (Ithaca, NY: Cornell University Press, 2004).

Kothari, S. & S. Singh, *The Narmada Valley Project – A Critique* (New Delhi: Kalpavriksh, 1988).

Morse, B. & Berger, T.R., *Sardar Sarovar – Report of the Independent Review* (Ottawa: Resource Futures International, 1992).

Padaria, R.N., Singh, R.P. & Y.P. Singh, *Big Dams Dilemma* (New Delhi: A.P.H. Publishing, 2000).

Paranjape, S. & K.J. Joy, *Sustainable Technology – Making the Sardar Sarovar Project Viable* – A Comprehensive Proposal to Modify the Project for Greater Equity and Ecological Sustainability (Ahmedabad: Centre for Environment Education, 1995).

Paranjpye, V., *High Dams on the Narmada: A Holistic Analysis of the River Valley Projects* (New Delhi: Indian National Trust for Art and Cultural Heritage, 1990).

Parasuraman, S., *The Anti-Narmada Movement in India: Can the Resettlement and Rehabilitation Policy Gains be Translated into a National Policy* (The Hague: Institute of Social Studies, 1993).

Pathak, M.T. ed., *Sardar Sarovar Project: A Promise for Plenty* (New Delhi: Oxford and India Book House, 1991).

Ram, R.N., *Muddy Waters: A Critical Assessment of the Benefits of the Sardar Sarovar Project* (New Delhi: Kalpavriksh, 1993).

Roy, A., *The Greater Common Good* (Bombay: India Book Distributors, 1999).

Sah, D.C., *Involuntary Migration: Evidence from Sardar Sarovar Project* (Jaipur: Rawat Publications, 2003).

Sangvai, S., *The River and Life: People's Struggles in the Narmada Valley* (Mumbai: Earthcare Books, 2002).

Santhi, S., *Sardar Sarovar: The Issue of Developing River Narmada* (New Delhi: Indian National Trust for Art and Cultural Heritage, 1994).

Shah, T., *Framing the Rules of the Game: Preparing for the First Irrigation Season in the Sardar Sarovar Project* (Anand: International Water Management Institute, 2003).

Sheth, P., *Narmada Project – Politics of Eco-Development* (New Delhi: Har Anand Publications, 1994).

Singh, S. & Pranab Banerji eds, *Large Dams in India : Environmental, Social & Economic Impacts* (New Delhi, Indian Institute of Public Administration, 2002).

John R. Wood, *The Politics of Water Resource Development in India – The Narmada Dams Controversy* (New Delhi: Sage, 2007).

Articles

Amte, B., 'Narmada Project: The Case Against and an Alternative Perspective', 25/16 *EPW* 811 (1990).

Appa, G., 'Narmada Projects Without World Bank Backing', 27/48 *EPW* 2577 (1992).

Baxi, U., 'What Happens Next is up to You: Human Rights at Risk in Dams and Development', 16 *American University International Law Review* 1507 (2001).

Berger, T., 'The World Bank's Independent Review of India's Sardar Sarovar Projects', 9 *American University International Law Review* 33 (1993).

Cullet, P., 'Human Rights and Displacement – The Indian Supreme Court Decision on Sardar Sarovar in International Perspective', 50 *International & Comparative Law Quarterly* 973 (2001).

Dharmadhikary, S., 'Hydropower from Sardar Sarovar: Need, Justification and Viability', 28/48 *EPW* 2584 (1993).

Dharmadhikary, S., 'Implementing the Report of the World Commission on Dams: A Case Study of the Narmada Valley in India', 16 *American University International Law Review* 1591 (2001).

Dwivedi, R., 'Displacement, Risks and Resistance: Local Perceptions and Actions in the Sardar Sarovar', 30/1 *Development and Change* 43 (1999).

Dwivedi, R., 'Resisting Dams and 'Development': Contemporary Significance of the Campaign against the Narmada Projects in India', 10/2 *European Journal of Development Research* 135 (1998).

Iyer, R.R., 'A Judgment of Grave Import', 35/45 *EPW* 3913 (2000).

Jayaraman, K.S., 'Narmada Valley Irrigation a Test Case for the World Bank', 335 *Nature* 583 (1988).

John, M., 'Interpreting the Narmada Judgment', 36/32 *EPW* 3030 (2001).

Kothari, A., 'Environmental Aspects of Narmada Valley Project', 35/3 *Indian Journal of Public Administration* 480 (1989).

Levien, M., 'Narmada and the Myth of Rehabilitation', 41/33 *EPW* 3581 (2006).

Mehta, L., 'The Manufacture of Popular Perceptions of Scarcity: Dams and Water-related Narratives in Gujarat', 29/12 *World Development* 2025 (2001).

Menon, M.S., 'Sardar Sarovar Project: Another Perspective', 38/39 *EPW* 4095 (2003).

Modi, R., 'Sardar Sarovar Oustees: Coping with Displacement', 39/11 *EPW* 1123 (2004).

Oza, N., 'Marginalisation, Protest and Political Actions: Tribals and the Sardar Sarovar Project', 32/29 *EPW* 1790 (1997).

Palit, C., 'Monsoon Risings – Mega-dam Resistance in the Narmada Valley', 21 *New Left Review* 81 (2003).

Paranjape, S. & K.J. Joy, 'Alternative Restructuring of the Sardar Sarovar: Breaking the Deadlock', 41/7 *EPW* 601 (2006).

Patel, J., 'Is National Interest Being Served by Narmada Project?', 29/30 *EPW* 1957 (1994).

Prajapati, R., 'Narmada, the Judiciary and Parliament', 32/14 *EPW* 693 (1997).

Raina, V., 'Sardar Sarovar: Case for Lowering Dam Height', 29/12 *EPW* 774 (1994).

Rajagopal, B., 'The Role of Law in Counter-hegemonic Globalization and Global Legal Pluralism: Lessons from the Narmada Valley Struggle in India', 18 *Leiden Journal of International Law* 345 (2005).

Ranade, R. & M. Dinesh Kumar, 'Narmada Water for Groundwater Recharge in North Gujarat – Conjunctive Management in Large Irrigation Projects', 39/31 *EPW* 3510 (2004).

Ruitenbeek, H.J. & C.M. Cartier, 'Evaluation of the Narmada Project: An Ecological Economics Perspective', 30/34 *EPW* 2138 (1995).

Sangvai, S., 'Reopening the Sardar Sarovar Issue: Significant Gain of the Narmada Struggle', 30/11 *EPW* 542 (1995).

Shah, U., 'Lowering Height of Sardar Sarovar Dam – What Purpose Does it Serve?', 29/10 *EPW* 667 (1994).

Srinivasan, B., 'Repression in Narmada Valley', 28/49 *EPW* 2640 (1993).

Talati, J. & T. Shah, 'Institutional Vacuum in Sardar Sarovar Project – Framing "Rules-of-the-Game"', 39/31 *EPW* 3504 (2004).

Thakar, H., 'Can Sardar Sarovar Project Ever be Financed?', 28/42 *EPW* 2262 (1993).

Tata Institute of Social Sciences, 'Sardar Sarovar Project: Review of Resettlement and Rehabilitation in Maharashtra', 28/34 *EPW* 1705 (1993).

Turanga, U., 'Damming Waters and Wisdom: Protest in the Narmada River Valley', 22/2 *Technology in Society* 237 (2000).

Upadhyaya, H., 'Narmada Project: Concerns over Command Area Environment', 39/19 *EPW* 1879 (2004).

Upadhyay, V., 'Relocating the Narmada Judgment', 36/39 *EPW* 3791 (2001).

Visvanathan, S., 'Supreme Court Constructs a Dam', 35/48 *EPW* 4176 (2000).

Whitehead, J., 'Space, Place and Primitive Accumulation in the Narmada Valley and Beyond', 38/40 *EPW* 4224 (2003).

Selected Internet Sites

Government and other official websites

Central Water Commission: http://cwc.nic.in

Government of Maharashtra, Relief and Rehabilitation (SSP page): http://mdmu.maharashtra.gov.in/pages/projectrelatedrehab/projectrelatedShow1.php

Narmada Control Authority: http://nca.gov.in/

Narmada Valley Development Authority: http://www.nvda.nic.in/

Sardar Sarovar Construction Advisory Committee (supervisory function over the construction of SSP): http://www.sscac.gov.in

Sardar Sarovar Narmada Nigam Limited: http://www.sardarsarovardam.org/

Other websites

Friends of the river Narmada (support and solidarity network for the NBA): http://www.narmada.org/

Friends of River Narmada and Association for India's Development in solidarity with the NBA: http://narmada.aidindia.org/

International Environmental Law Research Centre (IELRC) (legal documents related to SSP): http://www.ielrc.org/india/narmada.php

International Rivers Network (India programme): http://www.irn.org/programs/india

Support Narmadadam (a site in support of the SSP): http://www.supportnarmadadam.org

South Asia Network on Dams, Rivers and People (SANDRP): http://www.sandrp.in/

Maps

Map 1. Area of the Sardar Sarovar Project

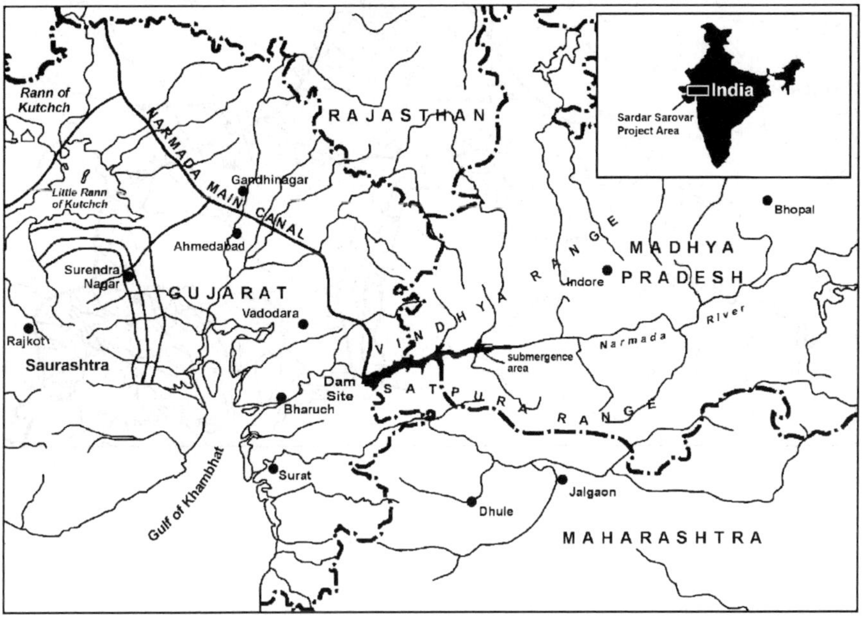

Map 2. Proposed Command Area of the Sardar Sarovar Project

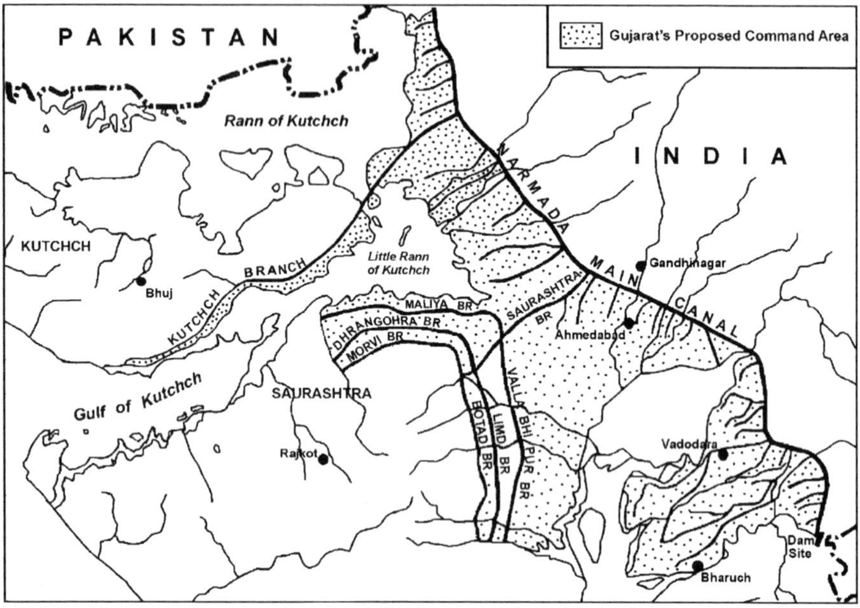

Map 3. Proposed Coverage of Narmada Pipeline Water Supply Project

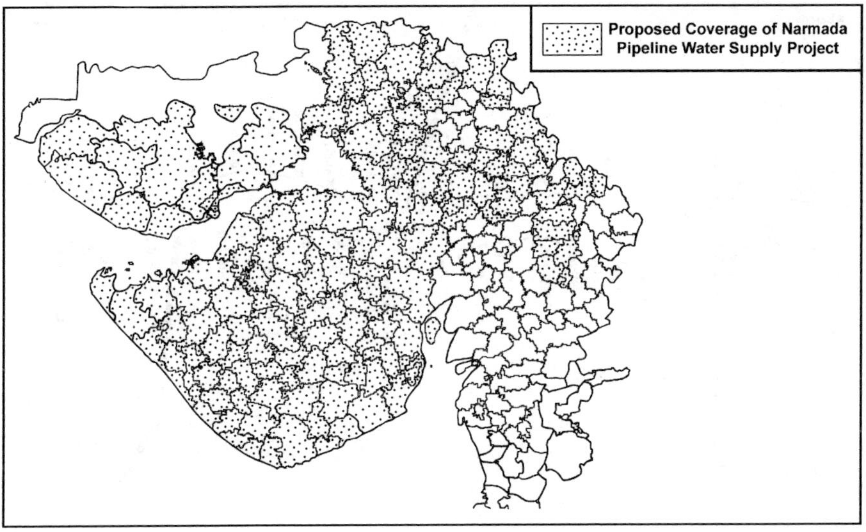

Map 4. Sardar Sarovar Project Submergence Area

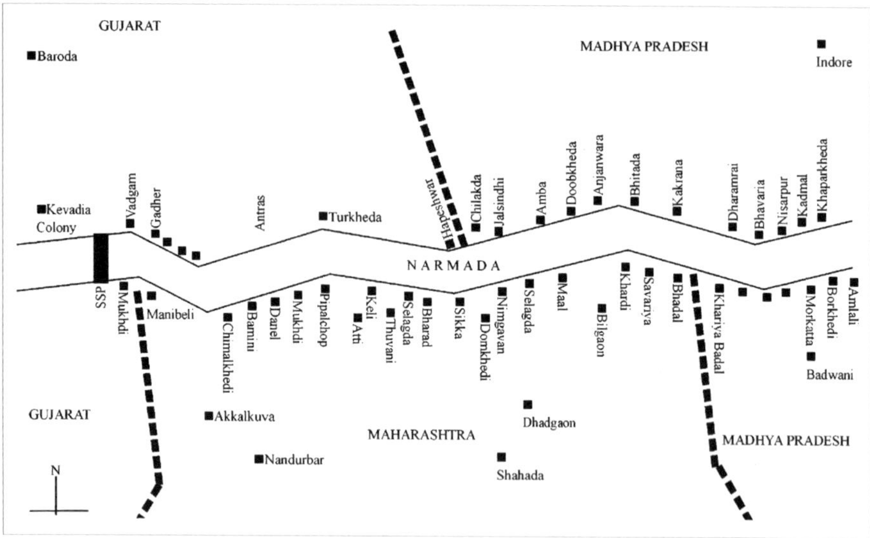

Index